Automatic Test Equipment:
Hardware, Software, and Management

Edited by
Fred Liguori, P.E.

Automatic Test Equipment Branch Head
Naval Air Engineering Center

A volume in the IEEE PRESS Selected Reprint Series, prepared under the sponsorship of the IEEE Aerospace and Electronic Systems Society.

The Institute of Electrical and Electronics Engineers, Inc. New York

International Standard Book Numbers: Clothbound: 0-87942-049-9
Paperbound: 0-87942-050-2

Library of Congress Catalog Card Number 74-18892

PRINTED IN THE UNITED STATES OF AMERICA

Automatic Test Equipment:
Hardware, Software, and Management

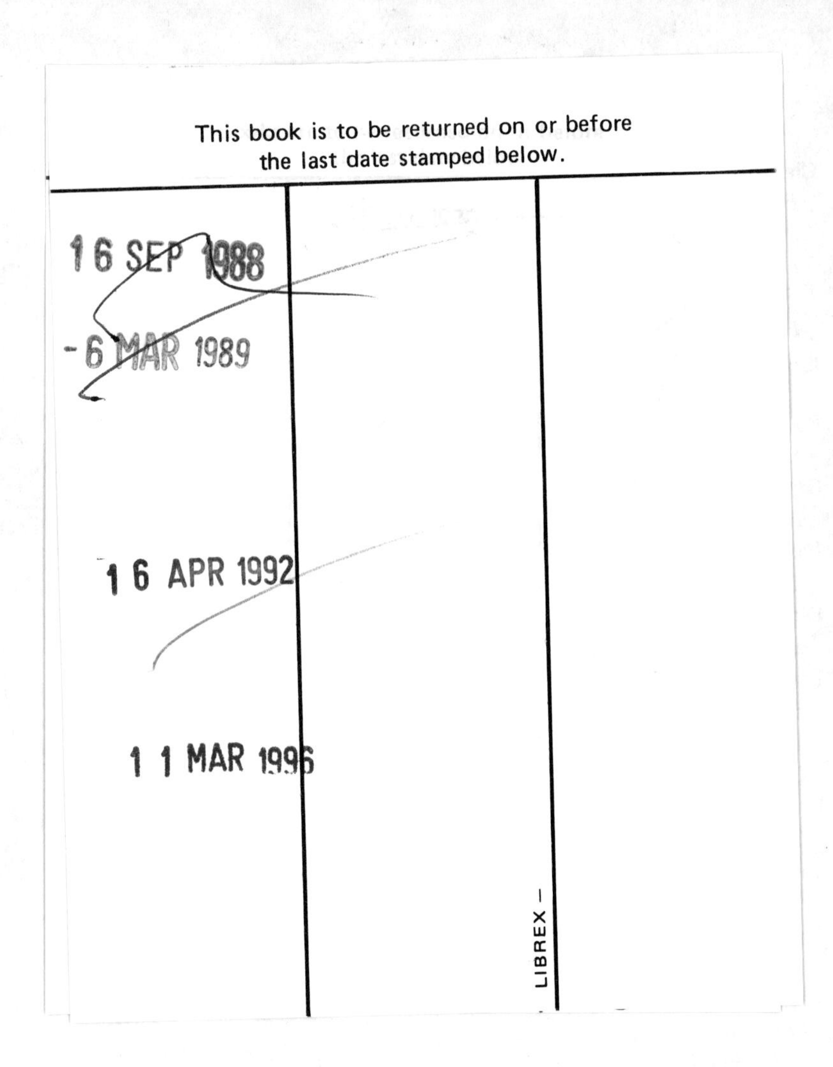

OTHER IEEE PRESS BOOKS

Clearing the Air: The Impact of the Clean Air Act on Technology
Edited by John C. Redmond, John C. Cook, and A. A. J. Hoffman
Active Inductorless Filters
Edited by Sanjit K. Mitra
A Practical Guide to Minicomputer Applications
Edited by Fred F. Coury
Power Semiconductor Applications, Volume I: General Considerations
Edited by John D. Harnden, Jr. and Forest B. Golden
Power Semiconductor Applications, Volume II: Equipment and Systems
Edited by John D. Harnden, Jr. and Forest B. Golden
Semiconductor Memories
Edited by David A. Hodges
Minicomputers: Hardware, Software, and Applications
Edited by James D. Schoeffler and Ronald H. Temple
Digital Signal Processing
Edited by Lawrence R. Rabiner and Charles M. Rader
Laser Theory
Edited by Frank S. Barnes
Integrated Optics
Edited by Dietrich Marcuse
Literature in Digital Signal Processing: Terminology and Permuted Title Index
Edited by Howard D. Helms and Lawrence R. Rabiner
Laser Devices and Applications
Edited by Ivan P. Kaminow and A. E. Siegman
Computer-Aided Filter Design
Edited by George Szentirmai
Key Papers in the Development of Information Theory
Edited by David Slepian
Technology and Social Institutions
Edited by Kan Chen
Key Papers in the Development of Coding Theory
Edited by E. R. Berlekamp

Contents

Automatic Test Equipment:
Hardware, Software, and Management

Introduction

In the mid-1950's those responsible for the maintenance of military electronics were facing some formidable problems.

1) Avionics and missiles were becoming too complex for the typical military technician to maintain.

2) The better trained and more capable technicians were leaving the service at the end of their first tour for more lucrative civilian jobs.

3) The special-purpose test systems required to maintain tactical systems were proliferating in quantity and complexity.

4) The time required to test systems was becoming excessive.

5) System designs were changing too fast to keep maintenance documents current.

6) Peace-time budgets were demanding a more economical approach to maintenance.

Against this background emerged the concept of multipurpose automatic test equipment (ATE), which promised testing at computer speeds, fully automatic operation by low-skill operators, the virtual elimination of maintenance documents, and universal designs adaptable to any test problem through the flexibility of programming. Is there any wonder that the military saw ATE as the solution to their maintenance problems? Is there any wonder that such bountiful promises might fall short of the mark?

Although many military applications of ATE have been moderately successful, they have not fulfilled the promises. Putting the computer in control has not resulted in testing at computer speeds, because systems were not designed to respond to such speeds. Operator skill requirements were reduced, but a need developed for a new highly skilled technician called a test programmer. While the conventional troubleshooting manual was replaced by the program, several new documents were required in support of ATE, such as unique operating instructions for executing each test program and program listings. The universal design approach caused ATE systems to be so large and flexible that they were extremely complex, and instrument performance was degraded by complex routing systems. And finally, the flexibility of easily changeable programs made test program development almost impossible to control and extremely expensive. Perhaps even more traumatic than these technical problems was the impact on traditional procurement methods. The old approach of using special-purpose test equipment allowed each new weapon manager to buy and apply support equipment in a vertical orientation to support only his weapon system. Using a multipurpose ATE to support several weapon systems requires unprecedented coordination and cost sharing by all potential users. There was no vehicle for implementing a horizontal support concept. Cooperation among weapon managers required coordination at the flag rank in military organizations. Finally, the support equipment acquisition managers were asked to pay the enormous cost of test program development when the real payback for test programs is to reduce the long-term logistics support costs rather than save procurement dollars. Thus even though ATE could be shown to have long-term cost effectiveness, the method of appropriating and disbursing military funds was not amenable to trading off higher short-term costs with long-term cost savings.

With all those problems, it is a wonder that ATE has enjoyed continued sponsorship by the military. The reason it has is simply because mission effectiveness demands ATE. The complexity of modern weapons and pace of modern warfare is such that the military can no longer depend on the "on-the-spot" mental troubleshooting process because it is too time consuming and error prone. Hence, ATE has earned a permanent place in military maintenance, not because it fulfilled all its glorious promises, but because in spite of its high cost and complexity it provides a capability to support missions otherwise unrealizable.

In the past few years, commercial applications of automatic testing have blossomed at a much greater pace than military applications. At the current rate of growth, commercial investment in ATE will soon, if it has not already, outstrip the high military investment. There is every reason to expect that the success of commercial applications will far exceed the military. Commercial applications have the following advantages over the military.

1) ATE was adopted in the commercial environment not with a false hope of achieving dramatic cost reductions, but rather to relieve the bottleneck of production, which is testing.

2) The commercial manufacturing environment is more adaptable to new techniques than the maintenance world.

3) With enlightened management and good planning, it is possible to trade off the large initial investment cost of ATE against reduced long-term operating costs.

4) The essential skills of test design and programming can be nurtured in the industrial environment.

5) The commercial user can learn from early military mistakes to avoid such approaches as "general-purpose" designs that are cumbersome and costly.

6) The manufacturing environment, involving large quantities of identical or similar products, is much more cost effective for ATE than the wide-mix short-run requirements of maintenance testing.

7) The manufacturer has full control of the product to be tested and can design testability into it and perform testing at optimum stages during product development and production.

With all those advantages going for ATE in the commercial environment, there is little doubt that automatic testing will soon dominate manufacturing technology. It is quite probable that any manufacturer who does not incorporate automatic testing in his production line in the near future may find his operation unable to compete.

After close to twenty years of military involvement in automatic testing and a phenomenal explosion of commercial applications for several years, it is somewhat amazing that no book has been published on this subject. It is hoped that this book will begin to fill the void. But much has yet to be done. The papers presented in this book emphasize the application aspects of ATE, for there are many more users and potential users than designers. Yet there is a genuine need for a book on ATE design.

The purpose of this book is to provide guidance in preparing for ATE, understanding its basic concepts, and managing ATE. A substantial number of the papers included here are devoted to test design and software preparation, because these tasks represent the most costly and least understood aspects of ATE. Much of the material is taken from the records of the annual Automated Support Systems Symposium sponsored by the IEEE. For years this was virtually the only source of documentation on the subject except for some seminars sponsored by the Secretariat for Electronic Test Equipment, Project (SETE). More recently, other conferences such as WESCON and IEEE INTERCON have been devoting sessions to ATE. And finally, there now is a short professional course on ATE offered by the Center for Professional Advancement, Somerville, N.J. Hopefully, this book will help give ATE the exposure it deserves and spark its transition from a rather obscure empiricism to a mature technology.

PART I: Planning, Preparing, and Staffing for ATE

In the 1960's many "experts" in automatic testing technology were forecasting an almost complete transition from manual test systems to automatic test systems. That transition is underway but is far from nearly complete. It has been and continues to be a turbulent transition because the impact on traditional ways of doing business has been far greater than most people imagined. In fact, there are those who believe that if top management had known the extent of impact that automatic testing would have on traditional means of maintenance support, the military would not have pursued multipurpose automatic testing as the vehicle for testing so many of the military electronics systems. Commercial applications of ATE have so far been less revolutionary because they have been introduced more gradually into the production environment. Thus there has been more opportunity for adjustment to the differences introduced by ATE.

There are many reasons why the introduction of ATE in the maintenance environment has been in most cases a traumatic experience. One reason is that in many instances insufficient attention was given to planning and preparing for the introduction of ATE. This is not to be taken as a criticism, for had one wished to plan extensively for ATE it would have been virtually impossible without having the prior experience to make such planning practical. Now that many systems have been introduced into the maintenance environment, many factors pertinent to good planning have been revealed.

Certainly now there is little excuse for not planning future applications of ATE for both maintenance and production testing. The four papers included in this part address four key areas of planning for new applications of ATE. These are

1) designing for automatic testability;
2) preparing for effective procurement;
3) determining the testing requirements;
4) appropriate staffing.

The paper entitled "Designing Equipment for Automatic Testability" is more than eight years old, yet the concepts offered are still relevant. The testing philosophy to be used in testing a unit with ATE cannot be an afterthought independent of the unit under test (UUT) design. This message has been heard and acted upon by at least some ATE users. The Navy, for example, has imposed a specification for testability (see Part VII-C, [2]) on certain contracts involving the design and manufacture of avionics to be tested by ATE.

So far, design for automatic testability has been somewhat less of a problem in commercial applications because manufacturers have already learned to consider manufacturing tests as an integral part of the fabrication function and have in-house control of the design of units they must test. Also, the typical commercial units of the 1960's have been analog circuit modules with fair access for testability either through the module connector or by pressure-point contacts. Furthermore, manufacturing testing is done at various levels of assembly where maximum accessibility can be achieved.

With the introduction of LSI, however, testability is proving to be the limiting factor of production. Access for testing of logic elements is not normally provided in the chip design. Thus it is incumbent on the designer to provide for access leads to internal elements for test purposes. The ability to initialize logic networks is imperative where memory elements such as flip-flops or latches are involved. This topic is deserving of a paper itself. Unfortunately, no suitable paper was available for inclusion in this book. It is a key area of concern for the future of LSI and ATE. Investigative and innovative work is needed in this area.

Preparing to procure an ATE in an effort which, properly done, will reap its reward many fold. The paper entitled "A Philosophy for Automatic Test Equipment Procurement" offers some advice which proved sound in a military application procurement. Many of the points made are worthy of consideration for any procurement, however. If there is one thought to be borne in mind regarding any ATE procurement it is this: Do it carefully and enlist the help of some reputable user who has seen the impact first hand. There is too much to consider and to be careful about to expect someone with only general test experience to appreciate the special considerations attendant to ATE and its dependence on software (see Parts V and VI).

A test requirement analysis (TRA) is a detailed analysis of what stimulus and measurement capability is needed to test a family of UUT's. The classical approach is to look at each UUT design carefully and fill out appropriate forms describing each test function. These are then integrated into an envelope of testing requirements for the test problem in total. Based on such an analysis, one may determine basic needs of a tester. Such an analysis will not generally tell how well these requirements are met by any proposed tester. In other words, it is concerned basically with capability and not with efficiency. The paper entitled "A Test Requirement Analysis for a TRA" provides an insight to the important task of performing a TRA. Again, there is no substitute for experience to do a TRA well.

To perform any of the preceding preparations for effective use of ATE as well as the follow-on applications work requires a staff of technically oriented self-disciplined individuals that are not easy to come by. Their background must include a combination of test design, programming, logic design, systems engineering, and technical writing. The paper, "ATE Test Designer—Programmer or Systems Engineer?" discusses the attributes to be sought in individuals for assignment in ATE. One would think that such versatile people as described in the paper would command higher salaries than design engineers. Unfortunately, industry has not seen fit to pay higher salaries for such talent. In fact, since most ATE people are on test design or applications assignments they do not even fare as well as design engineers. Thus there is a natural tendency for people with ATE backgrounds to migrate into more lucrative positions. Until this problem is resolved by making ATE work more financially rewarding, there will always be a problem attracting and keeping highly qualified people in this discipline.

DESIGNING EQUIPMENT FOR AUTOMATIC TESTABILITY

F. Liguori
Sr. Member, Technical Staff
Radio Corporation of America
Defense Electronic Products
Aerospace Systems Division
Burlington, Massachusetts

Summary

Enough has been accomplished in the field of general purpose automatic testing to prove its feasibility. Substantial progress has been made in adapting equipment designed for bench testing to automatic checkout and fault isolation. The next significant advance in automatic testing must come about by designing automatic testability into the units under test (UUT's).

This paper discusses the incorporation of testability into equipment designs by two means:

(1) General design features based on an awareness of automatic testing requirements

(2) Specific design features imposed by a specification tailored to a specific Automatic Test Equipment (ATE) or family of ATE systems.

General design features discussed include test point assignment, functional packaging, the use of manually-controlled adjustments and circuit design considerations. These subjects are discussed qualitatively because of the wide scope of UUT's that may be involved. The approach adopted is to present a basic philosophy for providing testability rather than attempting to develop rigid rules.

Specific design features for automatic testability are a function of the particular ATE system to be used in the maintenance function. Since even "general-purpose" test systems are not truly universal, the approach taken is to discuss the role of an ATE test capability specification in equipment (UUT) procurement rather than attempting to define such a specification.

Introduction

Automatic checkout and fault isolation of electronic and electro-mechanical equipment by programmable, general-purpose Automatic Test Equipment (ATE) is a reality. The Aerospace Systems Division of RCA alone has designed, built, programmed and fielded several generations of ATE systems and is actively producing another.[1] The practicability of employing ATE systems for both checkout and fault-isolation of units under test (UUT's) ranging from a plug-in circuit card to a complete transceiver has been demonstrated.[2]

Independent efforts to improve reliability and maintainability have advanced the state-of-the-art in systems packaging to where most military equipments are basically compatible with automatic testing. Advances in testing techniques and engineering-oriented programming languages have contributed considerably toward bridging the gap between the general purpose ATE and the special purpose UUT's. Studies are in progress which promise ever greater advances in testing and programming techniques. These efforts, however meritorious and successful, are nevertheless restricted to schemes for coping with testing problems in existence by virtue of the UUT design characteristics.

Many of the characteristics least compatible to automatic testing are not fundamental to the UUT design and could be appropriately modified during the design phase. Impetus for the next significant advance in automatic testing must, therefore, come about by designing automatic testability into UUT's. Just as surely as it has been learned that true reliability must be designed into a system, so too must true testability.

General Design Features Favorable To Automatic Testing

No feature is more important for automatic checkout than the availability of adequate test points, compatible with the lowest level of fault isolation dictated by the maintenance philosophy. Quantity alone is not the answer to test point requirements; other important considerations are accessibility, distribution, types of test points available, and parameters sampled.

---

[1] Digital Evaluation Equipment (DEE), first model installed at Letterkenny Arsenal and second at Tobyhanna Depot; Multipurpose Test Equipment (MTE) accepted by Army Missile Command in 1964; Depot Installed Maintenance Automatic Test Equipment (DIMATE) installed at Tobyhanna; Land Combat Support System (LCSS) under development.

[2] At RCA facilities and Tobyhanna on both DEE and DIMATE.

Reprinted from *IEEE Auto. Support Systems Symp. Rec.*, June 1965.

Quantities. Basically, the number of test points required to fault isolate a given UUT, using direct measurement techniques, is simply a sum of the number of independent interface lines between the modular elements, as defined by the required level of fault isolation. An independent interface is a tie point at the input or output of the UUT or between modular elements of the UUT, which is not tied in common with any other tie point. For example, if two modules of the same UUT are interconnected by a single conductor, this would represent only one independent interface.

For purposes of this discussion, direct measurements are those performed under normal conditions of circuit operation. Conversely, indirect measurements are those obtained by varying an input signal so as to obtain independent equations revealing operational characteristics of the circuit which must then be evaluated by mathematical techniques. Indirect measuring techniques can reduce the number of test points required. However, except for component fault isolation problems, the transfer functions involved are usually too complex for practical testing applications.

Accessibility and Distribution. Certainly any test point not accessible may as well not exist. No conscientious designer would provide inaccessible test points; however, his design might well be improved with proper guidance. For automatic testing, test points should be grouped together in as few different connectors as possible. The preferable way is to have a single connector containing all test points for the modular unit. This test connector should be in a position readily accessible for the mating connector.

Types. For manual testing, any well-defined point that can be readily reached with a probe might be considered an adequate test point. For automatic testing, however, the point must be of the attachable type, preferably a standard jack or connector.

Parameters. Each point selected must first provide a meaningful test parameter indicative of the associated circuit performance. Usually a choice exists between two or more measurable parameters capable of providing the same information. DC levels should be used as universally as possible because such signals are least susceptible to transmission problems. In certain cases where loading problems or isolation requirements exist, simple detection devices may be required. In choosing, preference should be given to the higher signal level point such as the plate of a vacuum tube rather than the grid. Higher level signals are not only easier to measure but are less prone to influence from the test apparatus. In general, parameters preferred for conventional laboratory measurements are also best for automatic testing criteria.

Functional Packaging

Functional packaging refers to the modularization of a unit consistent with its lowest echelon of maintenance required. This requires not only that replaceable units be readily removable, but that there be a direct correspondence between physical modules and logically singular functions. If, for example, a UUT is an eight stage amplifier, all components for a given stage should be contained on a single module. Where complexity dictates that the function be broken up into several modules, each should encompass a relatively complete circuit or sub-function. Functional packaging simplifies the testing process because it enables a modular unit to be tested with the fewest possible input/output signals and the simplest UUT-ATE adapter.

Standardization. The number of different modules should be minimized, and circuits standardized. Circuit standardization facilitates test standardization and substantially simplifies the testing process. Even at the expense of circuit redundancy or superfluous components, standardization of modules performing similar functions is usually the most economical design approach and always minimizes the logistics problem.

Sectionalizing. Where modularization is impractical, such as with case-mounted components, the main frame should be sectionalized. That is, components should be functionally grouped on removable sections or plates on the main frame.

Removability. Any group of components removable without unsoldering may be considered modular, but this does not always provide simple removability. Care should be exercised in the choice of fasteners to select not only secure devices but those readily removable by hand or with simple hand tools. Attention should also be given to avoid the possibility of damage during replacement by providing simple, clearly defined means of extracting elements of a unit. Unit alignment guides should be provided and complex mechanical alignment problems avoided in establishing module configurations. Also, modules should be capable of being removed without damage while UUT power is applied.

Stray Parameter Minimization. All modules or subassemblies should contain their own shields whenever they are subject to parameter variation due to separation from the equipment case. The performance of each module should be independent of electro-static or magnetic environment. Circuit designs should not utilize the placement of components or unshielded wire as a means for obtaining desired electrical parameters such as stray capacity. Stray parameters should be minimized, and readily identifiable lumped elements used for required parameters.

Adjustments and Controls

Minimizing Variable Devices. It is generally good practice to minimize adjustments and controls required to operate, calibrate and maintain a system or unit. For automatic testing it is even more important because each adjustment requires the automatic procedure to be interrupted for manual action by the operator. This not only appreciably delays the test process but it introduces human error, the elimination of which is an objective of automatic testing. In some situations, where the variable parameter is not a basic operational requirement, it can be eliminated by using more precise components. For example, it may be possible to eliminate a trimmer capacitor if a better quality main tuner were used.

Automatic Control of Variable Parameters. In many cases, variable parameters are essential elements of a system and they cannot be eliminated. Often, however, they may be electrically controlled rather than manually. Servo systems can be used for complete tuning operations, logic matrices can sometimes replace wafer switches, and relays can replace toggle switches. In general, electrical control should be favored over mechanical even though the initial cost will increase. In communication equipment, for example, the greatest single obstacle to automatic checkout of isolated assemblies is the existence of complex gear trains and tuning shafts. These mechanisms not only complicate disassembly procedures but they often force testing of several assemblies to be performed in the main frame of the unit, thereby complicating the testing process.

Selection of Connectors

Standardization. Standardization of connectors and pin number assignments is generally realized to be desirable for spares maintenance. For an automatic test facility, however, the need is even greater. Every UUT must somehow be adapted to the general purpose test system through cables and connectors. Not only is the cost of cables high, but without standardization, the problem of storage and retrieval becomes disproportionately great. With reasonable connector standardization, standard cables can be built to fulfill adapter requirements, thereby reducing the number of cables required.

Types. Good quality, floating and self-aligning connectors should be used as a general practice for modular designs because such equipment can expect a great deal of handling, assembly and disassembly. The connectors must be appropriately rugged and jam proof.

Circuit Design

One area in which significant contributions to automatic testing compatibility may be made is in circuit design. For example, testability may be improved by straightforward, simple circuit designs, the use of standard circuits, and the preference of digital techniques over analog. A few special requirements for automatic testing compatibility are discussed here.

Modular Specification of Units. Often a unit is designed to an overall specification dictating its performance. For effective modularization of the unit, however, each removable assembly and subassembly should be built and tested to its own specification as well as conforming to the overall specifications. This not only assures complete interchangeability of assemblies, but it provides test criteria for testing assemblies independent of the associated unit.

Warm-up Time. Circuit warm-up time requirements should be minimized. Often manufacturers will specify an unnecessarily long warm-up time, such as one minute when actually 45 seconds would suffice. The reasoning is that a few seconds make little difference in the utility of the device. In automatic testing, however, the warm-up time can be a significant percentage of testing time required to execute the remainder of the program comprising hundreds of tests. In mass testing operations, a few seconds lost per UUT may mean days lost in a complete run.

Sensitivity. The use of unusually sensitive components and/or circuit configurations should be avoided. For example, components that might be damaged by ohmmeter currents should not be used. Circuits with magnetic storage should be protected against data contamination. Circuits should not be designed so critically as to cause performance to degenerate due to the influence of the internal impedance of the measurement device.

Specific Equipment Design Requirements for Automatic Testing

Test System Specification

General design features favorable to automatic testing, as discussed earlier, deal with the philosophy and approach to making a system compatible to testing with a general purpose automatic test system. However, it is often difficult to translate philosophy into reality. Eventually a specific test system must be defined for the equipment designer to use as a guide in making the day-by-day decisions on testability implementation. In short, a complete test capability specification of a particular proven "general-purpose" automatic test system must be provided to the equipment designer whose units must be compatible to automatic testing.

Implementation of Automatic Test Capability

Once realized and fully defined by specification, the general purpose test system becomes the standard of testability requirements. A testability specification based on a specific ATE system is then imposed upon equipment manufacturers in the same manner as a maintainability specification such as MIL-S-23603 (WEP) is incorporated into development contracts. Design features of an equipment requiring test capabilities other than those of the existing test system would then require approval of the contracting officer. This provides a control which would enforce adherence to the ideal except in the cases where the contracting officer agrees that deviation cannot be avoided. As deviation requirements recur, appropriate modifications are made to the general purpose test system. Its specification is then modified and the evolution cycle completed. Figure 1 illustrates a possible flow process for controlling testability features of new equipment.

References

1. "A Tabulation, Automatic and Semi-automatic Checkout and Test Equipment", Ground Support Equipment, May/June 1964.

2. "Automatic Test Equipment" Op. Cit.

3. "A Survey of Programming Aspects of Computer-Controlled Automatic Test Equipment", Vol. I and II, Published by Project SETE, N. Y. University, June 1964.

4. F. Everhard, "DEE.... Digital Evaluation Equipment.... For Multi-Missile Checkout", Systems Support, Published by Radio Corporation of America, DEP, Camden, N.J.

5. D.B. Dobson, L. L. Wolff, "The DEE Concept.... A Family of Test Systems", Op. Cit.

6. D.B. Dobson, L. L. Wolff, "Automatic Test Equipment Checks Missile Systems", Electronics, 15 July 1960.

7. MIL-S-23603 (WEP) 1 March 1963 "System Readiness/Maintainability; Avionic Systems Design, General Specification For", Military Specification Published by U.S. Government Printing Office, Washington, D. C.

Acknowledgements

The author wishes to express appreciation for the valuable assistance and criticism of this paper rendered by the following members of the RCA Burlington Technical Staff:

F.U. Everhard
H.L. Fischer
E.B. Galton
P.M. Toscano

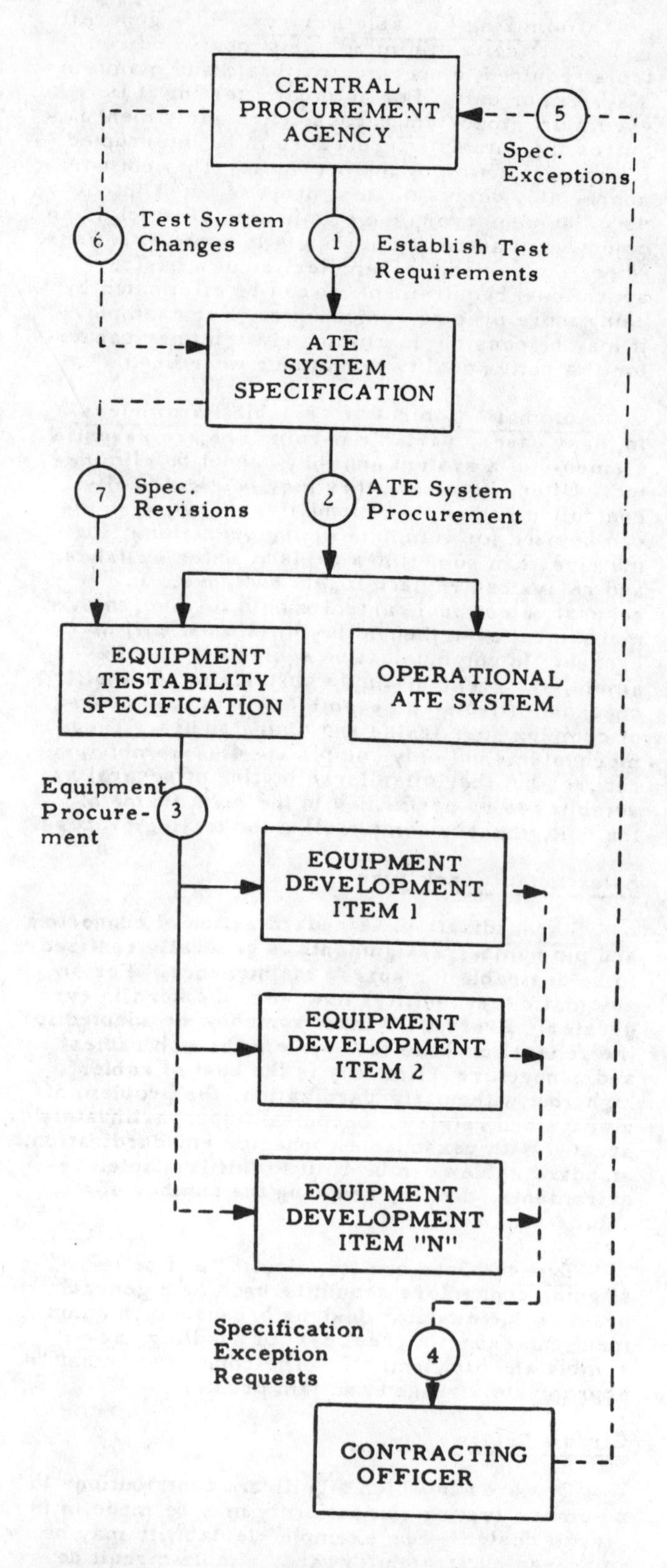

Figure 1 New Equipment Testability Control Process

A PHILOSOPHY FOR AUTOMATIC TEST EQUIPMENT PROCUREMENT

By Squadron Leader C.G.H. Frank, BSc, C Eng, MIEE, AFRAeS, RAF*

Summary:
Automatic test equipment design may be unsatisfactory to the customer unless the initial requirement is very carefully defined. The interdependance of the design of automatic test equipment and its test subjects, and the exact specification of all interfaces with the test equipment are discussed. Some preparations for introducing an automatic test equipment into service are also considered.

1. Introduction

The performance and effectiveness of automatic test equipments in service sometimes falls short of the customer's original expectations. This can be due to incorrect planning of the original requirement or to subsequent mis-use because of misunderstanding of the designer's original intentions. Certain aspects of this situation are discussed insofar as they affect automatic test equipment design, together with some other design factors.

2. Pre-procurement Planning Factors

2.1 Customer's Intention

The first requirement is a clear statement of the customer's intentions, followed by his reasons for requiring an automatic test equipment. A different approach to the design of the equipment may be necessary according to whether the prime aim is specific savings in men, materials, or time, to improve accuracy, integrity, repeatability, or some combination of these.

After determining his basic intention the customer must define closely what he requires, in terms of the original aim. This should start with the planned life of the test equipment, coupled to a statement of the degree of stretch likely to be required. Whether the test equipment is to be special-to-type or general purpose, and the degree of manual intervention envisaged must also be clearly stated, to clarify the requirement.

Definition of the whole system, of which the particular automatic test equipment will form a part, must embrace the main equipment (which may be an assembly of systems such as an aircraft, computer, series of control complexes or a production line) and include fault diagnosis, identification of the faulty unit, dismantling, as well as testing, repair and re-testing, re-assembly and overall test/checkout. Test subject repair policy is a factor and is in turn affected by delivery date; contract incentives should be planned as a defence against delivery delays. Finally, requirements for siting should be stated: data on floor bearing weights, door sizes, lighting standards, and on power supply availability, standby power, hydraulics, pneumatics and the tolerances on them, are essential.

* Royal Air Force Central Servicing Development Establishment.

Reprinted with permission from *IERE Joint Conf. on Auto. Test Systems Rec.*, April 1970.

2.2 Test Subjects

2.2.1 New Test Subjects - General Design Requirements

Current design of many electronics systems (particularly replaceable units for aircraft) tends to concentrate on performance, with reliability as a subsidiary: seldom are units designed to be tested. In addition to normal criteria, test parameters and test-access must be designed-in with the same degree of importance as performance and reliability: the means of testing must be established during design, to ensure that only a minimum of critical parameters need to be examined. Design must include a specification of performance which is stated in terms of a minimum number of critical parameters which, ideally, are also the test parameters.

For test subjects that are being developed at the same time as the automatic test equipment for them, the grouping of the individual circuits that go to make up an electronic system may have a considerable effect on the fault diagnosis. Of course the total system design is also a crucial factor, but it is no use disposing electronic circuits about, say an aircraft, or a large automatic plant, in a manner suitable to its functional performance alone, when so doing virtually ensures a hefty maintenance penalty.

2.2.2 Existing Test Subjects - General Design Considerations

When design of test subjects is outside the control of the automatic test equipment designer (which is too often the case) the customer must make available more information than normal about each test subject unit. For instance, the interdependance between two units as part of a larger system may make it uneconomic to design an automatic means to determine which of the two is faulty. This cannot be decided until design of the test equipment is in hand, and the decision requires more than the simple definitions of the two units to ensure success. It is likely that test subject run-up time and likely times-to-repair will have a big part to play in the design of the test equipment for utilisation.

2.2.3 Customer's Repair Policy

The customer must also state fully his test subject repair policy, the extent of component/sub-assembly replacement envisaged during test; also the expected frequency of and time allowed for changing between test subject types.

2.3 Automatic Test Equipment General Design Requirements

Customer requirements must be stated for automatic test equipment availability, permissible servicing times, mean-time-between-failures and all general aspects of the design over which control is required. When the range of test subjects is known, it will be possible to specify in advance acceptable process times expected, change-over times envisaged, and test equipment dead- or connection-times required. Test subject mean-times-between-maintenance-actions must be clearly stated (together with degree of confidence) and the test equipment total load in terms of numbers of each different type of unit in service, and the numbers of spares in the total system. Much depends on the customer's intended pattern of use of the equipment as to how the self-check part of test programmes is to be integrated into the test programme.

2.4 Contractor/Contractor Liaison

Particularly when the test subjects have been designed remotely from the test equipment, good liaison between the designers is essential. Test equipment modifications may accumulate and significantly upset the automatic test equipment design, and even when the definitions and specifications appear adequate, frequent incompatibilities may arise as the detailed work on the test equipment takes place: these can only be eliminated by formalised and careful contractor/contractor liaison.

3. Digression on Test Specifications

Production Test Specifications are essential for defining the state of products as they leave the manufacturer, but that state is determined to allow for a variety of factors which will bring about deteriorations in performance during use: the degree of rigor of the test specification being set to ensure survival of the product for its designed life before failure. However, after repair, or for periodic check, other tests are applied to determine the fitness of the equipment for further use: they are probably applied in in-use conditions and reflect expected in-use standards. Most often test schedules for these checks are derived arbitrarily by 'opening out' the Production Test Specification limits. What would appear to be required are proper test schedules for the purpose, written as part of the design process, which state simply those critical parameters, with tolerances, which alone are essential to proving that the unit is 'fit for use', at two different levels. The first schedule would contain only those items necessary to determine continued fitness for use, in situ, of the then-presumed-serviceable component, unit or system under test; the second would perform the same function more stringently for components or units after repair or during workshop test. The important factor is that the specifications should be much more intimately tied in with the design of the equipment, reflect the standard for which it was designed in use, and have the absolute minimum number of test parameters - this last has to be very carefully thought out.

Anarchy prevails in the layout, language, interpretation and tolerances used in test specifications at the moment - not surprisingly. It is thus becoming difficult to be sure that one is designing a piece of test equipment with the right range of stimuli and sensors and the correct circuit characteristics because of anomalies in the way specifications of the different subject equipments are written. What is required is that every user test specification offered by a manufacturer of electronic equipment should be written in a common language - ARINC ATLAS which the RAF is in process of adopting, is a good example. Wider acceptance of something along these lines will lead to much improved standardisation of documents because of the discipline imposed by the necessity to translate from English to a lower level of precise and formalised language, and agreement on compatibility of bands for test parameters and of impedances at test points may then follow.

4. Interface

The test subject/test equipment interface may present difficulties when test subject and test equipment are not designed together. In addition an automatic test equipment may have to be adapted, post-design, to cater for additional test subjects: the ultimate in this situation is the test equipment which is ordered after its set of test subjects has been designed.

One possible solution to this problem is a standard interface matrix of pin connections. Each pin would be allocated to an agreed standard range of stimulus or service, for attachment to which both test equipment and test subject manufacturers could design their equipments: there appears to be considerably less difficulty likely in defining this piece of 'neutral territory' than in standardising the contentious matter of primary plug and socket connections. The idea would allow a set range of test parameters to be developed, and facilitate adaptation of test equipments and potential subjects, because the development of the adaptor between subject and matrix would remain in the test subject manufacturer's hands.

5. Introduction and Commissioning

In order to ensure proper integration of the automatic test equipment within the customer's organisation, it is essential to prepare Management by careful briefing before the equipment is installed. Emphasis must be made of the degree to which the maintenance system had been orientated to function with the test equipment. Automatic test equipment represents an extremely high concentration of resources, and can prove to be an Achilles heel unless integration into the maintenance plans is adequate. A full repair philosophy for the automatic test equipment has to be worked out, and then backed by a properly planned and negotiated spares postion with adequate contractor support. Down-time due to unserviceability during working periods must be minimised by work study methods, and there must be inducement for the contractor to retain personnel skilled on the equipment for as long as possible, and to give advance notice before this service ceases.

Workshop and stores layouts need to be carefully re-planned to ensure that there will be no hold-up due to shortage of test subjects or inadequate test subject spares flow: if repair is away from the test equipment the method of removal/replacement of test subjects must be rapid. Facilities for rapid changes of automatic test equipment interface equipment must be planned, and a comprehensive system arranged for handling the bulk of input equipment and test program material that may have to be interchanged each time. A system will be required for continuously assessing the queueing problem and determining accurately the correct application of the equipment for a given state of stock of test subjects.

6. Contractor/Customer Liaison

Much of what has been said has emphasised the need for good customer/contractor relations from the inception of the project. Sufficient agreed authority should be vested in the customers representatives to permit progressive checks of test programs as they are written, verified and proved - ideally, the final check-out inspection might then become the commissioning demonstration as well. The final demonstration of the test equipment must include a physical check of every test with test subjects of known state; it should ascertain that tolerance bands are correct and check that reaction to faults is as required.

7. Modifications to the Automatic Test Equipment

One of the many advantages which may accrue from having customer liaison personnel at the contractor's works is in the field of test equipment modifications. Continuous vigilance is necessary to ensure that modifications to the automatic test equipment for improving the design, standardisation, modernisation of the design, do not impair the capability of testing the full range of test subjects originally accepted, because a return to manual test equipment on any such test subject, is very costly in equipment and labour. Care must also be exercised to ensure that the modification does not introduce different pass standards for different subjects or from different test equipments.

8. Program Updating

Great care is also necessary during the programming to ensure that the automatic test equipment will be able to deal with test subjects of different modifications standards. It is essential to demonstrate the facility to alter test control inputs, stimuli, and sensors to newly modified subjects during hand-over. The ideal is that an automatic test equipment (or one version of it if several are delivered) should be capable of self programming (in the computer sense) so that the customer may adapt test routines to changes in test subjects as they occur.

9. Conclusion

In considering some aspects of successful procurement and commissioning of an automatic test equipment, the important factors which emerge are primarily concerned with individual attention to detail concerning fundamental information about the customer's requirement and the test subjects.

A TECHNICAL REQUIREMENTS ANALYSIS FOR A TRA

A. M. GREENSPAN

AAI CORPORATION
P. O. Box 6767, Baltimore, Maryland

Abstract

The ultimate success or failure of a support plan for large scale systems designed around an automatic test and monitoring philosophy depends in large part upon the depth and quality of a Technical Requirements Analysis (TRA). However, little has been written to describe the technical considerations, management responsibilities and timing requirements critical to a successful TRA.

It is the intention of this paper to identify these functions with particular emphasis placed upon such critical factors as timing, documentation, controls and judgement criteria for assessing quality and progress.

INTRODUCTION

In order to systematically implement an automated monitoring concept for any large system; it is necessary that the process through which information will be gathered, developed and implemented as work progresses be firmly established. This must be done as a first step in implementing the system segment and sub-system TRA. The TRA will concentrate on descriptions of test methods, determination and location of test points, and creation of documentation necessary to support and describe the testing philosophy. The documentation must provide a detailed description of the methods and procedures for bottom up design implementation and development in accordance with mission requirements. A logical sequence of data development and the manner in which data should be reviewed, verified and utilized must also be stipulated. The following functional flow diagrams and descriptions recommend an approach for providing the direction and specific support and have been adapted from previous successful efforts at AAI. They describe the logical interrelationships and iterations which are required for smooth system development, integration, avoided inconsistencies and/or voids in the data base. Following these guidelines will help avoid lost time and rework due to problems discovered in later stages of a program.

The approach presented in this paper is not suggested to be an absolute or mandatory means for performing a Technical Requirements Analysis (TRA) leading to implementation of an automated support concept. Rather it is presented to serve as a check list and guide of tasks which should be performed as part of the TRA for any large system regardless of its specific purpose.

SYSTEM DEVELOPMENT OF DESIGN GOALS AND MONITORING OBJECTIVES

Figure 1 is a simplified block diagram of the means for developing a system plan for establishing the goals and guidelines necessary to implement an automatic support concept. The following will describe each block and explain how they should fit into a TRA plan. When finished these tasks should provide top down guidance for the implementation of a systems concept.

Items 1-5 represent documentation and policy guidelines which lay the foundation for system development.

(1) Specification

This is the system specification which will describe the system its goals and objectives. All subsequent data developed must be measured in respect to this specification to be certain the product developed will perform as described by its requirements.

(2) Monitoring Policy

The monitoring policy is developed to describe how, when and why monitoring must take place to support the requirements of the system specification. This policy should also include monitoring priorities if they exist, as well as test philosophy such as "Most-test-with-the-first-test" concept.

(3) Guidance Meetings

Guidance meetings on the level suggested by figure 1 should be held on a continuing basis to allow all participating entities involved with the systems development to remain informed of progress, problems and/or changes effecting the system. These meetings should take place on a management level and involve participants from the various subcontracting organizations as well as the customer and prime contractor. The purpose of these meetings is to insure that the various segments of the systems are compatible and thus will be able to be integrated with a minimum of

Reprinted from *IEEE Auto. Support Systems Symp. Rec.*, Nov. 1972.

problems. The meetings should provide a timely and smooth transfer of information between participating organizations as regards schedule, technical development, and data. The meetings should also be used to ascertain that the systems integrity is not being compromised and that changes in the development and implementation plans for various segments are promulgated at an early date.

(4) System Block Diagrams

System block diagrams should be provided as early as possible in the system development. The block diagram is used to show participants how their segments fit into the systems plan. The block diagram will also serve to indicate the interface requirements that will be necessary for systems integration.

(5) System/Function Matrix

This diagram becomes necessary for any complex system since the system configuration is subject to change based upon the systems operating mode and the use to which the system is being put. This document serves a similar function to that of the systems block diagram except that it provides more detail and takes into consideration the systems operation plans.

(6) System Monitoring Philosophy

The first step in developing a systems management and integration TRA for automatically monitored systems must be directed toward establishing a general monitoring philosophy. It is necessary to develop a heirachy of priorities based upon the system specification and engineering analysis. The monitoring philosophy will indicate system functions and parameters most critical for monitoring. It will also lay down ground rules for monitoring priorities and establish a matrix indicating how these priorities are effected by and related to the system modes of operation. Desirable variations or options should also be specified for the monitoring function as well as analysis of how the monitored parameters should be displayed and utilized.

(7) System Scanning

Once the system monitoring philosophy has been established an analysis for implementation is required. The analysis will first set up the criteria that insures system operability. Monitoring possible on a continuous basis and monitoring which requires special simulation or stimuli must be considered. The analysis should also determine the optimum manner for insuring maximal test over a minimum time; taking into consideration the various modes and options which had been established when setting down the monitoring philosophy. The scanning concept will be shaped by constraints involving ease of implementation and systems functions criteria.

(8) Go-Chain Functional Analysis and Objectives

Once the scanning concept has been made firm it is possible to use it as a foundation for establishing a detailed implementation plan. This task is based upon system documentation describing the segment and subsystem requirements which comprise the system.

Major system parameters as described by the system specification must be linked to segment engineering documentation. Operational sequence diagrams describing segment interaction and signal interfaces must be examined. Functional subsystem relationships and linkages should be clearly identified and analyzed on the basis of operating modes and system functions. In addition signal conditioning procedures, requirements and criteria should be identified and optimized to allow utmost efficiency thru standardization.

The analysis described above will then allow a preliminary selection of tests or test sequences necessary for determination of system operability. Use of the data developed and described in the preceding paragraphs will allow the selection of tests and test sequences on the basis of most important functions.

The conditions under which signals are available for monitoring must be considered as well as the times necessary to obtain and analyze these signals. It is necessary to consider the reliability of the monitored function to establish the frequency with which monitoring must take place. Monitoring policy described as block 2 will be used to provide guidance in test selection as well as the availability of various monitoring options; adequancy of test to determine a "Go" status as well as overall system efficiency and test sequence linkages must also be considered.

The functional analysis and objectives task described above at the system level should not be confused with the detailed test design and test point selection done on the segment level. The function of block 8 is to provide goals and objectives to the engineer at the segment level to give him a guideline for detailed development.

(9) "No-Go" Functional Analysis and Objectives

The Go-Chain functional analysis (block 8) provides guidelines necessary to develop a testing sequence to determine a system "Go" status. It also provides the necessary information for developing testing sequences in the event an on line monitoring test fails.

Much of the data necessary for developing guidelines for the Go-Chain are also necessary in developing guidelines for the "No-Go" diagnostic test philosophy. In addition to this data it is necessary to utilize the system maintenance and logistic support policies. It is important that the segment test designer be aware of the level to which fault isolation is required. This criteria may be different for various system segments. In some cases isolations to a group of functions may be adequate in others isolation to a function may be necessary and in still other cases isolation to a module level may be mandatory.

The availability of and requirements for test points must first be considered at the system level. Availability will be based upon the nature of the monitoring system. Requirements will depend upon function reliability, number of elements involved and the necessity to eliminate ambiguities. Virtual test points should be utilized whenever possible as well as utilization of halving techniques and other fault isolation good practices.

Other considerations in developing "No-Go Diagnostic" guidelines are function criticality, mean time to repair, and other operational and maintenance variables.

(10) Management Committee

In order to develop, organize, transfer and implement the data necessary for system development and integration the committee composed of personnel responsible for management of the various system elements should be formed. The information will consist of technical as well as schedule data. The committee should set up task responsibility and schedule matrices, as well as assign responsibility to committee members acting as liason to the system developer from participating organizations. The liason will take the form of implementing system plans as well as reporting problems and/or progress as system development takes place.

(11) Task Responsibility + Schedule Matrix

This matrix is an important working tool for a management committee. It should clearly detail all system tasks and identify the dates when work should begin and end for each. The name of the organization and the person within the organization responsible for the task should be specified.

Once a matrix of this type is completed it can serve to guide the committee as to where its attention must be concentrated and who must be contacted to provide indications of problems or progress. In addition should redirection or additional information be necessary for any segment of the system, the matrix provides the committee with the name of the critical personnel to be contacted.

(12) Working Groups

Working groups are formed by the management committee as temporary organizations to perform specific tasks which help the management committee perform its function.

The end result of the work of the organizations and documentation development described above is establishment of system Design Goals and Monitoring Objectives which should be formalized, edited and distributed to all system groups for their guidance. Subsequently all work developed at the subsystem or segment level will be examined to see how well it meets the Designs Goals and Monitoring Objectives that have been established by the management committee. It is only thru setting up of Goals and Objectives early in the system development and carefully insuring that all system elements meet them, successful development of complex automatically monitored systems can be expected to succeed.

BASIC FUNCTION TEST DESIGN

The lowest common denominator in the design of a system is the basic function which is normally comprised of modular elements. It is at this lowest common denominator that detailed monitoring consideration must begin. For, if the least element in the system is properly monitored and the monitoring philosophy is consistently carried to the succeeding levels, it must follow that the overall system is properly monitored. If automatic monitoring and test is not considered until higher system levels, such as a rack or subsystem, the probability of missed components, and/or inconsistent results discovered later in the program becomes large. In addition review of the systems development is realistic and practical only if it is implemented in manageable steps. Starting at a rack or segment level will almost guarantee that the reviewer will not be able to ascertain if test points have been utilized adequately, efficiently, or if the monitoring sequence provides for all necessary testing contingencies. This lack of certainty must result in chaos when all the system racks are integrated to the subsystem, segment and then system levels.

The following method is suggested for test requirements analysis at the basic function level for automatically monitored and tested systems.

(1) Detailed Design of Functional Element

Figure 2 illustrates the start of a system development at its lowest common denominator, the function.

Block 1, illustrates the detailed design of some function residing in any given rack. The design of the function will be based upon some performance specification required for that function as well as some performance specifications of the module boards used or components for the function.

When the design of the basic function is complete, its supporting documentation should consist of a block diagram indicating the modular elements that the function is composed of. In addition, the diagram should indicate inter-rack and intra-rack interface for the function as well as wave shapes and timing diagrams where applicable. A theory of operation and failure mode analysis should be written for the function and any pertinent schematic information should also be included as part of the function documentation package.

(2) Testing Philosophy (methods and procedures)

During the design and documentation of a functional element it should be considered from the viewpoint of testability. The first step in analyzing a function for test is to establish its place in the philosophy or overall test plan. This test plan will consider the nature of the function, optimum ways to test functions of that type and methods for dividing or segmenting the function to enable a conclusive decision regarding operability. In addition, guidelines for testing established at the system level must also be adhered to.

(3) Qualification of "Go-Chain" Test Requirements

After the test approach has been analyzed it is necessary to conceptualize the manner in which a test plan should be implemented. This requires analysis of signal availability, for certain signals will be present constantly allowing test through synchronous monitoring. Other signals may require special stimulus and/or be monitorable only through simulation and synchronous timing with system.

Analysis of signal availability leads to preliminary selection of testing points and associated signal conditioners. The test point selection will be made to allow rapid and complete determination of function availability. That is the designer will select the test point or points necessary to verify most efficiently that the function is good or "Go". As support for the selection and location of the test point or points, a test description detailing the purpose of the test and the assumptions made and/or requirements necessary for the test point to be used should be provided.

(4) Fault Isolation or No "Go" Test Requirements

Once the designer has developed a concept for determining a functions availability he must next consider how function failures will manifest themselves and how to isolate the cause of a function failure. The considerations are must the same as those for determination of function availability except that a function "Go Chain" test verifies the quality of certain modular elements used to comprise the function. If the test fails, the designer must determine if additional tests should be made to provide a more specific identification of the faulty elements or if all of the elements comprising the failed function should be condemned.

It is possible that in some cases, the fault isolation requirements and analysis, described above would not be necessary, at the basic function level. Requirement for this additional analysis would depend not only upon the number but also upon the type of modules required to comprise the function.

The reader should not be confused by the name "Qualification Test" (3) or "No-Go Test" (4). This does not refer to the manner in which the test point will ultimately be used in the system. It is merely a means of describing what the designer hopes to establish for the function under consideration. A Go Test point for given function, when considered from a system basis, may be the final point monitored in a No-Go chain before condemning a group of modules. The use and functional names of test points will be constantly changing as the system grows and becomes integrated. What is of ultimate importance initially is seeing that an adequate number of test points have been provided. That is why it is important to begin consideration of test point emplacement at the basic function level, for if every basic function has a properly emplaced and adequate number of test points, the only remaining task as the system grows and develops is to utilize these points properly.

(5) Basic Function Review

The primary function of the review is to evaluate the proposed test plan and provide help and direction as required. A secondary, but also important function is to consider the impact of the test approach on spares and maintainability requirements.

Copies of the documentation should be made available to all participants of the review a number of days before the review takes place in order for all participants to familiarize themselves with the material. Thus when the review takes place the reviewers can have gathered material to help resolve problem areas which have been indicated, or to suggest subtle problems which may have been overlooked by the test designer. In addition alternative approaches may be suggested which could enhance the results obtained or result in more efficient utilization of test points and/or signal conditioners.

Numerous results are possible from a review; the designers' concept may be approved totally, approved generally with action in certain indicated areas, approved partially with numerous exceptions, or rejected and redirection given by the review team. In all cases except the last, the designer must take the necessary steps to clear any action items generated as a result of the review, and go on to implementing the design concept that has been developed.

The documentation supporting the test point location and signal conditioning requirements should not become official until approved by the concept review team.

The procedures shown in figure 2 and described above represent the first step in the implementation of the TRA established at the system level. The process should be repeated for every basic function residing within each rack. Once a rack is fully represented, i.e., all functions residing within a rack have been documented and approved the next step may be taken. This second step is shown in Figure 3.

RACK TEST DESIGN

The second step in the TRA Analysis takes place at the level immediately above that of the basic function.

(1) Analysis

Once all of the basic functions have been analyzed and documented as described in Figure 2 and compiled into packages representing given racks or major functions, the focus of the analysis must shift to the next level which can be assumed to be rack itself. In performing a rack analysis additional data must be developed to supplement that developed for the basic functions lying within the rack. Whereas the first analysis was concerned with the basic function, the rack analysis will be concerned with secondary functions which will be a result of combining primary functions within the rack and/or will be utilized as outputs to other racks or segments.

Each of the rack output signals should be described by a performance specification describing the signals, truth tables, nominal characteristics and/or tolerances. In cases where basic functions and rack output signals are identical it is only necessary to remove the documentation representing that function from the data package. In cases where the rack output is the result of combining a number of basic functions a new specification based on the effects of the resulting combination must be written.

To aid the rack analysis and the later review function a flow diagram should be made which describes the manner in which basic functions combine to form secondary functions. This intra-rack flow diagram will act as a road map when determining the required testing sequences for monitoring and fault isolation. In addition an inter-rack block diagram describing the destination of all secondary functions developed within the rack will prove useful as a check list to insure complete analysis and in developing test sequences at the next higher level.

Schematics and prints should be supplied as necessary. In most cases these items will only be used to help describe wave shapes or support a monitoring concept. Logic diagrams will also fall into this category.

Theory of operation of the secondary functions should describe the role that the function plays how it is used in the system what it is controlled by and what is controlled by it.

The failure mode analysis for the secondary functions should describe the manner in which failures are likely to manifest themselves within the function itself or relative to system functions.

Interconnection diagrams describing cable, plug and pin connections are also required in order to adequately analyze test point placement and to provide detailed repair procedures.

(2) Determine Test Requirements

Once all of the secondary functions have been identified for the rack under consideration an additional level of segregation is necessary. The secondary functions representing the rack outputs also represent the GO-Chain for that rack. Again the reader must remember that GO-Chain when used in the context of the rack does not imply a system level description or typing of a signal or its associated test point. It is nearly a convenient means of describing the purpose of the test; which is to prove operability at a given level, in this case the rack. Thus if all "GO-Chain" Tests for a rack pass it would merely indicate that the rack was operating properly and not necessarily that the system was good. In addition what is referred to as a "GO-Chain" Test for a rack may in fact be part of a "No GO" routine from a system point of view.

The segregation required that the analysis of the rack "GO Chain" signals is segregation of the signals into classes. That is, the signals will be classified into groups according to whether they can be monitored asynchronously or require synchronous timing with the system.

Of those signals identified as asynchronous additional considerations must be made to determine their manner or use. For example the asynchronous signals can be classified as continuously available, requiring stimulation, or mode dependent. Each of these three classes of signals should be grouped to facilitate preparation of test program and for ease of review by the review teams described earlier.

Those signals identified as synchronous must also be classified. They must be classified by the system signals that they are synchronous with and can be identified by as well as their mode dependency and/or stimulation requirements.

(3) Establish, Test Continuity to Basic Function

Identification and documentation of the secondary (rack) functions as specified in items 1 and 2 constitutes a complete documentation package of the "GO-Chain" signals representing the rack. Only one item more is required to allow a thorough review of the data. This is the documentation that indicates the relationship between the secondary functions and the basic functions. That is, failure of any secondary function being monitored should result in diagnosis and ultimate condemnation of the elements or modules that comprise some basic functions. These test linkages should be documented as well as a rack overview testing sequence which recommends the order in which the rack signals should be tested to verify operability. Creation of this document requires analysis of signal priorities within the rack as well as signal dependency and consideration of virtual test points.

(4) Rack Concept Review

The purpose of the rack concept review is to evaluate the documentation for the rack. The participants at this review as well as the purpose of the review are similar to that described for part 5 of figure 2. Successful completion of the review and correction and verification of any action items raised as a result of the review act as gating items for the next step in the system implementation process.

SUBSYSTEM TEST DESIGN

The next analysis required takes place at a level immediately above that of the rack. This is the subsystem level; since subsystems are made up of a number of racks tied together to provide necessary system functions. The necessary condition to begin analysis at this level is completion and approval of the data packages, described in figure 3, for the racks necessary to make up the subsystem.

(1) Subsystem Analysis

The data items required for subsystem test requirements analysis are similar in type to those required for rack analysis. The additional data that must be developed for the subsystem will supplement that developed for the racks. The subsystem analysis will be concerned with functions that are developed thru combining rack functions and/or functions that interface with other subsystems. In addition subsystem signals that serve as a prime system function will be of interest.

Each of the subsystem output signals should be described in a performance specification containing truth tables, nominal voltage levels and tolerances as required. In cases where a rack output and subsystem output are identical it is only necessary to transfer the necessary documentation from the rack grouping to the subsystem grouping.

As in figure 3 a block diagram should be made to aid subsystem analysis and the review function. This block diagram should describe the manner in which rack signals are interrelated to obtain the subsystem output signals. The inter-rack block diagram serves as a guide in determining the testing sequence required for the subsystem or for fault isolation to a rack within a subsystem. In addition the block diagram will help identify the signals which must be monitored at the system level.

Schematics and prints to help describe wave shapes and/or support monitoring concepts will again be necessary on the subsystem level.

A theory of operation describing the manner in which the subsystem interfaces with the system is required. This description should include the way in which the subsystem signals affect the operation of the system and the way in which the system modes effect the subsystem signal output. The interrelationship between subsystems must be clearly described as it is important in creating a monitoring sequence.

The failure mode analysis for the subsystem signals must describe the manner in which subsystem signal failures manifest themselves at the system level. The causal relationships described in the failure mode analysis will prove very important in establishing the branching requirements of a monitoring sequence.

Interconnection diagrams describing cable plug and pin connections between the subsystem and other system elements are required for test point placement analysis and to establish detailed repair procedures.

(2) Determining Test Requirements

The signals identified in step 1 of figure 4 as being representative of the subsystem and therefore requiring monitoring must be separated and classified as was necessary in figure 3. Again those signals continuously available will be separated from the signals which are available only during given specific periods. Within these two categories the signals will be classified as to their requirement for stimulation, mode dependency or relationship to some option or controlled at an operator's discretion.

Signals having synchronous requirements with the system timing to achieve monitoring must again be classified by the signals they are synchronous with and identifiable by.

The documentation package for the subsystem should also include a rack linkage diagram showing signal traceability between racks until the signal appears as a subsystem output.

Finally system operator options and their effect on subsystem signals must be described as necessary to make the documentation package complete.

(3) Establish Test Continuity to Rack Level

The final requirement for subsystem documentation is to establish testing continuity on the subsystem level. This means that the order in which subsystem tests are performed must be documented. Normally there will be a number of sequences each representing the monitoring of the various signal classifications described earlier. Each of these test sequences will show the order of testing as well as the branching required if a test failure occurs. That is; should a subsystem test fail it will lead to a basic function test. The continuity between rack and basic function tests will already have been documented as described in figure 3 step 3.

In addition, test sequences and branching documentation should also be provided to establish continuity between subsystem operation. This means the action of operators and their ability to effect monitoring sequences must be considered.

In addition displays may suggest special sequences if they provide additional information which can make monitoring or fault isolation more efficient.

(4) Subsystem Review

A review should be held to evaluate the documentation for the subsystem. The participants remain the same as do the purposes of this review. At least one week is required to allow the reviewers to familiarize themselves with the documentation and to avail themselves of supplementary material and record suggestions or comments. Each reviewer will examine the material from the viewpoint provided by his area of expertise and often will be able to provide help in problem areas which have been indicated by the documentation. It is important that the review be recognized as a time for providing help as well as a time for monitoring the quality of the output.

Approval by the review team or clearance of action items resulting from this review are gating items for system implementation.

CONCLUSION

Completion of the steps described in figure 4 represents completion of a cycle. The system has been broken down to its basic element, the function, and has had testing requirements implemented at this lowest level. The correctness of implementation was verified thru review by experts in a number of cognizant disciplines allowing continuation on to the next system level on the rack. Testing is then considered at the rack level and the correctness of test implementation for the rack is again reviewed and verified. This leads to analysis and implementation at the subsystem level. Successful completion of the review at the subsystem level brings about analysis and verification that design goals and monitoring objectives are being met for the system. Establishment of those requirements has been described in Figure 1 and thus TRA documentation required to support the automatic test and monitoring implementation process is complete, compatable and consistent; possessing traceability through application of standardized documentation and management control practices.

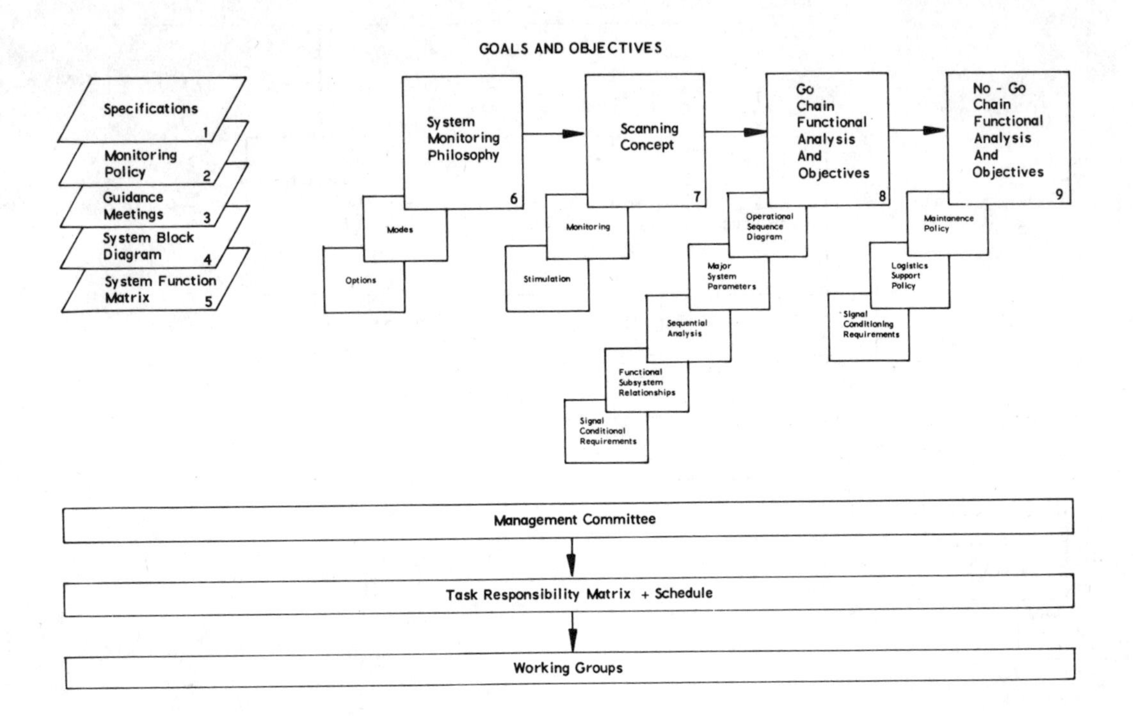

Figure 1

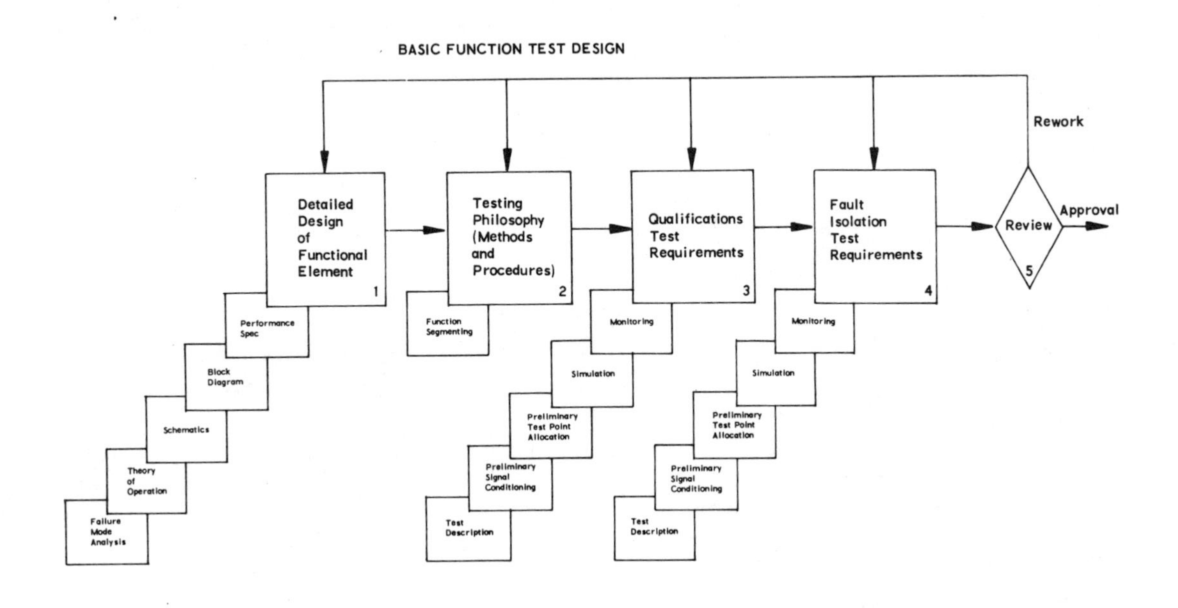

Figure 2

RACK TEST DESIGN

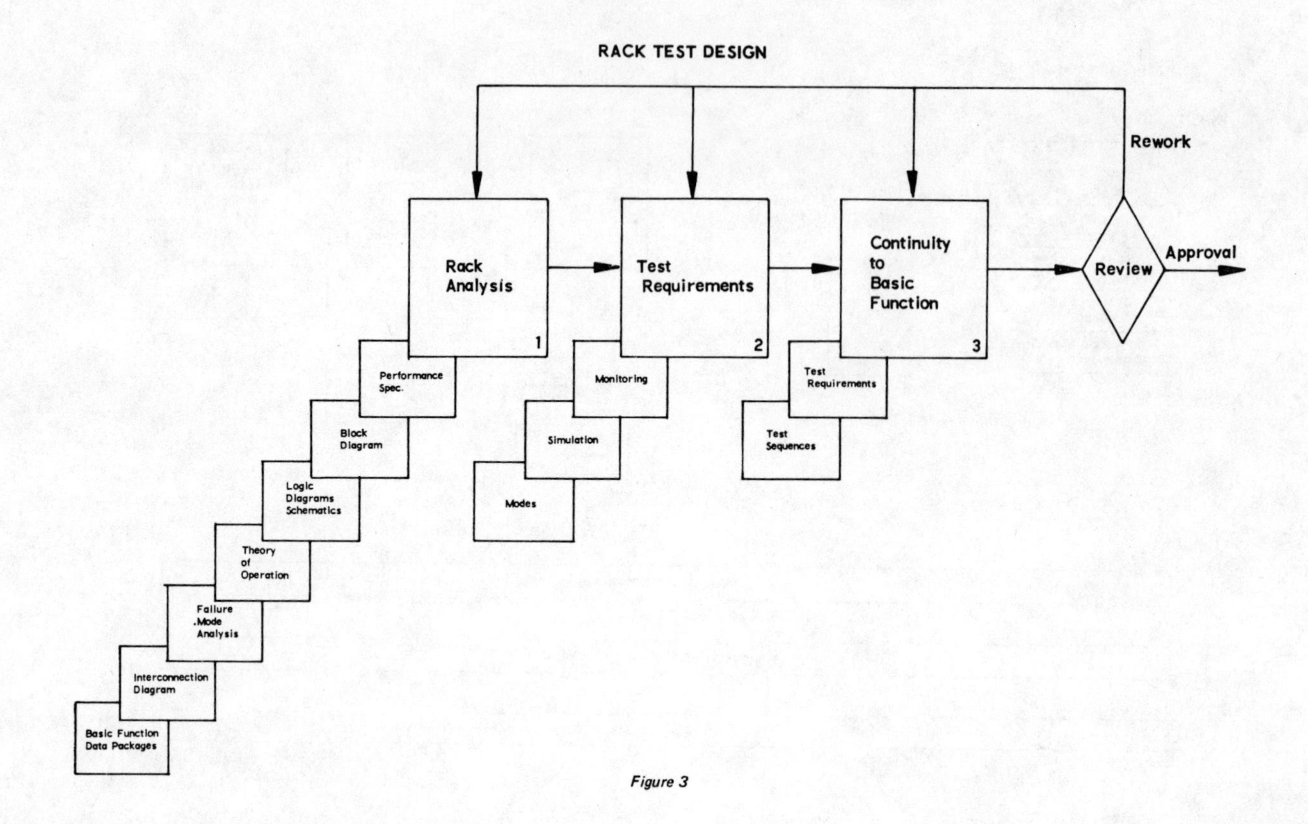

Figure 3

SUBSYSTEM TEST DESIGN

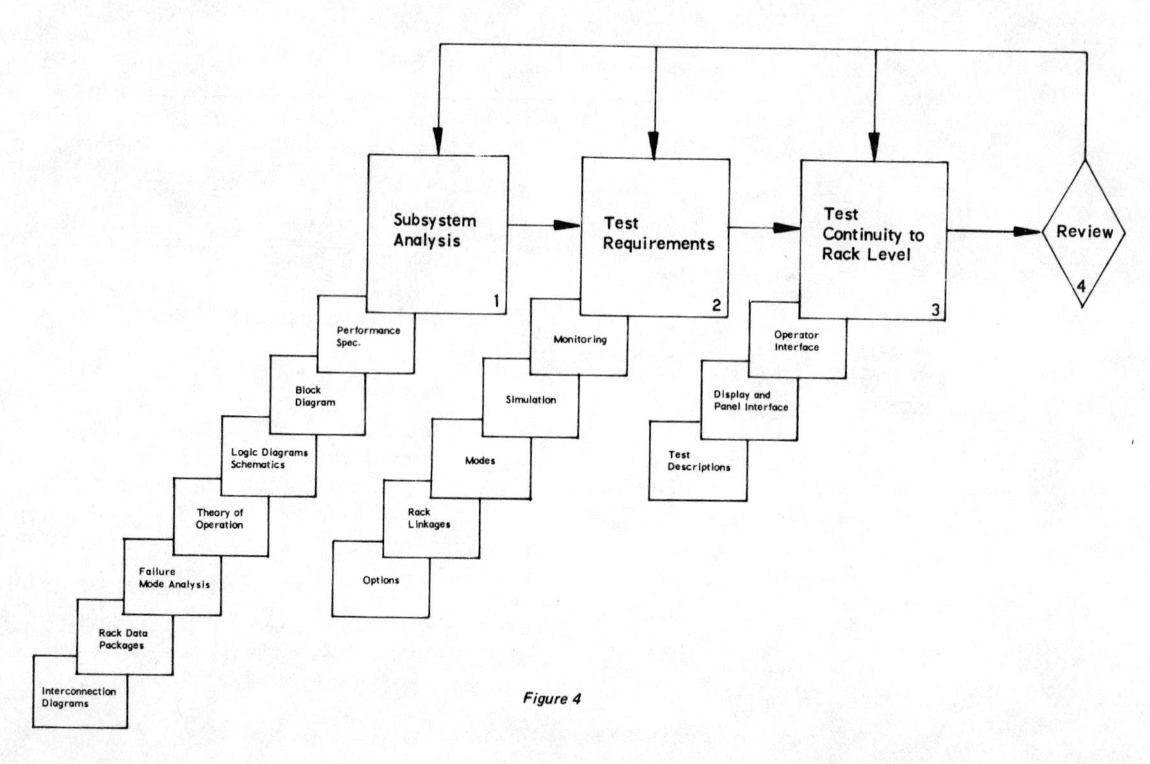

Figure 4

ATE TEST DESIGNER: PROGRAMMER OR SYSTEM ENGINEER?

by

F. Liguori, Sr. Project Member
Radio Corporation of America
Burlington, Massachusetts

Designing automatic test equipment calls for some fairly sophisticated equipment designers to conceive and develop systems which include nearly all aspects of the electronics field from power supplies to microwave and digital computer technology. Nevertheless the types of talent required are fairly well definable and clearly fall in the category of electronic design engineering. Defining the role and qualifications of individuals responsible for preparing test programs for such systems is not so clear cut.

During the experimental phases of ATE, it was found easier to teach engineers the rudiments of programming than to train programmer/mathematicians in the principles of test engineering. Hence it has been customary to use engineers in the task of ATE programming. However, the skills and work habits which the test designer should have are not widely found.

The test designer-programmer must be a multidisciplined individual who can bridge test requirements and ATE capability with a test procedure composed of engineering ingenuity and attention to detail.

For an insight into the technical requirements of a test design programmer, it is appropriate to look at the nature of his assignments. For each aspect of his assignments there is a corresponding need for aptitude, education and experience. Salient technical requirements and desirable personal characteristics of the ATE test designer are developed in this paper.

Introduction

The process involved in generating and proving out programs for testing units on an Automatic Test Equipment (ATE) can be defined by a functional block diagram containing any number of "functions", depending on the level of detail desired. Basically the process reduces to four major functional areas:

1. Test requirements analysis
2. Test programming and interface design
3. Program production and verification
4. Program validation on the ATE.

These four functions of the test design and programming process represent the bulk of the effort and cost in the preparation of programs for units under test (UUTs). The functions are generally performed in the sequence given; how ever it is not always clear where one operation begins and another ends. The validation task, for example, can require substantial rework involving the other three functions.

The individual responsible for these tasks is variously called a programmer, test engineer, test designer, systems engineer, and so on. Each title is apt but none is exactly correct. He is all of these and perhaps a little more. A brief look at his assigned tasks should help define his role if not his title.

Test Requirements Analysis (TRA)

This function might also be called preliminary test design because, generally, a test philosophy and plan are generated concurrently with the analysis of the test requirements for the UUT. The test plan is usually based on a particular specification, most commonly a design or test specification. For UUTs with substantial field history, test specifications generally exist. For newer equipment, the design specification may be the only authoritative source for test data. Generally such a document contains a section on test requirements and/or maintainability. Sometimes field and depot maintenance manuals are provided for basic information but these are never adequate in themselves for test requirements analysis and planning.

The first task the test designer-programmer faces is that of interpreting and evaluating a wide variety of documentation to extract the important test requirements for a UUT he may never even have seen. To extract essential test criteria from documents prepared by many people for many different objectives (none of which is ATE-oriented) requires a man of many talents. He must be the classical expert who must be allowed to judiciously "break the rules". What makes his task particularly demanding is that each UUT may

Reprinted from *IEEE Auto. Support Systems Symp. Rec.*, Nov. 1967.

involve technology quite different from his major area of experience. Thus, one UUT may be an audio amplifier, another a radar IF strip, and still another a digital control network. UUT complexity may vary from a disposable module to a complete transceiver requiring fault isolation to the replaceable piece part.

From this one can conclude that the test designer-programmer must be a person of broad technical background. But how diversified can one be and still be knowledgeable enough to make detailed technical judgments? Rather than expecting detailed knowledge of many technical areas it is more practical to seek a mature individual who can ferret out information and quickly assimilate knowledge in a technology not already familiar to him.

A Person of Broad Technical Background

The test requirements analysis evolves into a plan for testing the UUT in accordance with its performance criteria and a philosophy of maintenance, hopefully well defined by other documents. Unfortunately, no matter how well-documented the maintenance philosophies, there are necessarily a number of grey areas requiring sound engineering judgment. Judgment implies people and people imply subjective and often divergent decisions. To avoid this, the test design-programmer must adhere to certain additional ground rules which evolve through experience in the application of ATE to units and procedures never intended to cope with the special problem of automation. Hence, while the test programmer must be resourceful and must exercise judgment, he must not be a "loner" who neither communicates with the group nor adheres to rules based on another's design. This is a talent in great demand, certainly not limited to ATE technology.

In Great Demand

As the test design plan is formulated it must be given sanction just as any sound design, through a design review (or two, or three). At this point the test designer has already dealt with many people of divergent disciplines, including the customer, in his efforts of gathering information during the TRA. But now he must face some formidable opponents: ATE hardware designers, customer representatives and a host of other experts on the "battlefield" of the design review. Up to now he has simply been a somewhat annoying information seeker. Now he must present a logical design approach and "sell" many interests. He must defend the design even where specifications must be compromised or hardware designs modified. In the tradeoff between sound test programs and conflicting specifications, schedules and budgets, the test designer-programmer is often discriminated against since he is last in the long chain of events leading to a complete tactical equipment with maintenance support.

Test Programming and Interface Design

Upon completing the design review cycle, the test designer returns to the quiet, even-paced function of test programming and interface unit design. Having established a test approach and resolved the most serious specification conflicts, he must formalize the test flow chart. Each test must now be fully defined in terms of stimuli inputs, routing, UUT operating conditions, signal conditioning, parameter measurements and decision criteria for automatic flow. Here the test designer must complement his basic engineering skill with programming logic and good documentation habits. It is not sufficient that the tests follow logical sequences; the consequences of each alternative in the logic flow must be correctly predicted and accounted for. This is equally true whether programming in machine language, in assembly language, or in a problem-oriented compiler language. Now that compilers are widely used in program generation much of the tedious bookkeeping work is handled automatically, but the need for precision remains.

.. I don't CARE!
I say, it's the
PROGRAM

Must Face Some Formidable Opponents

Often Discriminated Against

Sometimes Confused

The outcome of the test designer's efforts in this phase of programming is a problem-oriented, pseudo-English language flow chart. This flow chart is the product of a systems analysis of the UUT, a thorough familiarity with the test system capabilities, programming rules and sometimes confused test philosophy. Perseverance and logical consistency are perhaps the most important attributes to add to the test designers' characteristics in order to achieve a good test flow diagram.

Concurrently with the flow chart design, a design must be generated for interfacing the UUT with the ATE, so that testing can be carried out without adversely affecting UUT performance or requiring excessive manual interventions (or program interrupts). The UUT designed to be tested automatically has yet to be found, so interfacing is no simple task. This is truly a systems engineering problem when all interfaces are considered. Hence the test designer-programmer must also be a competent systems engineer, capable of overcoming seemingly insurmountable obstacles.

Capable of Overcoming
Seemingly Insurmountable Obstacles

Program Production and Verification

Once the test process has been defined by a detailed flow chart and interface diagram, a program must be generated for operation on the ATE. The coding task can be handled by engineering aids who translate the flow chart information into series of statements written on coding forms. The extent of knowledge required is dependent on the "software aids" available to translate and/or compile programs written in problem language. In any case, the test designer-programmer must monitor the coder's work to ensure faithful translation. This requires close attention to detail, for the test designer is ultimately responsible for all aspects of the program.

Requires Close Attention to Detail

Program Validation on the ATE

All of the steps involved in producing a test program ultimately lead to trying the program on the ATE with the UUT connected. This validation effort is undoubtedly the most exciting aspect of the test designer's job. Up to this point the task has involved paperwork alone. Now the test programmer operates the ATE system for substantial periods of time. There is a great deal of pressure because of the great demand for ATE system operating time. The test designer must remain calm despite the schedule pressures, interruptions for demonstration, equipment failures, UUT problems, interface adapter problems and software aid debugging. Again the systems aspect of the job prevails in that the test programmer must resolve problems among the many possible trouble areas. And, of course, his own program contributes to the problems for he is not infallible, either.

He is Not Infallible

Finally, the test designer must be a capable technical writer who practices good writing habits throughout the test programming process. Clear, precise, writing is essential in all engineering endeavors but especially so in test programming, for there the product line is paperwork. Although the test designer is actively involved with hardware when analyzing a UUT, operating the ATE and designing interface devices, his major product is a program recorded in a test design document. It is not sufficient that the program perform well. Clear and complete documentation is required to allow others to continue his efforts, evaluate or modify the program. A good test design document cannot be hastily written at the end of the project. The test designer must make writing an everyday habit so that it becomes a natural means for guiding his daily work and communicating results to others. Thus he will develop his test design document through each phase of the test design and programming process until it becomes a finished and tested product like the validated program.

The Product Line is Paperwork

In summary, the test designer-programmer can be considered basically a systems-oriented test engineer, with some programming experience and technical writing ability. He must also have certain personal characteristics such as self-direction, persistence, attention to detail, adaptability to new problem situations and above all a capability for achieving some admirable accomplishments under severe handicaps. Many engineers cannot take this type of work for long, whereas others thrive on it. Test programming is a very real engineering task even though it is practiced by a select cult. To accomplish its tasks effectively this group must first consist of competent engineers and must enjoy the full support of managers who appreciate the importance of the test design and programming task.

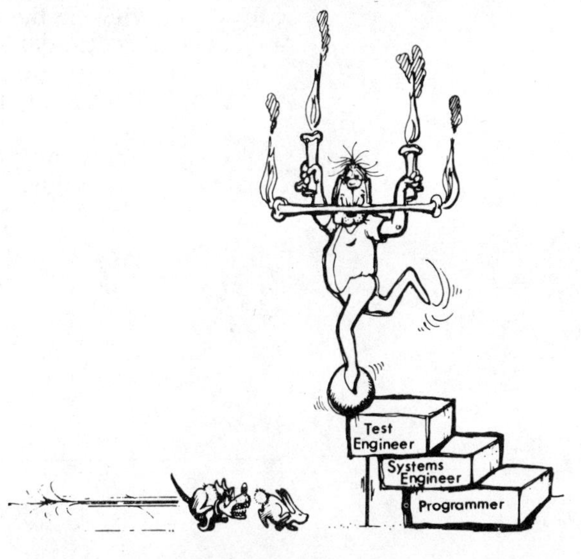

A Capability for
Achieving Some Admirable Accomplishments

A Select Cult

Acknowledgment

The author wishes to express special appreciation to the following for assistance in the preparation of this paper.

- Members of the RCA technical staff involved in reviewing the material, particularly members of Section 326.
- J.D. McCready for the artwork.
- L. Tritter for editorial services and constructive criticism.

PART II: Test System Concepts and Design

To adequately cover the subject of automatic test system design would require several volumes dedicated to the topic. ATE spans the spectrum of electronics from direct current to microwaves and even includes fields such as infrared technology, hydraulics, pneumatics, and prime movers. Each of these fields has developed its own specialized technologies and is fairly well documented in the literature. ATE simply adopts elements of all pertinent technologies to a given testing mission. By introducing a programmable control system (consisting of either a specially designed logic network or an adopted computer) into the test system, some measure of automation is achieved. Except for recent developments implemented in the so-called third-generation ATE, typical systems have made little use of the control system except to interpret commands from the stored program that operates the system in the automatic mode. It is the automation of the testing process that makes ATE different from conventional manual testing technology. This sounds like a small difference but it has significant impact. The purpose of the papers included in this section is to introduce the concept of ATE and emphasize the differences introduced by automation. It is intended to provide user information rather than hardware design guidelines.

The first paper, "A Look at Automatic Testing," is a survey article which introduces anyone unfamiliar with ATE to some basic concepts and typical systems. The paper also explains some of the economic considerations of concern if ATE is to prove cost effective for a given application.

The second paper, entitled "An Automatic Test System Utilizing Standard Instruments and an Abbreviated English Language," gives a more detailed description of a typical commercial ATE. It describes a system comprised of Hewlett-Packard 9500 system building blocks, a product line that has enjoyed substantial application in a variety of electronic testing missions. It is representative of the building block concept which so far has dominated ATE system designs. The concept utilizes stimulus and measurement instruments of standard design, together with a computer and routing system to achieve a system design tailored to a specific need. The concept has proven successful in many applications. It does, however, mislead most first users into believing that a system is equal to the sum of its parts. Experience shows that once a collection of instruments is integrated into a system, the system takes on a new character with different overall performance characteristics. Once this is recognized, and an in-depth evaluation and documentation of the system capability at its interface to the outside world is completed, one may begin to program the system effectively. As fundamental as this sounds, few of the first users learned this requirement before initiating lengthy futile attempts to apply the system before its performance capability was fully defined.

The ultimate requirement to make a system out of a collection of integrated building blocks is the development of the test programs that must control the system. The second paper also introduces some fundamentals of test design and programming. That topic is developed more fully in later sections of this book.

The third paper, "Automatic Test Systems Dedicated or Integrated," questions the soundness of the general-purpose building-block approach to ATE design so often taken for granted in early ATE systems. It points out that a dedicated system (one designed from more basic elements than "off-the-shelf" building blocks) can achieve certain advantages providing one has a well-defined mission for the tester. In fact, experience has borne out that ATE design, like anything else, is a compromise situation where advantages and disadvantages of a design approach are relative to the needs of the user. Some of the advantages and disadvantages of integrated (building-block) and dedicated (specially designed) systems approaches are discussed.

The fourth paper, though simple in its title, "Automatic Test Equipment Design," takes a giant step in the more effective use of a computer in ATE. Most ATE systems, until very recently, have relegated the computer to the mundane task of interpreting building-block control commands. The test programs for such systems tend to automate manual test procedures rather than taking a fresh look at how the computer could really help in the testing function. The system described in the paper utilizes the computer as an integral part of the measurement function. By implementing sampled-measurement theory, all measurement functions are reduced to a series of fundamental instantaneous measurements of amplitude and frequency. This approach greatly reduces the hardware requirements over the more conventional method of using a collection of stand-alone measurement devices. The equations and theory are not new. What is new is the practical implementation of these equations which have long been taught in engineering school. Until recently they were only of theoretical value because the instrumentation and computation circuitry to utilize such concepts was not generally developed. This approach of using the computer as an integral part of the measurement function represents a major breakthrough in the art of ATE design. It is destined to supplant many of the existing instrument design techniques.

The fifth paper, entitled "EQUATE—New Concepts in Automatic Testing," describes a system typical of the third generation variety. In addition to using the sample measurement technique for its measurement subystem, it also utilizes the computer as an integral element of the stimulus generation subsystem. In this approach, complex waveforms are synthesized by generating a series of narrow pulses of varying amplitude which are then converted to analog form and filtered to produce continuous waveforms of the programmed variety. This process, very similar to the mathematical integration function, can also be considered to be the reverse of sample measurement techniques. As with the previous paper, the techniques are classically mathematical but now reduced to practical implementation by using basic pulse generators, filters, and the power of the computer. Unlike the previous paper, this one does not develop the theory of equations involved. No suitable paper was found to cover that aspect of stimulus generation. The implementations to date are few and very new. New literature should be forthcoming shortly on the concept.

AUTOMATIC circuit-testing system for digital integrated circuits. (Courtesy Teradyne)

An IEEE SPECTRUM applications report

A look at automatic testing

Automatic testing of electronic devices has been a major factor not only in the overall improvement of product quality and reliability, but also in the dramatic lowering of product costs

Harold T. McAleer General Radio Company

This survey reviews the current state of automation in the testing of electronic components, networks, and circuits. Elements and characteristics of typical test systems, both hardware-controlled and computer-controlled, are described. Paths to be followed and pitfalls to be avoided in achieving automation are discussed in an effort to help the reader toward a better understanding of the subject and its broad applications. ''How to automate successfully,'' a major theme in the report, places emphasis on economic justification.

Reprinted from *IEEE Spectrum*, May 1971.

The need for and desirability of automatic testing have been well established and require little justification. The testing function itself, which used to be considered somewhat of a poor relation, a necessary evil, has gradually emerged as a well-recognized and essential part of any manufacturing operation—particularly in electronic manufacturing. This testing function appears in many areas and in many different forms. It may be called incoming inspection, line inspection, final test, or quality control, and it may involve sampling tests or 100 percent inspection. From the standpoint of product quality, 100 percent inspection at every recognizable step in production would be ideal. This ideal was seldom possible because of the high costs involved, both in time and money. As the inspection process becomes automated, however, 100 percent inspection becomes economically feasible and an attainable goal, especially when one realizes how rapidly the cost of repair rises as a faulty item gets buried in the final product.

Experience with transistors offers an example. It has been calculated[1] that the cost of *not* subjecting transistors to 100 percent testing as received from the manufacturer, but rather sample testing them to an acceptable quality level (AQL) of 0.65 percent, can amount to 8½ cents per transistor purchased. For a user of a million transistors per year, this amounts to an annual *loss* of $87 500!

Integrated circuits provide another example. Back in 1969 when the price of integrated circuits varied from 85 cents to $3.50 each, depending on complexity, the typical failure rate, as received from the IC manufacturer, was around 1.5 percent. Now that the price is down to 20–85 cents each for the same units, the failure rate is up to 2–3 percent.* With an average IC count of 20 to 30 per board and an average time to diagnose and repair of 20 to 30 minutes, we can see the importance of culling out bad units as early in the manufacturing process as possible.

Since electronic equipment is used more and more in products involving safety considerations—spaceships, aircraft, automobiles, medical instrumentation—and the cost of failures may be measured not only in dollars but in human lives, the need for thorough testing (as well as sound design) becomes even more important.

Levels of testing

In general, five distinct levels of testing can be recognized, one in the engineering-design phase of a product, three in the manufacturing phase, and one in the postshipment phase. In the design phase, testing is required not only to prove out or modify the design, but also to evaluate and select the materials, components, or techniques used in the product. In the manufacturing phase, testing is required at the input to the production process to qualify the materials used in the product (incoming inspection). It is also required during the process to assure that all the steps have been successful (line inspection) and at the end of the process to qualify the final product (final inspection). In the postshipment phase, (service) testing is required, to establish that operation is proper or to diagnose failures and aid in repair.

* Typical user experience. However, for a premium of 5–10 cents each the customer can buy units with a guaranteed failure rate of less than 0.1 percent; i.e., the IC manufacturer will test them more thoroughly.

Different test information is required at these different levels and different types of test equipment are often used. Engineering and incoming-inspection tests may require the acquisition and analysis of detailed measurement data, whereas line-inspection and final-inspection tests are more concerned with "go" or "no-go" decisions to maximize throughput.[2] Testing and test equipment used during the production process is sometimes separated into two types—equipment and techniques for the rapid sorting of items into "good" and "bad" categories and equipment and techniques for the slower diagnosis and repair of the bad items. Diagnosis and repair equipment is also used in the servicing of delivered products.

Distinctions between laboratory, production, and servicing equipment are disappearing, however, as design engineers are required to design and specify not only the product but also the test equipment and techniques.

Measuring systems

The block diagram of Fig. 1 shows the variety of items or functions to be found in any measuring system, manual or automatic. The major functions are within the blocks, and the words around the blocks show examples of hardware items that accomplish the functions. Although not perfect, the diagram provides a useful basis for discussion. The tinted blocks show the essential items: a device to be tested, a stimulus source to provide a signal for the device, a measurement instrument to quantify the response, and an operator to make the whole thing go.

The other blocks show the functions that can be added to ease the operator's task and automate the process. As equipment for these functions is added, the operator acts less as a mechanical part of the measurement process and can devote his attention to the results rather than the details of the measurement. If the results are used for automatic control of other equipment operating on the device under test, a process-control system can be achieved.

Device under test. It is useful to categorize the device under test according to the hierarchy shown in Fig. 1. Thus, a component is made out of materials, a network is made out of components, a circuit is made out of networks, etc. The categories can be further subdivided: components and networks can be linear or nonlinear, active or passive; circuits can be analog or digital, discrete or integrated. The hierarchy can be shown to be imperfect (for example, is a transformer a component, a network, or a circuit?) but in general any electronic device can be fitted into one of the categories.

Adaptor. The adaptor block is used to define items that interface the device under test into the measurement system. This category includes test boards, fixtures, or sockets that contact the leads or terminals of a component or network, or a multipoint prober that contacts the nodes of a microcircuit.

Input. Input equipment facilitates the insertion of the device under test into the adaptor. Examples are vibratory-bowl component feeders, rotary and linear transports, and other mechanical devices. Such items are usually associated with a sorting or binning device to deposit the tested unit into a useful location.

Condition. Conditioning equipment applies to the device under test a secondary stimulus, such as power,

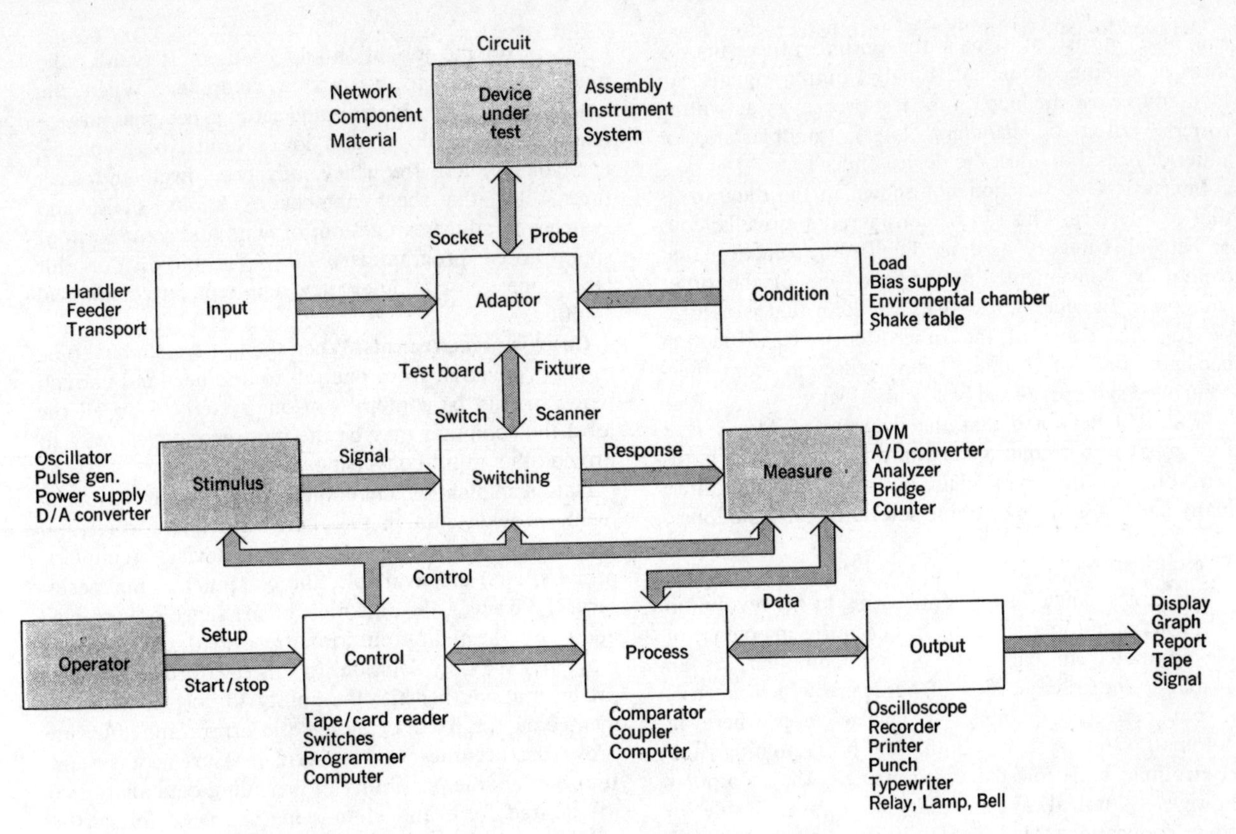

FIGURE 1. Functional block diagram of a manual or automatic measuring system.

bias, heat, cold, or shock, or sets up terminal load conditions.

Switching. Switching equipment is used to connect the device or device adaptor to the test system and to vary the connections of the device terminals. Examples are solid-state scanners (multiplexers), relay trees, crossbar scanners, and reed-relay scanners.

Stimulus. Stimulus sources for measurement systems are legion, and include dc supplies, oscillators, synthesizers, function generators, ramp, pulse, burst, and word generators, and D/A converters, among others. For use in automatic test systems such sources are often required to be programmable; that is, all their functions should be controllable by electric signals instead of (or in addition to) manual controls.

Measure. Measuring instruments are likewise legion; they include voltmeters, current meters, phase meters, impedance bridges, frequency counters, A/D converters, etc. These instruments should also be programmable. Data output in digital form is a useful, but not always necessary, characteristic of such instruments. The signal and measure functions are often combined in a single instrument, as for example in an automatic capacitance bridge or a digital ohmmeter.

Control. The control function can be accomplished by a variety of equipment in a variety of ways. Card readers and tape readers are often used to program and sequence automatic equipment with the actual program determined by the holes in the cards or tape. Patchboards, plug boards, and switches can be used as "memories" to set up test conditions and sequences. Cam-operated motor-driven switches are used as well as relay networks. Specialized digital-logic circuits are often applied to this function. The most versatile programmer/controller of all, the computer, will be discussed in a later section.

Process. Data from the measurement instruments must often be "massaged" or processed in some way. One example of processing is limit comparison. The comparator may be an analog or digital circuit. It compares the measured data with preset or programmed limits and determines whether the data are above, below, or within the limits and provides appropriate output signals (lamp illumination, contact closures, digital-logic signals, etc.). Another form of processing is coupling. The coupler may store and convert parallel measurement data (all digits at once) to serial data (one digit at a time) to operate card punches or tape punches or strip printers, or it may perform the opposite function (serial-to-parallel conversion) to operate gang punches or line printers or displays. Yet another form of processing is computation or data processing. The processor may perform arithmetic computations on the data to provide corrections or conversions or to somehow convert the data to more useful form.

Output. The output function of an automatic test system is also performed by a variety of equipment and appears in a variety of ways. One typical output function is that of display. Measurement results or test conditions may be displayed by means of oscilloscopes, panel meters, indicator lamps, digital indicators (visual display), or by ringing a bell (auditory display). Hard-copy output can be provided in the form of graphs, printed lists, or typed reports by means of recorders, printers, or typewriters. Output can consist of condensed data in the form of punched cards, punched tape, or magnetic tape for off-line processing by other equipment. Output can also consist of signals for on-line processing by local or remote equipment. Finally, output signals can be used to close the loop on the test process by activating sorters, automatic handlers, lasers, sandblasters, or other devices that affect the device under test.

Interface. One function not shown in the diagram is that of interface. This can be considered a miscellaneous or catchall category used to dignify any function that cannot be conveniently forced into one of the other categories. In general, it defines any item that is used to tie together some of the other items. This function becomes one of extreme importance in computer-controlled systems, as will be discussed later.

It should be noted that the diagram of Fig. 1 is a functional rather than a hardware diagram. The hardware diagram may be similar, but often it is not, since many hardware items combine several of the functions.

The computer

The largest single factor contributing to the revolution in the art of automatic testing has been the emergence of the computer, in particular the minicomputer, as an available and reliable item of hardware with immense, if not perfect, capability. It appears everywhere in automatic test systems, both for programming and controlling and for data processing. The computer however, is not always necessary, or even desirable, in some applications. How do you know when to use one? The following paragraphs discuss some considerations.

Operational flexibility. If the job of the test system is fixed, and well defined—for example, the sorting of resistors into prescribed categories—and the test system will be used for that one specific task forever, a computer may not be required. If simple and economical hardware—timers, programmers, bridges, comparators, etc.—exists to do the job, it should be used. It won't take much, however, to warrant a computer. When the number of possible controls and settings becomes large—nominal values, upper and lower limits, bias voltage, test voltage, test frequency, advance, dwell and soak times, etc.—the sheer number of knobs, dials, and switches to adjust can get out of hand and some form of sophisticated programmer will be required. For this task alone, the computer may represent an economical solution.

Changing requirements. When the test system has to be rapidly converted from one job to another (and enough hardware can be combined in one system to do all the jobs) the computer may be the most economical way to provide this rapid conversion.

Data manipulation and output. If the measuring instruments provide data in the right form and format and processing equipment and output devices (couplers, printers, etc.) are available, the computer is not necessary. If, however, the data must be massaged or corrected and a complicated output format provided, the computer may offer the only solution. In this regard the computer provides a side benefit: the ability to correct measurement data for fixed or systematic errors and thus improve the accuracy of the basic measurement system. It also offers the possibility of providing data analysis in the desired form and eliminating the need for further off-line processing.

Computer functions. The computer can perform two of the basic functions shown in Fig. 1, the control and process functions. It can perform the control function by means of a program stored in its memory and the process function by means of its central processor. To make use of these capabilities, however, two other elements are required: an interface system and software.[3]

FIGURE 2. Simplified block diagram of a typical computer-controlled component-test system emphasizing the role of the interface system.

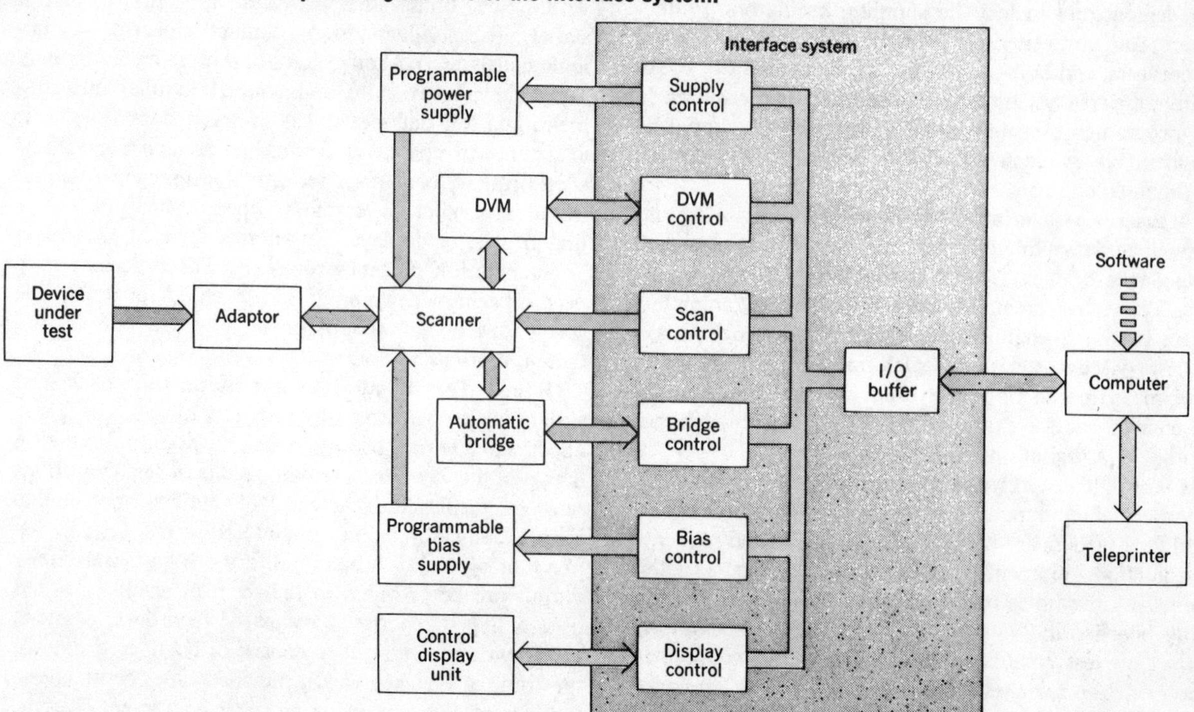

Interface system. Figure 2 is a simplified block diagram of a typical computer-controlled test system drawn purposely to emphasize the role of the interface system. The test system includes several programmable elements to perform the functions of Fig. 1. The interface system is required to wed these elements to the computer, since at this stage of the art neither computers nor test instruments are directly compatible.[4] Cost considerations in the design of general-purpose test instruments do not allow the inclusion of expensive interface circuits. As we grope our way toward standardization, however, test instruments are becoming available that make the interfacing task, if not unnecessary, at least easier. Similarly, the design of computers cannot anticipate all the possible interface requirements. In general, the interface circuits perform the functions of I/O (input/output) bus expansion, level conversion, decoding, storage, transfer, and control. By storing data signals, both from and to the computer, the interface system allows the (fast) computer to keep busy while the (slow) test instruments are doing their job.[5] When the test instruments are ready, either to receive instructions or give up their data, their interface control circuits can provide an "interrupt" signal and an identifying signal (flag) to the computer to request its attention.

Some minicomputers make provision for plug-in interface-circuit cards within their packages; others require an external unit. General Radio uses its 1761 Interface System in computer-controlled test systems. Hewlett-Packard provides plug-in interface cards for its computers and offers its 6936A Multiprogrammer System, a main-frame unit and a series of plug-in interface cards for its test instruments. Digital Equipment offers card-cage hardware and digital-circuit modules that can be used to construct interface systems, and other manufacturers offer similar items.

Expanding role of the computer. As the art progresses, it is becoming increasingly difficult to separate the elements of computer, interface, stimulus source, and response measurement, as well as to decide where one function leaves off and the other begins. Many automatic test sets of the future (and some of the present) will consist of card cages or main frames to house plug-in functional modules—memories, arithmetic elements, storage registers, stimulus sources, measurement modules, and peripheral-control circuits for input, control, and output devices.

Software. To make most efficient use of the limited core memory of the typical minicomputer, operating programs are often written in assembly language. (Such programming efforts are not trivial; it can take 4 to 6 man-months to produce 4000 words worth of debugged software and can cost $10 000 to $20 000, depending on how you value a man-month.) This approach works fine if the test set will be used with a few infrequently changed programs. If not, more memory, and perhaps the use of a high-level test-set language, may be needed.

Computer languages. Much has been written and said about computer languages,[6,7] the terminology of communication between the programmer and the computer, and the level of such languages. There are no strict definitions of language level. In general, high-level languages allow the programmer to write program statements in terms that are familiar to him, such as English words or mathematical equations, and with few rules of order or format, so that he can ignore the particular characteristics of the computer used to run the program. Low-level languages, on the other hand, force the programmer to adapt his problem to the characteristics of the machine to be used. They may use mnemonics or octal numbers and are related to the operation of the computer and its subsections. The use of higher-level languages may require several intermediate steps before a final operating program is prepared or run—that is, editing, debugging, punching tapes, loading tapes, etc. Given sufficient computer capability, however, the steps can be handled automatically (with little programmer action) by the use of a sophisticated operating-system program, and programs can be prepared and run at essentially the same time.

Table I shows some common types of programming languages. For further details, see Appendix A.

Test-set languages. One also can speak of programming and operating languages for automatic test systems. Test-set languages can be high level or low level, depending on their relationship to the application (problem) or to the test set (equipment). When you set the function knob of a digital counter to FREQUENCY, you are using a high-level language, related strongly to the application but of little flexibility. Similarly, setting the power switch to OFF uses a lower-level language. This concept has given rise to a host of test-set languages, Pol (Problem-Oriented Language), Tool (Test-Oriented Operator Language), Atlas (Abbreviated Test Language for Avionics Systems), and many others that have no names but do their jobs just as well. If the test set is dedicated to a specific application (capacitor testing, transformer testing, logic-circuit testing), the language can be made high level and use terms relating to the device under test (nominal value, transformer terminals, logic-drive pattern). If the test set is for more universal use, its language should be lower level and use terms relating to the instruments within the set (function, range, connection).

The program of a dedicated test set can provide two levels of language for the users of the set, a low-level language for the programmer, which will enable him to change certain operating constants or test sequences, and a higher-level language for the operator, which will enable him to control the set for a specific test. A partic-

I. Common programming languages

Type	Example	Typical Statement	Comment
Machine	PDP-8	100011101101	Machine code, object code, run code
Assembler	Pal III	HLT	Can make optimum use of memory and minimize run time. Error prone. Expensive. Great variation between programmers. Difficult to document
Compiler	Fortran Algol	D = A + B + C	Off-line compiling. Object program requires more memory and runs slower. Less variation between programmers. Easier to document. Compiler program may be expensive
Interpreter	Basic	IF E > 2 PRINT GO	Interactive for easy changes. On-line run. Requires even more memory and runs even slower

ularly useful mode of operation with such languages is an interactive or conversational mode, wherein programming questions are automatically displayed, printed, or typed by the set and replies are typed by the operator. The main resident program—the operating system—interprets the operator's instructions and runs the test.

Typical test systems

To illustrate the concepts described earlier, the following paragraphs discuss a few examples of commercially available automatic test systems. Excellent surveys are available in trade publications and commercial reports.[8-10] The examples will be discussed in order of the device under test in the hierarchy previously described.

Component tests. Although threatened by film technology, classical discrete components (resistors, capacitors, inductors, transistors, diodes, etc.) remain with us. Their form is changing, but their function remains the same. Many automatic systems are used in the manufacture of these components—some for measurement only and some for process control.

Diode test. Figure 3 shows a hardware-controlled system for the automatic sorting of semiconductor diodes. The system includes a handler/sorter, which includes a vibratory-bowl feeder, a rotary test table with three test positions, and a five-bin sorting mechanism. The instrumentation includes three test instruments; a capacitance comparator, which compares the junction capacitance of the diode to a preset limit; a stored-charge (switching-time) detector; and a diode classifier, which tests for short and open circuits, reversed polarity, peak inverse voltage, reverse current, and forward voltage drop. The results from the three instruments are combined in the diode classifier to provide bin-sort signals to the handler. These systems can test and sort up to 10 000 diodes per hour.

Capacitor tests. Figure 4 shows an automatic system for the testing and sorting of capacitors. The input, adaptor, and switching functions are performed by a manually loadéd transport unit that grips the capacitor terminals and connects them to various soaking, measurement, and discharge busses. The transport is stepped by signals from a control unit, which also houses other interface circuits. Signals from the control unit also enable the transport to perform an output function: dropping the tested capacitors into sort bins. Programmable power supplies provide stimulus signals for the measurement of dielectric strength and leakage current, and also provide a conditioning bias voltage for capacitance measurement. An automatic bridge provides both stimulus and measure functions for the measurement of capacitance and loss, and a digital voltmeter measures leakage current. The digital output data of the automatic bridge and digital voltmeter are processed by digital limit comparators to provide sort decisions. The instruments are controlled by a program unit, which includes a card reader to interpret holes in a prepunched card to program voltages, soak time, instrument ranges, and sort limits, and an overriding series of thumbwheel switches to allow manual programming. A further output function is provided by a series of electromechanical counters, which display a running tally of the sort decisions. This system provides a good example of a hardware-controlled system for a specific inflexible purpose.

Another capacitor-testing system, shown in Fig. 5, is used for evaluation and life studies rather than physical sorting. Components are loaded on test boards, which are inserted into a test socket. A reed-relay scanner is used to connect the components to the stimulus and measurement instruments. A minicomputer and interface system control the stimulus and response instruments, process the measurement data, and provide a typewritten output report on a teletypewriter. The system software provides an initial test sequence that produces a printout of parameter values and indicates out-of-limit units, and a final test sequence that provides parameter values, deviations from initial values, and a statistical summary.

Network tests. One passive network of extreme importance is the connection network, which has the

FIGURE 3. Automatic diode-sorting system. (Courtesy Teradyne)

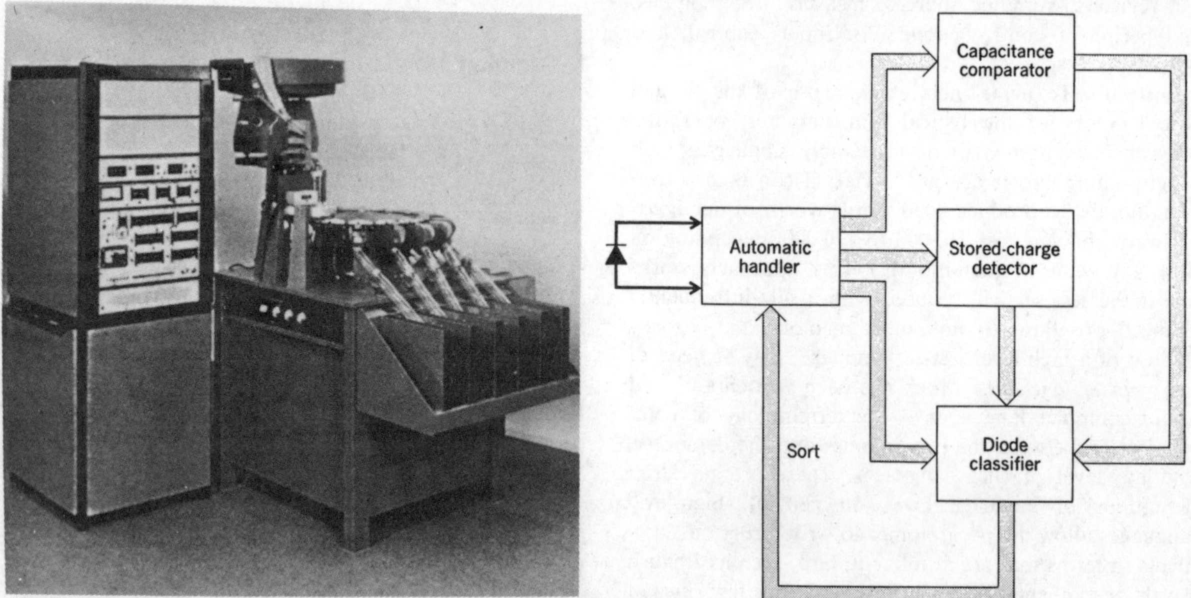

mundane task of connecting some circuit nodes together and isolating others. This type of network appears in many forms—a cable, a wiring harness, a wired backplane or motherboard, a bare printed-circuit board, or a hybrid-circuit substrate before the components are attached. These networks are perhaps the most critical elements in electronic hardware and the ones most prone to error in manufacture, and therefore considerable effort has gone into the design of automatic test equipment to check them out. The stimulus and measurement units are relatively simple, since the measurements required are usually dc continuity, dc or ac dielectric strength (hipot), and dc leakage resistance; however, the control and switching units can get quite complicated. The less expensive systems ($20 000) use paper-tape readers and bar relays and can handle networks with up to 500 nodes. The more expensive systems ($100 000) use computers and random-access mercury-reed scanners and can handle networks with up to 100 000 nodes.

Another important passive network is the multipair communications cable. Figure 6 shows two versions of a computer-controlled test set for the measurement and evaluation of the capacitance parameters of such cables.[11] (Most of the important transmission characteristics are determined by these parameters.) A functional block diagram is shown in Fig. 7. The input function is provided by a large "fanning fixture" and the switching function is performed by a special-purpose reed-relay scanner. Stimulus and measurement functions are performed by an automatic capacitance bridge and the control and process functions are provided by an interface system and a minicomputer. A teletypewriter provides the output report. The set uses a high-level language to enable an interactive dialogue with the operator to determine test limits and desired output data. Although individual measurements can be printed, in the usual mode of operation all the data are condensed into a compact statistical summary. A typical output report is shown in Fig. 8.

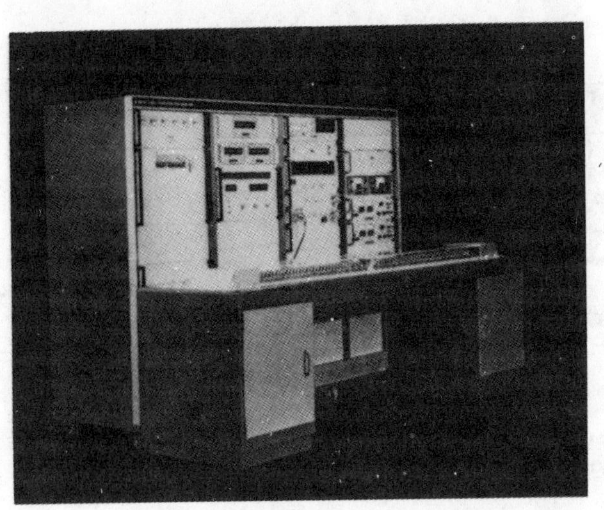

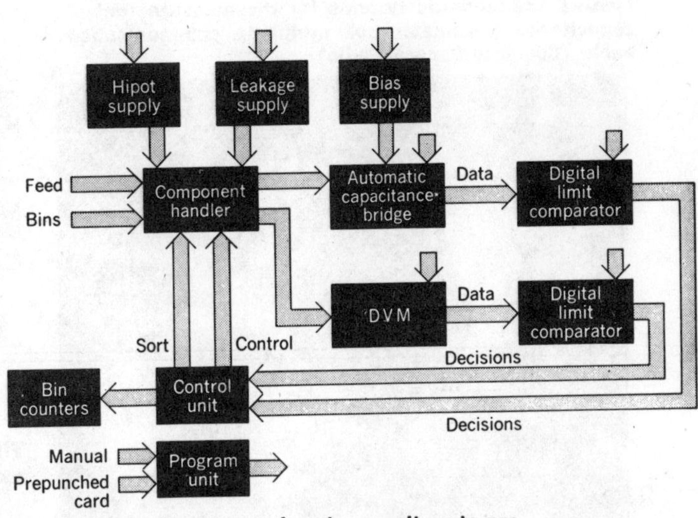

FIGURE 4. Automatic capacitor-sorting system, which measures and sorts capacitors to preprogrammed specifications. (Courtesy General Radio)

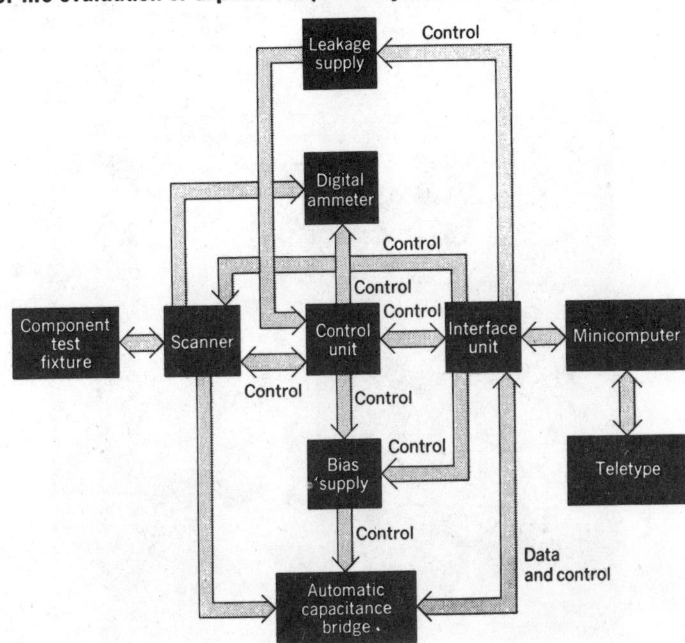

FIGURE 5. Computer-controlled system for life evaluation of capacitors. (Courtesy General Radio)

Circuit tests. Circuit testers usually fall into three categories, depending on the type of circuit to be tested—that is, digital, analog, or hybrid (combination).

Digital circuits. The explosion in the use of digital circuits, both discrete and integrated, has created an entire industry in test equipment with its own evolving terminology and jargon.[12] Digital-logic engineers speak of three basic types of testing[10]:

1. Functional testing, to check truth-table or "Boolean" operation of digital circuits by applying and sensing patterns of 1's and 0's on the pins of the circuit.
2. Parametric testing, to check dc or static parameters, such as forward and reverse currents, saturation and offset voltages, etc.
3. Dynamic testing, to check time-domain parameters, such as rise and fall times, propagation delay, etc.

FIGURE 6. Automatic systems for the measurement of capacitance parameters of multipair communication cable. (Courtesy General Radio)

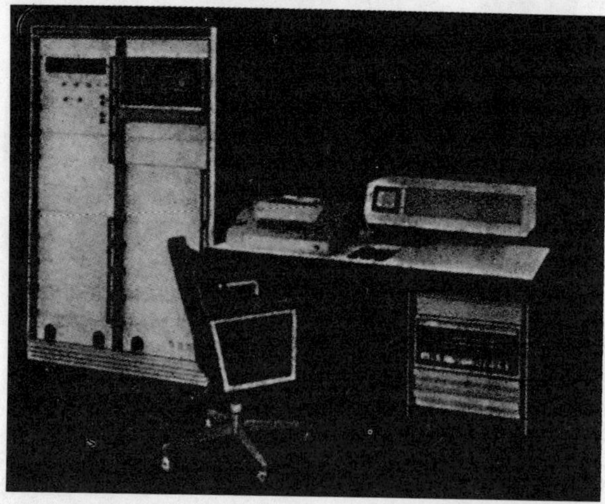

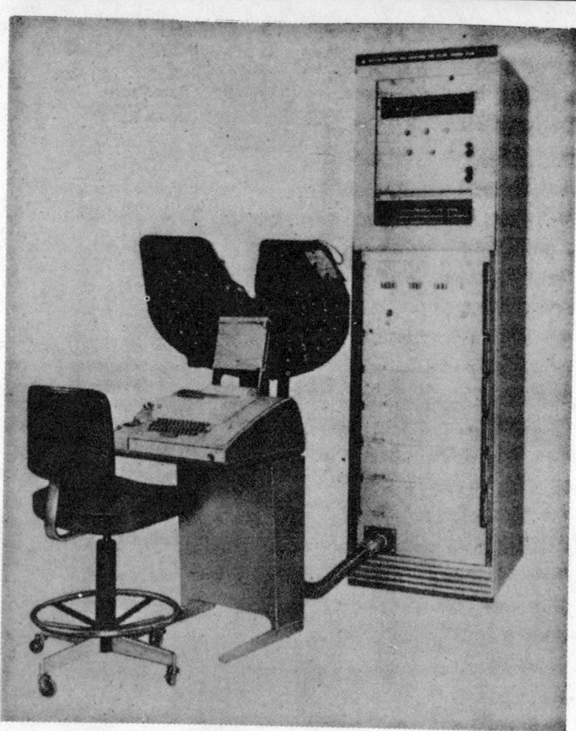

Inexpensive ($10 000) functional testers now available apply a sequence of bit patterns to the input pins of both a known good circuit and the circuit under test and compare the output patterns. Only one input signal is altered between consecutive tests to avoid indeterminate states. As one author points out, however,[13] since it takes at least 2^n tests to provide a complete input sequence for a unit with n interdependent inputs, it might take 40 000 years to test a 60-input device, even at a 2.5-MHz test rate! The answer, of course, is to bring out internal logic nodes as test points to reduce the number of interdependent points to ten or so and the test time to milliseconds. Such testers are useful in the rapid sorting of bad from good units but are of little help in diagnostic repair.

The more expensive ($20 000 to $100 000) testers use tape readers or computers to exercise a predetermined test program for both sorting and diagnostic information.

Figure 9(A) is a photograph of the General Radio 1790 Logic-Circuit Analyzer, a high-speed functional tester for digital circuits.[14] This system, shown in block-diagram

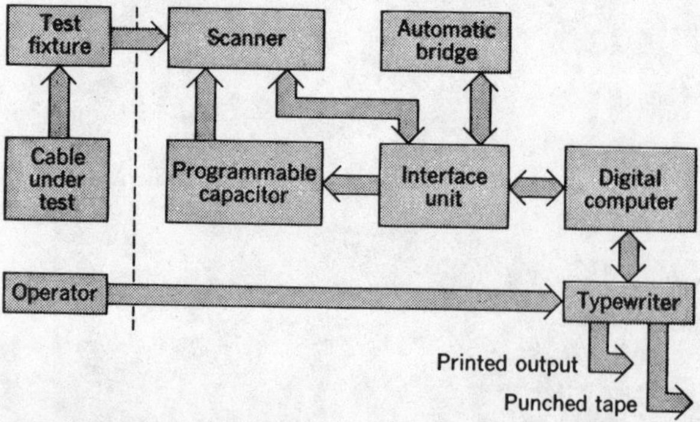

FIGURE 7. Block diagram of cable capacitance-measuring system shown in Fig. 6.

FIGURE 8. Statistical summary of results from cable capacitance-measuring system of Fig. 6.

```
IDENTIFICATION INFORMATION:
135382 ADMC-1500 L2 RD-WH UNIT #4  6/20/69

PAIRS:   100
LENGTH (FEET):   750

SELF TEST:  OK

MUTUAL
ERROR THRESHOLD:   3.020 NF
MUTUAL DISTRIBUTION - CELL NOS ARE NF/MI +/-1

 L     71    73    75    77    79    81    83    85    87    89
 0      0     0     0     0     1    13    16    29    13    19

 91    93    95    97    99   101   103   105   107   109    H
  5     2     1     1     0     0     0     0     0     0    0

NUMBER OF PAIRS TESTED:   100        COEFF. OF DEVIATION:  3.987%
AVERAGE:   85.777  NF/MI
STD. DEVIATION:   3.32  NF/MI        LIMITS:  PULP XXX  PIC XXX

UNBALANCE TO GROUND DISTRIBUTION - CELLS ARE 100 PF/1500 FT WIDE

  1    2    3    4    5    6    7    8    9    10    11    12
 21   17   17   13    8    6    8    5    3     0     2     0

 13   14   15   16   17   18   19   20
  0    0    0    0    0    0    0    0

STD. DEVIATION FROM BAL.:  430 PF/1500 FT
AVERAGE UNBAL.:   372 PF/1500 FT
PAIR 32:   +1910 PF/1500 FT   +980 PF
PAIR 47:   +1960 PF/1500 FT   +915 PF

PAIR-TO-PAIR UNBALANCE
CAP. UNBAL. LIMIT:  124.207 PF    SKIP LIMIT:   13.799 PF   (11.11%)
CAP. UNBAL. DISTRIBUTION--CELLNOS ARE CELL MAX. % OF CAP. UNBAL. LIMIT

  2.5    5    7.5   10   12.5   15   17.5    20   22.5   25
4597   135   79     38   36     18   17       9    6      6

 27.5   30   32.5   35   37.5   40   42.5    45   47.5   50   H
  2      3    2      0    2      0    0       0    0      0   0

STD. DEVIATION FROM BAL.:   3.644% of CAP. UNBAL. LIMIT
NUMBER OF TESTS:   4950    COMBINATIONS SKIPPED:   4358

SELF TEST:    OK
END OF TEST
```

form in Fig. 9(B), is an example of a test set developed by a manufacturer out of necessity for in-house use and then offered for sale after prove-in. The set includes a minicomputer for control and processing, a control panel, tape reader, display oscilloscope, and teletypewriter for operator communication and printout, a device adaptor for connecting the circuit under test, and an interface system that provides logic-level drive and sense circuitry for digital circuits with up to 240 pins.

The system communicates with the operator in four ways. The operator control panel contains control buttons and switches to control the system operation, and lamps to indicate the go/no-go results of a test. A display oscilloscope serves as alphanumeric output of test conditions and results, and is also used to display instructions to the system operator. A teletypewriter and paper-tape reader provide keyboard input, paper-tape input and output, and printed output where a permanent record of test results is desired.

A key feature of the system is the use of a high-level test-set language that allows test programs to be prepared by technicians familiar with the terms of digital-logic circuits. The language contains English-like commands, which perform functions such as the following:

1. Set specific inputs or outputs high or low.
2. Check specified outputs against output patterns specified in the program.
3. Generate program "loops" to allow selective repetition of part of a program.
4. Branch to another place in the program if specified output patterns exist.
5. Print instructions or error comments to the operator.
6. Insert specified delays in the program.

A typical program is given in Fig. 9(C). Two types of programs may be prepared: a simple truth-table program to allow quick sorting of good and bad circuits or a more complicated diagnostic program to facilitate troubleshooting and repair. Programs can be prepared either on line or off line on paper tape. An interactive interpretive mode provided ("combined interactive system") allows generation, translation, and execution of a test program on line. An "autoprogramming translator" permits the use of a known good circuit to ease program preparation by automatically recording output states.

LSI circuits. Continuous evolution in the technology of integrated circuits, particularly in the manufacture of large-scale integrated arrays, has sparked much interest in automatic test systems, as well as some controversy.[15,16] The development of the devices to be tested is running somewhat ahead of the development of the test equipment.

The title illustration shows a typical test system used in this area, the Teradyne J283 Circuit Test System, called the "slot machine" because of its application in the testing of sequential logic. The system consists of three free-standing racks or "kiosks," a teletypewriter, and one or more test decks. One rack contains a minicomputer and magnetic-tape transport. Another contains a control

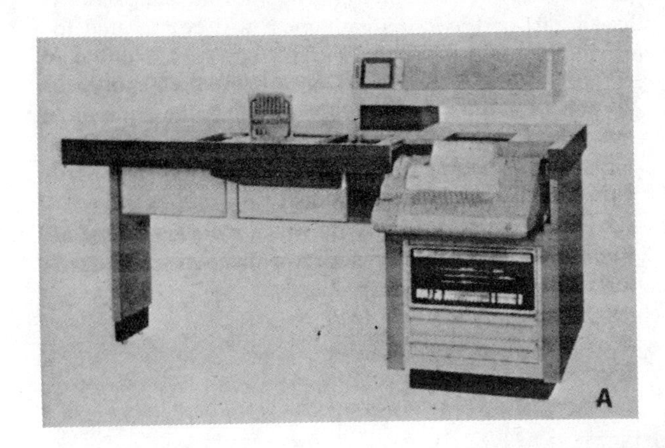

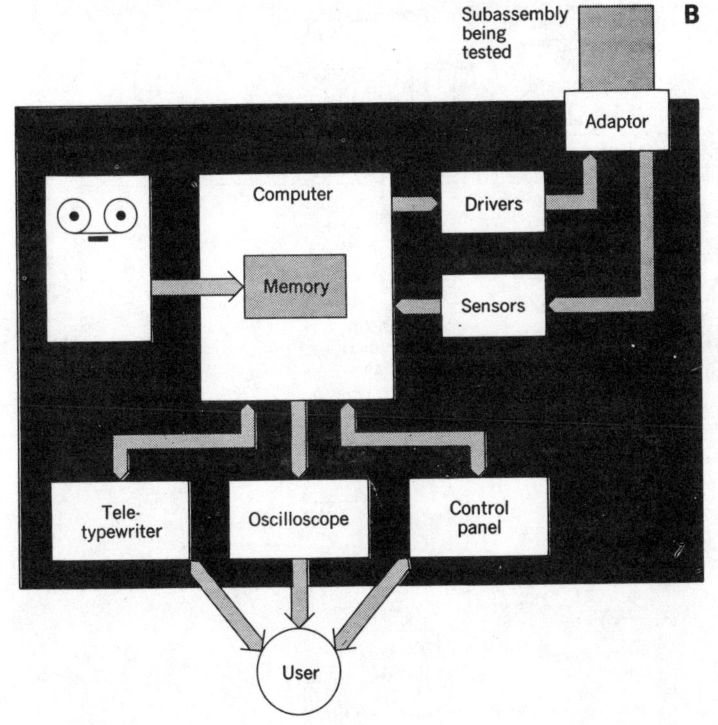

FIGURE 9. Automatic system for functional testing of digital logic circuits. A—System. B—Block diagram. C—Typical program. (Courtesy General Radio)

C

```
/BOARD NUMBER 1790-4140
/USE ADAPTOR 4
/PROGRAM REVISION 1
*I (1, 2, 14, 15, 17)
*0 (20, 21, 22, 23, 30)
1; IH(#) OH (22, 23)
2; IL (14, 2)
PRINT MOVE S1 TO POSITION 5
AND PRESS CONTINUE
PAUSE 1
DO 5, 30
3; IH (2) IL (17) $
4; IL (1) OH (30)
5; IH (1) IL (2) OL (30)
DELAY 100
IGNORE (21, 22)
6; OL (20)
IF (23) 7
PRINT PROBE TEST POINT 1 AND IF HIGH,
THEN 1C8 IS BAD
OTHERWISE CHECK 1C6 !
PAUSE 2
.....................................
.....................................
END
```

unit, a CRO display unit, and a dc parametric measurement unit. The third rack contains multiplex relays, buffer amplifiers, and the drive and sense circuits for functional testing.

The system can perform functional and parametric tests on digital circuits with up to 120 input/output terminals over a range of ±30 volts. The functional test rate is 50 kHz. A clock-rate tester is available in a separate kiosk to permit functional testing of dynamic MOS circuits with clock rates of 10 Hz to 8 MHz. An independent dynamic test system can also be incorporated to provide measurements of propagation delay and rise and fall times on 48-pin devices. Programs are prepared on magnetic tape, using the teletypewriter and display unit and a high-level test-set language.

Analog circuits. Figure 10 shows an example of the Hewlett-Packard 9500 Series Automatic Test System,[17] used to test analog circuits. A system is configured from a selection of building-block modules, including computers, stimulus sources, measurement units, processing units (called converting and conditioning units), switching units, and output units—most of which are commercial test instruments. Also shown is a block diagram of one of the simpler configurations. Figure 11 gives a program used to test the gain versus frequency of an audio amplifier. The program language is HP Basic, a modified version of the "Beginners All-Purpose Symbolic Instruction Code" developed at Dartmouth College in the mid-sixties. The software includes a 5500-word interpretive compiler and "software-driver" routines for the hardware elements.

Figure 12 shows the Instrumentation Engineering System 390, a computer-controlled system for the testing of printed-circuit boards—blank, digital, analog, or hybrid—or other forms of circuits. The system is another example of a modular building-block approach that allows a system to be configured from a selection of stimulus, measurement, switching, control, and output units. The system software uses Atlas language and can be used in a compile-before-run mode or in an interpretive direct-run mode.

Military systems. As is true of almost all our technology, many significant advances in automatic test systems have been spurred by the requirements (and the money) of the military services. Many military contractors (Northrop, LTV, General Dynamics, etc.) have furthered the art of automatic testing. Although this article is not directed specifically to such equipment or such applications, a brief description of one program may be useful.

The VAST (Versatile Avionics Shop Test) system[18,19] has been developed by PRD Electronics over the past ten years or so under several Navy contracts. About $100 million has been spent to date.

The system is intended for maintenance tests, both aboard ship and at supporting shore sites, on the avionic equipment in Navy aircraft, particularly the new Lockheed S-3A, the Grumman E-2C and F-14A, and the LTV A-7E. This project influences not only the testing of the avionic equipment but also the basic design and packaging of the equipment itself, since the suppliers of the aircraft and avionics equipment will be required to provide interface devices (device adaptors) and compatible test programs, and demonstrate that the equipment can indeed be tested and diagnosed on the VAST system.

Figure 13 shows a typical VAST test station. The hardware contains three basic sections:

1. The computer subsystem, which includes a general-purpose digital computer and two magnetic-tape transports.

FIGURE 10. Automatic test system for analog circuits. (Courtesy Hewlett-Packard)

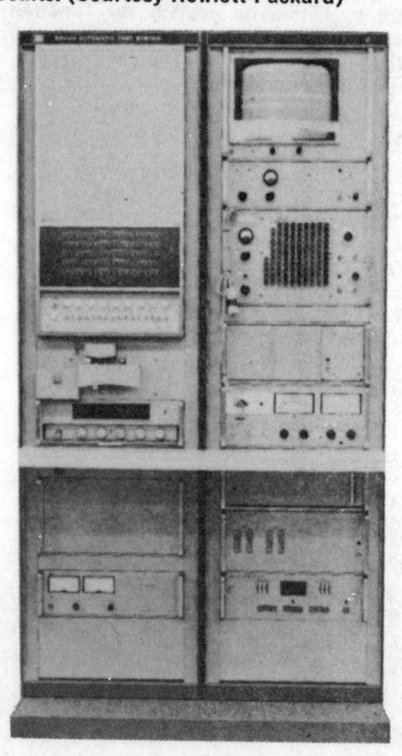

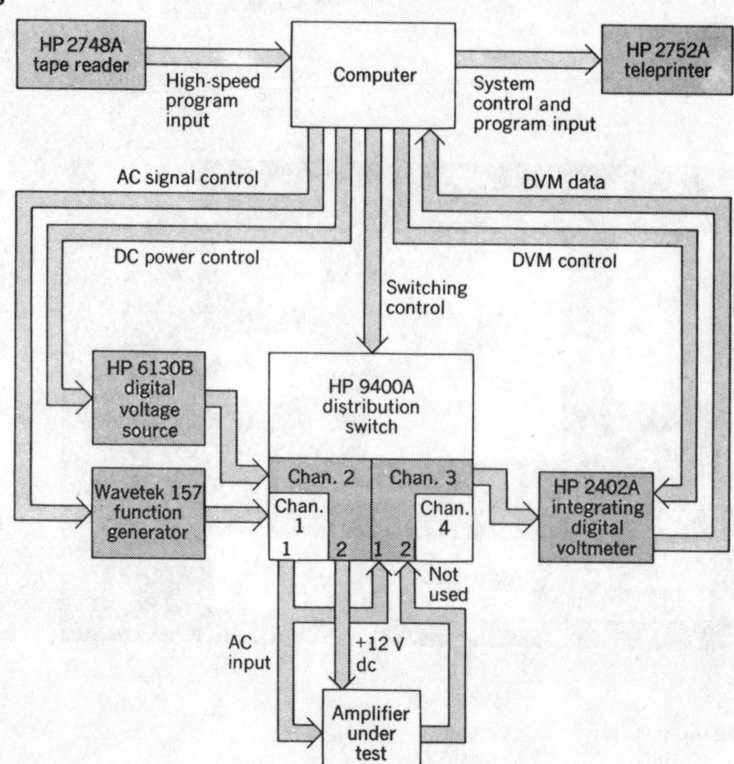

2. The data transfer unit, which represents the operator/machine interface. This unit includes a control panel with mode-select buttons, status indicators, and an ASCII keyboard; a display panel that provides a cathode-ray tube presentation of test data and instructions; and a maintenance panel to monitor autocheck results.

3. The stimulus and measurement section, which includes the basic building blocks used to configure a particular test station. These blocks are chosen from a list of about 50, which are divided into three categories: stimulus, measurement, and operational. The stimulus blocks include dc and ac power supplies, signal generators, function generators, pulse generators, delay generators, word generators, etc. The measurement blocks include digital voltmeters, frequency meters, power meters, impedance meters, spectrum analyzers, etc. The operational blocks include switching and interface modules.

Several sophisticated software techniques have been developed for the VAST system.[20] A high-level language, Vital (Vast Interface Test Application Language), is an extension of Atlas, which was developed in 1968 by Aeronautical Radio, Inc. The new language includes several sublanguages, or "dialects," to obtain the ease of programming of a high-level language but still retain the flexibility of harder-to-use low-level languages.

Programs are prepared off line. Several compiler programs are used to convert the dialect programs into VOSC (Vast Operating System Code) at a compiling station, which includes peripheral items and a PDP-8 minicomputer. The compiling station is in turn linked to a remote Univac 1108 computer. Source programs are entered on punched cards and are compiled into an object program (in VOSC) on magnetic tape. This object program is entered into the VAST test station along with an operating-system program that provides control of test-station operation and interprets the instructions of the object program during a test run.

Test complex. Most of the systems just described are "stand-alone" test systems. Some can include a few multiplexed test stations but in the main they are autonomous units. Another concept being considered, and in some cases implemented, is the test complex consisting of a central control station and remote test centers. Each test center can include a master test station and several "satellite" or "slave" test stations. The approach is intended to minimize cost by "putting all the computer in one place and developing the software once and for all." In practice, however, it has been found that the minicomputer itself is often the most economical item to use in a remote test center to communicate back to a central station. It also allows some degree of autonomy at the test center in case the central unit goes down.

Such a test complex is in operation at the Raytheon plant in Andover, Mass.[21] It is called IHATS (Improved Hawk Automatic Test System) and is used for factory test of missile components. Each test station includes a

FIGURE 11. Programs for system shown in Fig. 10. A—Test result. B—Program example.

A

```
AMPLIFIER FREQUENCY RESPONSE TEST
SPECIFIED GAIN IS 1Ø, TOLERANCE 1.7%

CONNECT AMPLIFIER
SERIAL NUMBER IS?12

FREQUENCY        GAIN          % ERROR
 1ØØØ             9.8424       -1.576
 2ØØØ             9.85134      -1.48661
 3ØØØ             9.85929      -1.4Ø711
 4ØØØ             9.83948      -1.6Ø521
 5ØØØ             9.84841      -1.5159
 6ØØØ             9.81771      -1.82291      OUT OF SPEC
 7ØØØ             9.82665      -1.73349      OUT OF SPEC
 8ØØØ             9.82665      -1.73349      OUT OF SPEC
 9ØØØ             9.8247       -1.753Ø1      OUT OF SPEC
 1ØØØØ            9.83363      -1.6637
```

B

```
1Ø PRINT "AMPLIFIER FREQUENCY RESPONSE TEST"
2Ø PRINT "SPECIFIED GAIN IS 1Ø, TOLERANCE 1.7%"
3Ø PRINT
4Ø G = 1Ø
5Ø DVS (1, Ø, 1Ø)
6Ø PRINT "CONNECT AMPLIFIER"
7Ø PRINT "SERIAL NUMBER IS";
8Ø INPUT S
9Ø PRINT
1ØØ PRINT
11Ø SWH (1, 2, 1, Ø)
12Ø DVS (1, 12, 1ØØ)
13Ø PRINT "FREQUENCY", "GAIN", "% ERROR"
14Ø FOR F = 1ØØØ to 1ØØØØ STEP 1ØØØ
15Ø SYN (1, 1, F, .283)
16Ø WAIT (5Ø)
17Ø DVM (2, 1, V1)
18Ø SWH (1, 2, 2, Ø)
19Ø DVM (2, 1Ø, V2)
2ØØ E = ((V2/V1-G)/G)*1ØØ
21Ø PRINT F, V2/V1, E
22Ø IF ABS(E)>1.7 PRINT "OUT OF SPEC"
23Ø PRINT
24Ø NEXT F
25Ø END
```

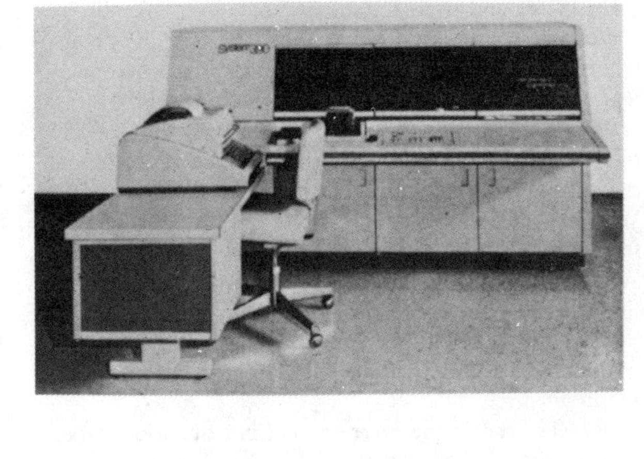

collection of stimulus, measurement, conditioning, switching, adaptor units, etc. (usually commercial items) plus specially designed interface circuits for control and communication with a "line computer" on a multiplexed basis. Several line computers in turn communicate with a central "collection computer" for data collection on magnetic tape. Programs are prepared and data are analyzed on an isolated "utility computer" and associated peripheral units. The system uses a special compiler language, TCOMP, based on Fortran IV, which allows program preparation by test engineers rather than programmers.

Cost. The prices of commercial automatic test systems seem to cluster about decade and half-decade values (1:3:10), with, of course, a continuum in between. Table II shows what to expect.

How to automate successfully

There are two basic approaches to achieving automation: the do-it-yourself approach and the buy-it-outside approach, with several gradations in between. In the do-it-yourself approach you attempt to do as much of the job in house as your time, talent, and budget allow. (Of course you have to buy something—from wire and solder to ICs or complete instruments or subsystems—but the basic responsibilities for design, procurement, fabrication, assembly, test, documentation, installation, training, and maintenance are yours.) In the buy-it-outside approach you attempt to contract for as many of these services as possible. Either approach can succeed or either approach can fail, depending on how you go about it. The following paragraphs discuss techniques that can make the second approach successful.

Getting started. First, study your task. Try the "man-from-Mars" technique and mentally step back as far as you can to visualize your operation as part of an overall system. Then gradually narrow your view to specific operations. Where does the device to be tested come from? Where does it go? What paper work goes with it? Why? What happens to it? What tests are to be performed? Why? What accuracies are required? Why? What reports are required? Why? This exercise can sometimes disclose unnecessary or redundant steps or operations no longer required because of changes in other parts of the organization. Or it may disclose new or impending requirements.

The chickens in the other yard. Find out what other people with similar problems have done. Read, study, telephone, visit. Present your problem informally to a few potential suppliers to get their ideas. Don't tell them how to do it; tell them the basic task. You can suggest ways but don't, at this stage, demand them. Others may have different ideas based on previous experience; you're not committed to follow them.

Put it in writing. Prepare a written specification for your automatic system. This exercise will clarify your thinking and provide an essential tool for you and the supplier. Again, specify what the system must do rather than how to do it. Separate your specifications into "must have" and "would like." In other words, don't demand 0.1 percent accuracy if 1 percent will do; don't demand four-week delivery if 12 weeks will do. You can get almost anything you want, but cost and time will increase exponentially as specifications get tougher.

Get competitive bids. Send your specification to as many vendors as you think are qualified to do the job. Give them time to absorb the request and invite them in to discuss it to make sure they understand the requirements. Keep prospective vendors separate unless you can handle a bidders' conference in a firm, unbiased manner.

In some ideal utopian world a prospective buyer would get a group of three or four prospective vendors together and say, "Look, fellows, I've got $37 500 and this is the job I'd like done. What can you give me?" The vendors would leave and come back later with their best proposals—all priced at about $35 000 (with optional extras to $45 000).

Or, better yet, he might say, "If I can get a machine to do these things in four seconds it's worth $37 500 to me. If you can get it down to two seconds it's worth $75 000!"

In the real world this does not happen often. A buyer is afraid that the vendors will combine their efforts to milk him of his money for a machine of lesser value. This, of course, is a small danger, since most vendors are engaged in a competitive life-or-death struggle and have no intention of cooperating with each other.

Get it in writing. Request a written proposal from your prospective suppliers in addition to a quotation. The proposal should list and describe the equipment to be provided, both electrically and physically. It should define the specifications of the system as a whole. It should describe the operation of the system from the operator's point of view (rather than the designer's). It should describe the other services provided, such as installation, training, warranty, and repair. It should include a list of recommended spare parts. Get as many competitive bids as possible.

Evaluate the bids. When you open the bids you'll be surprised. The prices will be higher than you expected, the accuracies will be poorer, the speed will be slower, you'll get less equipment, the equipment will be harder to use, and you'll get fewer services. Don't be dismayed; the approach can still succeed.

There are four basic considerations in evaluating the bids: the equipment, the supplier, the economics, and yourself.

The equipment. Lean toward standard off-the-shelf items whenever possible. Such items are usually well debugged and conservatively specified, and will have a good backup of instruction and maintenance manuals, available spare parts, and trained service technicians. You may require some modified or specially designed items, but it is important to appreciate the differences between these items and standard items. In this area you are highly dependent on the integrity of the supplier.

The supplier. You should thoroughly evaluate the

II. Commercial automatic test systems

Price Range, dollars	Description
10 000–30 000	Hardware systems of dedicated capability. Sometimes with tape readers, patchboards, or other forms of semiautomatic control. Printed or machine-readable output
30 000–100 000	Computer-controlled systems. One or more test stations and processed output
100 000–300 000	Computer-controlled with several test stations. Flexible capability and completely processed output
300 000–1 000 000	Multicomputer installations. Central control with remote test centers and satellite test stations

supplier. Where is he? What are his local sales and service facilities? What's his reputation? What else has he done like this? How did it work out? If you are contemplating a major expenditure, visit the suppliers' facilities. Talk to the people who will be doing your job.

The economics. Although you have probably made some form of economic analysis to establish your budget, do it again based on the actual bids and the actual throughput promised. Appendix B describes some methods used to compare investments. All methods require an estimate of the initial investment and the first-year savings.

The cost of the initial investment includes not only the billed price of the equipment but also the additional costs of installation, training, program preparation, spare parts, etc. As a rule of thumb, these costs can add 10 to 15 percent to the cost of the equipment.

The computation of operational savings requires an estimate of the cost of doing a job with the new equipment versus doing it the old way. The following factors are involved:

1. Number of units to be tested.
2. Number of types to be tested.
3. Average failure rate.
4. Old labor costs, including overhead.
5. New labor costs, including overhead.
6. Preparation time for test procedures.
7. Sorting good units from bad.
8. Troubleshooting and repair of bad units.
9. Tooling and maintenance.

Table III shows a savings calculation for the testing of logic-circuit boards. The comparison is between the use of manual test equipment and an automatic system. The results illustrate two things. First, the automatic equipment saves $20 742 in doing a year's work; second, it uses anly 398 hours in doing it. This reduction in time is, of caurse, the reason for the cost savings, but it also shows that the machine will be sitting idle most of the time. It will be underutilized and has the capacity to provide greater savings during the year if the work volume increases.

Armed with such an estimate, you can find an accounting expert who can calculate the payback period or discounted rate of return or net present value. You will usually find that you are losing money every day you are without the system—even a high-bid one. Amazingly enough, this is generally true (at least in the type of investment discussed here).

Yourself. You should evaluate yourself along with the bids. Can you assign a system "father" who will learn and master the system and nurture it as his own? (If it's foisted upon him by the "front office," it will never work.) Do you have, or can you develop, the capability to get the most out of the system and keep it running? Will you train new people as the present people get promoted or leave?

Follow-through. Once you've made up your mind and placed the order, get behind it solidly. Keep in touch with the supplier and work with him. Agree on an acceptance-test procedure—the simpler the better. If possible, visit his plant to look at and try the equipment before it is shipped; last-minute modifications are accomplished much easier at his plant than at yours.

Pilot operation. When your new system is installed, don't put it into full critical use at once; drive it slowly for the first "1000 miles." You'll find bugs, either in your operation of the system (cockpit errors), or in the set itself—early-life failures or some combination of switch positions that the designer never thought of. Work these problems out patiently and smoothly and get your operators used to the system. Then you can put it into use with full confidence.

Service and maintenance. If the system is really to do a job for you, keeping it running is of prime importance. Most vendors provide free service at your plant for 90 days. This usually works well since most of your problems will occur early and, if the system is in trial use, the few days (or weeks) it takes them won't hurt you much. Beyond the 90 days, service at your plant will take time and will cost you—unless you've signed up for a service contract (for about ½ percent of the system price per month) or unless you're willing to diagnose the difficulty and send the defective item back to the vendor. If the system is in critical use and down time is disastrous, the best approach is to fix it yourself. This requires people who know the system inside out and a good stock of spare parts. The spare parts can range all the way from a few relays to spare circuit boards to spare instruments.

Conclusion

This article has attempted to survey the state of automation in electronic testing and to discuss some of the characteristics of current approaches. The message is simple: automation is desirable and everyone should have it. Get it any way you can. Rent it, lease it, buy it, make it. It's not as easy to achieve as it may seem (nothing worthwhile is), but neither is it as hard. The rewards can be startling and will be in direct proportion to the effort put in.

III. Operating savings analysis, logic-circuit boards, annual cost

	Manual	Automatic
Preparation:		
Number of types	10	10
× hours per type	× 40	× 16
Total hours	400	160
× cost per hour	× 15	× 15
Cost	$6 000	$2 400
Sorting:		
Number of boards	5 000	5 000
× minutes per board	× 10	× 0.25
× hours per minute	× 0.017	× 0.017
Total hours	833	20.8
× cost per hour	× 15	× 10
Cost	$12 500	$208
Troubleshooting and repair:		
Number of failures	1 000	1 000
× minutes per failure	× 30	× 10
× hours per minute	× 0.017	× 0.017
Total hours	500	167
× cost per hour	× 15	× 15
Cost	$7 500	$2 500
Tooling and maintenance:		
Total hours	100	50
× cost per hour	× 10	× 15
Labor cost	$1 000	$650
+ material cost	500	1 000
Cost	$1 500	$1 650
Total cost	$27 500	$6 758
Total hours	1 833	398
Cost savings		$20 742

Appendix A.
Software languages; some definitions*

Figure 14 is a generic block diagram applicable to all computer systems. Any system can be considered as a subsystem consisting of some of the elements in the diagram. Let us consider the function of each element, beginning with the lowest.

Base machine. The base machine consists of registers and decision-making circuits that perform the basic data-processing operations. This machine is controlled by signals for opening or closing gates to allow the data to flow between the machine registers. These signals are sometimes called the *microprogram* of the machine, and the language that describes this microprogram is called the *register-transfer language*. Programs for basic machines are very complex and are usually determined only by the computer manufacturer; recently, however, machines that can be microprogrammed by the user have been introduced.

Emulator. The emulator block usually consists of hardware decoding circuits or a read-only memory, which decodes certain bit patterns, called the *machine code*, to operate the base machine. These circuits determine what is usually called the *instruction set* of the machine. Instructions consist of a pattern of bits in an instruction word. Some special machines use a string of ASCII characters as the instruction set, so they may be programmed directly from an I/O device, such as a teletypewriter. A few machines have been built whose machine language is a high-level language. The emulator and base machine together make up what can be called the *real machine*. This is what we generally buy from a computer manufacturer.

Interpreter. The interpreter block defines a program, sometimes called an *operating system*, or *simulator*, which can be thought of as a means for converting the real machine into a new machine, called a *pseudo-machine*, whose effective machine code is different from that provided by the real machine. The machine code for the pseudo-machine, called *interpretive code*, is usually chosen to make the pseudo-machine perform more directly those functions desired by the application. As a consequence, the program in interpretive code requires fewer bits of memory than the same program would require in machine code directly.

This efficiency in memory space is obtained at the expense of both speed and flexibility. The pseudo-machine usually executes the program at a slower rate than would occur with the same program coded in machine code. Also, the pseudo-machine will tend to be less general than the real machine. The interpreter is designed for the effective performance of certain functions needed by the class of programs for which it was intended. Consequently, it performs certain other functions poorly and, in fact, it may not allow the programmer to perform still other functions at all.

Program store. The *program store* block consists of the medium in which the program is stored. It may be either core memory or paper tape.

Translator. The *translator* takes the source code generated by the programmer, often performs some rudimentary checking for errors, and translates the programmer's statements into interpretive code. Note that translation is not part of the real-time operating environment. It is performed once when the program is written, either on the machine in question or some other machine of a different type, or even manually by the programmer. If the translation is performed mostly by machine, the source language that the programmer uses can use terms very close to the terms in which he thinks about the problem. If this is the case, he is using *high-level language*.

Programmer. If the machine translation process is relatively simple (such as a basic assembler), then the *programmer* must first translate his programs into the language needed by the translator program. The closer the terms of the source language are to the terms in which the programmer considers the problem, the higher the level of the language. The further the programmer has to translate his thoughts into the terms of the machine, the lower the level of the language.

Degenerate cases. There are many practical systems where various parts of this hierarchy seem to be missing. In these cases the functions are actually absorbed by some other block. Some examples may illustrate this.

Missing interpreter. In many systems the interpreter or operating system block is missing. This occurs with most of the software provided by minicomputer manufacturers with their systems, including the following:

1. The assembler program requires the programmer to translate his problem into a language very close to machine code. It further translates his assembly-language code to machine code. When writing in assembly language the programmer has access to all the machine functions provided by the machine designers. For this reason, good assembly-language programmers can write efficient programs that use the machine resources efficiently. In small-computer systems most of the programs are usually written in assembly language, since efficient use of the limited machine resources is very important.

2. A Fortran compiler translates a higher-level language, Fortran, into machine code. This enables the programmer of algebraic problems to think more in terms of the problem and leave most of the translation to the Fortran translator or compiler. Good compilers generally use machine resources more efficiently than a poor assembly-language programmer and less efficiently than a good one.

Missing translator. There are several important software systems in which the translator appears to be missing. In these systems the program is stored directly in the source language of the programmer. Examples are most uses of Basic and Focal. In these systems the program will take much more memory space than a translated program, since the source language contains much redundant information. In addition, the system will usually run slower since each program statement must be processed each time it is executed. These systems have the advantage of being easy to change on line and are often more interactive than those in which translation is needed. There are several programming languages that are used in this way during program preparation and debugging. After prove-in, the program is translated into machine code for repetitive use.

Between these examples is a continuum of possibilities. The skill of the system designer determines which is the

* The thoughts in this Appendix were provided by R. G. Fulks of General Radio Company; the words were compiled by the author.

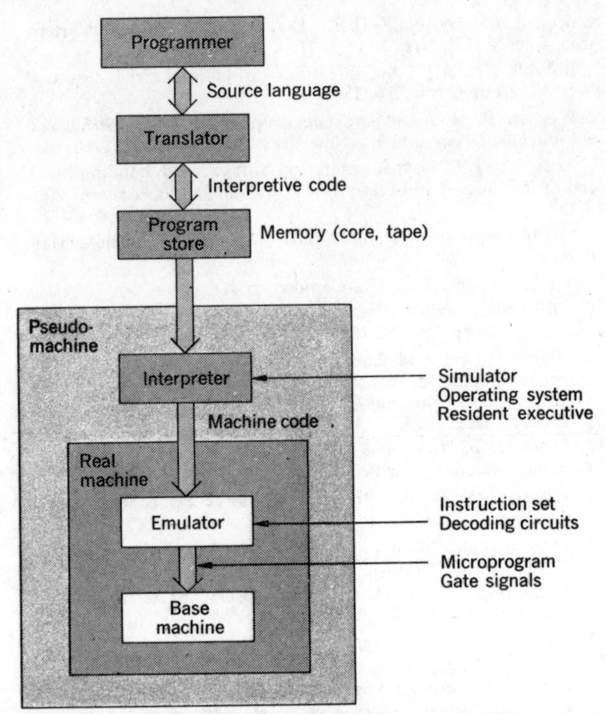

FIGURE 14. Block diagram of a computer system.

most appropriate for a given system problem. The right decision can result in the optimum use of program space and machine time.

Appendix B. Economic justification for automatic test equipment

Justifying an automatic test system from an economic standpoint is relatively easy. In fact, it can be said that automation, at almost any price, is worth it—or soon will be. Looking backward or forward ten years from any point of our lifetime we see a smooth increase (with time) of the cost per unit of production using skilled labor only, and a decrease of the cost per unit using a combination of semiskilled labor and automatic equipment. This seems a paradox, since the cost of automatic equipment itself might seem to be related to the increasing cost of the labor that goes into the equipment. That it is not is due to the insatiable demand for the products of technology, which causes, in turn, an increasing demand for automatic equipment and brings to bear the forces of "the economies of scale"; that is, the more items you make, the less their unit cost. Although the cost of the equipment goes up (after the initial decline from the first few developmental models), the efficiency of the equipment goes up even faster. This is further illustrated by the paradox of the equipment cost/product cost ratio. When diodes sold for $1.50 each you could buy a diode tester for $300. Now that diodes sell for a few pennies apiece, the test equipment costs $100 000! Morever, the $100 000 testers have helped achieve the price reduction of the diodes.

The real problem in automatic testing is not in justifying automatic equipment, but in selecting which automatic equipment to buy. Even so, we are still called upon to justify such expenditures to our supervisors. The following paragraphs may help.

The steps in justifying the purchase or construction of an automatic test system are identical to those in justifying any capital investment. Basically you have to compare the money saved by the new machine with the cost of the machine, and establish that the use of the money required to buy the machine is better than any other use you can think of. The preceding sentence makes it sound easy, but a rigorous analysis can get quite complicated, for here we leave the comfortable world of volts and ohms and enter the strange world of our professional colleagues in the accounting department, with its tax shields and discounted cash flows. We quickly find that what we thought was a dollar isn't a dollar at all. It may be a half dollar, or 75 cents, or even $1.25! This is the result of two aspects of our economic system: the corporate income tax and the time value of money. If you're a profit-making corporation, you can view any additional dollar of operating expense or saving in its incremental effect on the profit of the corporation. If you "save" a dollar you must give half of it to the Federal Government as tax and you wind up with only half a dollar. Similarly, if you spend a dollar for operating expense, you reduce your tax by half of it and you're only out half a dollar. Thus some of the operating savings and expenses in the use of capital equipment are viewed at half their value (tax shield).

The time value of money, on the other hand, shows us that a dollar to be saved during the next year is not worth a dollar today. In fact, if we can invest money and earn 10 percent on it, that dollar return spread over next year is worth only 95 cents today. Therefore, the time distribution of savings and expenses must be considered.

Three basic methods are currently used to estimate the relative profitability of capital investments: return on investment, payback period, and discounted cash flow.[22,23]

Return on investment. Several methods are used to estimate the percentage rate of return for a capital investment (in order to compare it with other possible investments). A simplified version is the MAPI formula,[24] wherein you use the estimated savings in the first year of operation to determine an "urgency rating," the percentage return in the first year. The following calculation shows the urgency rating for a $30 000 test system that will provide savings of $15 000 a year (based on ten-year straight-line depreciation).

Operating savings:	$15 000
Less depreciation:	3 000
Net operating advantage:	12 000
Less added tax:	6 000
After-tax return:	$ 6 000

$$\text{Urgency rating} = \frac{6000}{30\,000} = 20 \text{ percent}$$

Payback period. The payback period is the time necessary for the savings (after taxes) to pay back or make up for the initial investment. The shorter the period, the better the investment. Rather than savings, total cash flow is used; that is, depreciation charges are added to the actual savings, since these charges are an additional source of cash. (Depreciation dollars never leave the company; you've already paid for the machine.) Table IV compares two investments on this basis, a $30 000 system that will save the company $15 000 a year

IV. Payback-period investment approach

Invest-ment	Operating Savings	Depreci-ation	Net Return After Tax	Cash Flow	Payback Period, years
$30 000	$15 000	$3 000	$6 000	$ 9 000	3
50 000	20 000	5 000	7 500	12 500	4

V. Discounted cash flow technique

Year	Savings	Depreci-ation (sum of years)	Net Cash Flow After Taxes	Discounted Value at 20 Percent Return	
				Factor	Amount
1	$15 000	$10 000	$12 500	0.909	$11 363
2	15 000	8 000	11 500	0.751	8 637
3	15 000	6 000	10 500	0.621	6 520
4	15 000	4 000	9 500	0.513	4 873
5	15 000	2 000	8 500	0.424	3 604
Total:	$75 000	$30 000	$52 500		$34 997

and a $50 000 system that will save $20 000 a year.

Discounted cash flow. The discounted cash flow technique takes into account the smaller present value of future returns. The usual method is to tabulate all the future savings, select a desired interest rate (opportunity cost), and reduce the values of the future returns to present values by factors read from a discount table. You then add up all the discounted (reduced) returns. If the sum is greater than the investment, you'll do better than the selected interest rate; if it's less, you won't. Table V shows this calculation for a $30 000 system that will save $15 000 a year for five years and a desired intrest rate of 20 percent (returns beyond five years and salvage value are ignored).

Since the total discounted value (net present value) of $34 997 is greater than the investment of $30 000, the actual rate of return is greater than the 20 percent selected. This technique can be refined by choosing different interest rates and interpolating until the net present value equals the investment, thus determining the actual interest rate.

The preceding methods are fairly straightforward and relatively simple. The procedure can be made a good deal more complicated, however, if all the fine points are accounted for, such as the growth of operating savings from year to year as the equipment is utilized more (more volume, more shifts) or as the cost of labor *not* paid for increases, the salvage value of the equipment if and when you sell it, the interest-compounding period to be used (continuous, monthly, annually), nondepreciated start-up costs, varying tax rates, etc. If your company requires such refined analysis you'd best consult a friend in the accounting department. Some equipment manufacturers provide forms, tables, charts, and even computer programs to simplify the analysis, but, if you use them, make sure you apply the same techniques to all choices.

REFERENCES

1. Lucey, R., and Newth, D., "When to inspect all incoming semiconductors," *Electron. Equipment Eng.*, Mar. 1968.

2. Holt, A., and Stoughton, A., "Guidelines for the purchase of memory testing equipment," *Evaluation Eng.*, July/Aug. 1970.

3. Bartik, J. J., "Common computer interfaces," IEEE Computer Society, Mideastern Area, Cherry Hill, N.J., Mar. 1971.

4. Litzinger, J., "Hardware/software characteristics of a computer-controlled test system," IEEE Computer Society, Mideastern Area, Cherry Hill, N.J., Mar. 1971.

5. Bobroff, D. A., "Avoid pitfalls in computerized testing," *Electron. Design*, Aug. 16, 1969.

6. Wegner, P., *Programming Languages, Information Structures, and Machine Organization.* New York: McGraw-Hill, 1968.

7. Ross, D. T., "Fourth-generation software—a building-block science replaces hand-crafted art," *Computer Decisions*, Apr. 1970.

8. "IC test equipment buyer's guide," *Solid State Technol.*, Mar. 1970.

9. A.T.E. Reports, Box 746, Camden, N.J.

10. "Electronic industry manufacturing markets report," Theta Technology Corp., Wethersfield, Conn.

11. Fulks, R. G., and Lamont, J., "An automatic computer-controlled system for the measurement of cable capacitance," *IEEE Trans. Instrumentation and Measurement*, vol. IM-17, pp. 299–303, Dec. 1968.

12. *Glossary of Integrated Circuit and Related Terminology.* Motorola Semiconductor Products Inc., Phoenix, Ariz.

13. Johnson, W. R., Jr., "Proving out large PC boards," *Electronics*, Mar. 15, 1971.

14. Fichtenbaum, M. L., "A computer-controlled system for testing digital logic circuits," *NEREM Rec.*, pp. 38–39, 1969.

15. Robinton, M. A., "A critique of MOS/LSI testing," *Electronics*, pp. 62–64, Feb. 1, 1971.

16. Curran, L., "Readers reply on MOS/LSI testing," *Electronics*, Mar. 1, 1971.

17. *Hewlett-Packard J.*, Aug. 1969.

18. Loughlin, R. G., and McCoy, F., "The introduction of VAST," presented at WESCON, Los Angeles, Calif., Aug. 1970.

19. Fuller, E. J., and Proctor, J. A., "The evolution and development of a digital test capability for VAST," presented at the Automatic Support Systems Symposium for Advanced Maintainability, St. Louis, Mo., Oct. 1970.

20. Ellis, M. T., "Vital: A general-purpose multiple-dialect automatic test language," presented at the Automatic Support Systems Symposium for Advanced Maintainability, St. Louis, Mo., Oct. 1970.

21. Gundal, R. K., and King, J. F., "A central time-sharing mini-computer—provides data collection and automatic control over a variety of remote test systems," *NEREM 70 Tech. Appl. Papers.*

22. Ammer, D. S., *Manufacturing Management and Control.* New York: Appleton, 1968.

23. Helfert, E. A., *Techniques of Financial Analysis.* Homewood, Ill.: Richard D. Irwin, Inc., 1967.

24. Terborgh, G., *Business Investment Policy*, Machinery and Allied Products Institute, Washington, D.C., 1958.

Harold T. McAleer (M) received the B.S. and M.S. degrees in electrical engineering from the Massachusetts Institute of Technology in 1953. While in college he was employed as a cooperative student at the General Radio Company working on the design of high-frequency measuring and recording instruments. After two years' service as an engineer with the U.S. Army Signal Corps at Fort Monmouth, N.J., he returned to GR as a development engineer, in which capacity he was engaged in the design of frequency counters and associated instruments. In 1966 he transferred to GR's Systems Group to work on the design of automatic measurement systems. In 1968 he became manager of custom products at General Radio and is responsible for marketing, design, and manufacture of custom-tailored test systems. He has written several technical articles and is a registered professional engineer and a member of Eta Kappa Nu, Tau Beta Pi, and Sigma Xi.

AN AUTOMATIC TEST SYSTEM UTILIZING STANDARD INSTRUMENTS AND ABBREVIATED ENGLISH LANGUAGE

Robert A. Grimm
Hewlett-Packard Company
Palo Alto, California

Summary

This system provides stimuli and measurement capability required for testing many dc, low frequency and RF circuits, modules and components. The system consists of standard, system-oriented instruments connected to an instrumentation computer.

This system, which has been built, is intended to illustrate the capability and ease of use that can be provided with standard instruments, a small computer, and a simple interpretive language.

Programmable stimuli provided include dc voltage, audio and video ac voltage (both amplitude and phase), resistance, and frequency.

Relay trees and crossbar switches provide routing of signals to instruments. Three- and four-wire switching provides switching of guard voltages on dc and ac to provide high common mode rejection. Remote sensing of voltage after relay distribution unit is provided to eliminate errors caused by contact resistance voltage drops.

The language used in programming the system is BASIC. A program consists of a series of numbered statements, each statement giving one instruction to the system. A single instruction, for example, will set a power supply voltage. A single instruction will program the digital voltmeter to the function and range desired, and instruct it to take a measurement.

Comparisons against limits, print-out of test results, branching to other tests, are all provided with simple English-like statements. Complete arithmetic capability plus square root, exponents and trigonometric functions are available.

System interprets and executes test line by line, so no compiling is required. Editing or modification of program may be made at any time, by typing statement number and revised statement.

The programming language can be learned quickly in about four to eight hours. This permits test technicians and engineers to write their own test programs simply and efficiently.

Introduction

Automatic testing of electronic and electrical equipment and components requires the same functional elements as in manual testing. These are:

Stimulus
Measurement
Switching
Control
Evaluation
Recording

In manual testing, the functions of switching, control, evaluation, and recording are usually performed by a test operator using general purpose electronic instruments. Increasing emphasis and interest in the advantages of automatic testing has resulted in the availability of general purpose instrumentation designed and adapted for use in automatic test systems. Control of the test sequence, evaluation and recording of test results may be accomplished using an instrumentation computer designed for ease of interfacing with test equipment.

This paper describes such a system recently built for testing radio receiver circuit components. The equipment used

Reprinted from *IEEE Auto. Support Systems Symp. Rec.*, Nov. 1968.

in the system is standard or slightly modified general purpose equipment. The programming language is simple, easy to learn, and powerful.

Hardware

For convenience in presentation and understanding, the stimulus and measurement capability of the system will be described in three sections: DC, AC and RF.

DC Stimulus

Four programmable dc power supplies are provided. (Refer to Figure 1.) Each can be programmed from 0 to ±10 volts in one millivolt steps, and up to ±50 volts in 10 millivolt steps. Output current capability of each supply is one ampere. Current limit may also be programmed in 8 steps from 20 to 1000 milliamperes. Each supply can slew from one voltage to another in 20 microseconds.

These power supplies may be used not only to supply operating power and bias voltages, but also as input signals for testing dc circuits or components.

The output of each power supply is connected to a relay tree distribution unit. Any of sixteen outputs may be called for. Four wires are switched through each relay tree: High, Low; and High, Low Remote Sense. By carrying through the relay tree the remote sense lines, the programmed voltage is regulated at the remote load terminals independent of voltage drop in the cables and contact resistance in relays.

The four relay trees are contained in a single 7 inch high, rack mounted unit.

DC Measurement

DC voltage measurement is provided by a high accuracy digital voltmeter. This unit can make measurements with one microvolt resolution on lowest range, and measure voltages to 1000 volts. Programmed range or automatic ranging is provided. Measurement time is 25 milliseconds.

A crossbar scanner is provided to allow three-wire voltage measurements to be made on up to 200 channels. The crossbar scanner and digital voltmeter were chosen to provide high common mode and superimposed noise rejection, together with low thermal offset in all circuit wiring and switching.

Current may be measured by the drop across a series resistor.

DC Resistance Measurement

Resistance measurements are provided by inclusion of a plug-in option in the digital voltmeter. Full scale resistance ranges may be programmed from 1 kilohm to 10 megohm, in five decade steps. Resolution on lowest range is .01 ohms.

This unit operates through the crossbar switch also. Provision is made for four-wire Kelvin resistance measurements to avoid errors that could be introduced by switch contact resistance.

Although most circuit testing is done with active tests, resistance tests are very useful for initial testing to assure correct type unit is being tested and to check for shorts. It is also very helpful in fault isolation.

AC Stimulus

The ac signal source (see Figure 2) is a low distortion oscillator with frequency programmable from 0.1 Hz to 99,900 Hz. The output is programmable from 0.01 to 9.99 volts rms in 10 millivolt steps.

An output buffer amplifier with a gain of one provides very low output impedance. (This gain may be manually changed to 2, 5, or 10, if desired.)

A relay tree switching unit, as on dc, provides programmable connection to any of 16 outputs.

AC Voltage Measurement

The same digital voltmeter and crossbar scanner are used for ac voltage measurement from 50 Hz to 100 kHz. Lowest full scale range is one volt rms, and measurements up to 750 volts peak may be made. Measurements again are three terminal, guarded.

AC Low Frequency Measurement

A plug-in option in the digital voltmeter provides frequency measurement capability from 5 Hz to 200 kHz. This measurement can be made on any of the 200 input channels to the crossbar switch.

DC TESTING

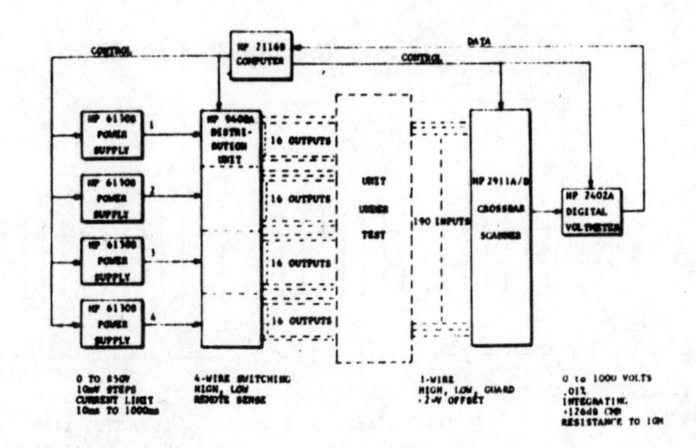

Figure 1.

DC Stimulus and

Measurement Block Diagram

AC TESTING

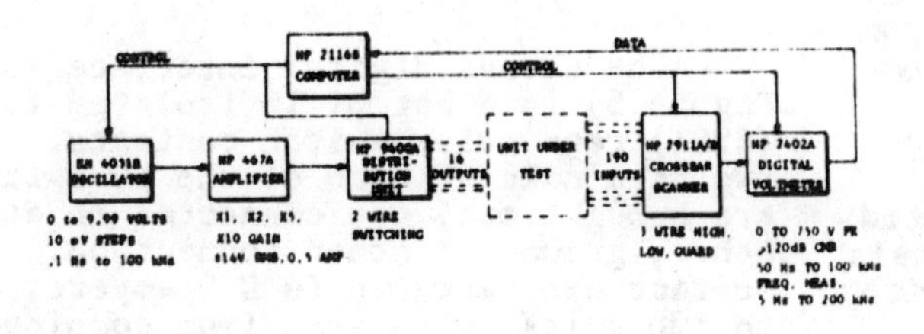

Figure 2.

AC (to 100 kHz) Stimulus

and Measurement Block Diagram

<u>RF TESTING</u>

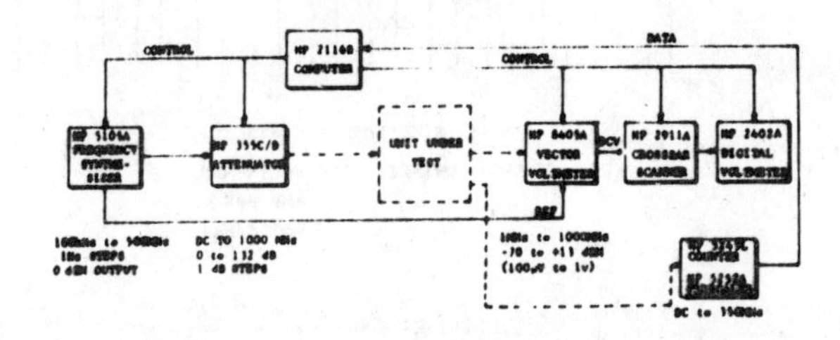

Figure 3.

RF (above 100 kHz) Stimulus

and Measurement Block Diagram

RF Stimulus

The stimulus for RF circuit testing (refer to Figure 3), such as RF amplifiers, IF amplifiers and filters, is a frequency synthesizer. This unit may be programmed to any frequency from 100 kHz to 500 MHz in 0.1 Hz steps. Its accuracy is better than one part in 100 million. The output is zero dbm, or 0.224 volts rms, into 50 ohms.

An attenuator which is flat from dc to 1000 MHz controls the output of the synthesizer. It may be programmed for attenuations from zero to 132 db, in one db steps. The output of the attenuator, then, is directly settable in minus dbm.

RF Voltage Measurement

Both amplitude and phase of RF voltages are measured with a vector voltmeter. This unit, utilizing high frequency sampling techniques, measures signals from one to 1000 MHz. The amplitude of signals that may be measured can be from +13 dbm (1 volt) to -80 dbm (22 microvolts). The voltmeter range and approximate frequency are programmed.

The outputs from the vector voltmeter are dc voltages which are connected in the system to two of the input channels of the crossbar scanner, and measured with the digital voltmeter.

Note in the block diagram that the reference input of the vector voltmeter is connected to the synthesizer output, so that phase shift through the unit under test may be measured.

RF Frequency Measurement

The frequency of signals, such as from the receiver local oscillator, are measured with a frequency counter with a high frequency prescaler plug-in. This provides frequency measurement capability from dc to 350 MHz without programming.

Digital Interface

Two sets of Input/Output digital data capability are provided by the computer in addition to the analog capability.

The first is a duplex register capability (see Figure 4) that provides an output 16 bit pattern on one set of 16 wires, and also the capability to read a digital logic pattern on another set of 16 wires. These I/O lines may be used to drive or read logic patterns, or read signals from various external devices.

DUPLEX REGISTER I/O CARD

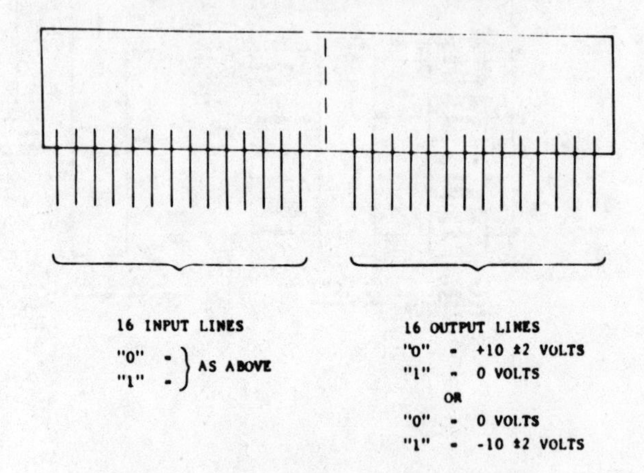

Figure 4
Digital Interface Utilizing
Duplex Register I/O Board

The second digital interface (see Figure 5) is a set of 16 isolated form A (SPST) reed relay switch contacts. The two terminals of each of the 16 switches are brought out, not connected to each other, ground or other connection. Each contact can carry up to 0.5 ampere, at up to 100 volts, with a maximum combined rating of 10 watts. For control of higher level signals, these contacts may be used to control external relays.

REED RELAY I/O CARD

16 ISOLATED FORM A CONTACTS
EACH CONTACT: 100 VOLTS MAX
.5 AMP MAX
10 WATTS MAX

Figure 5.
Digital Interface with Reed Relay Board

In this system, these contact closures are used to control an automatic circuit component handler.

Interface of Instruments with Computer Controller

The computer used in this system is a general purpose computer, designed specifically for instrumentation applications. The Input/Output interface in this unit consists of printed circuit boards with edge connectors on two opposite ends. (See Figure 6.) One end of the board plugs into the computer backplane, the opposite edge connector accepts a connector that through a cable goes directly to the programming connector or data connector of the instrument.

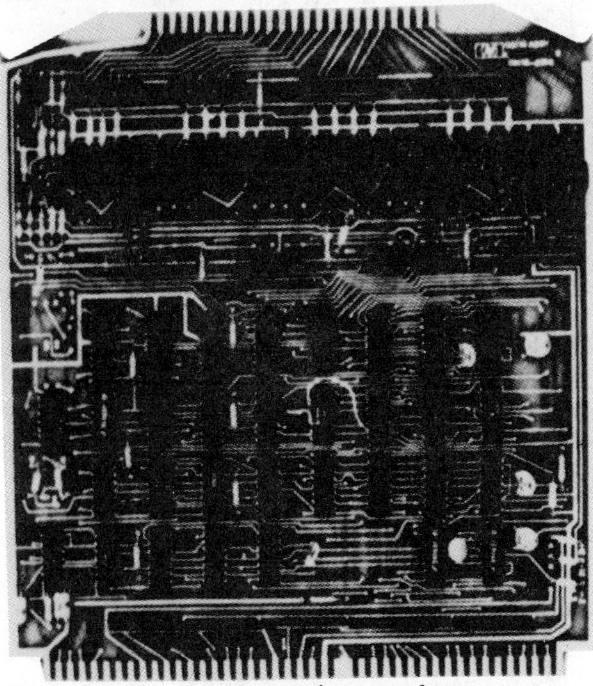

Figure 6 Typical I/O Board

For most instruments, one I/O card will provide programming, and one card will accept data. The number of I/O cards required for the instruments described above is shown in the following chart.

INSTRUMENT	NO. I/O CARDS
DC Power Supplies (all four)	1
DC Relay Tree Distribution Unit	1
Oscillator	2
AC Relay Tree Distribution Unit	1
Crossbar Scanner	1
Digital Voltmeter Control	1
Data	1
RF Frequency Synthesizer	1
RF Attenuator	1
Vector Voltmeter	2
Duplex Register	1
Reed Relay Register	1
	14
Computer Main Frame I/O Capacity	16

The computer used has 16 I/O slots into which the above cards plug. See Figure 7. In this system, two additional cards are used: one for a teletype, one for a high speed paper tape reader. For systems requiring more than 16 I/O cards, an I/O extender may be added that will hold an additional 32 cards.

Figure 7
Inside View of Computer

Computer and Peripherals

The computer in this system is a general purpose unit with a 16 bit word length and memory cycle time of 1.6 microseconds. 8192 words of core memory are provided.

This computer, being general purpose, will do all of the normal computer operations of arithmetic and more complex algebra and trigonometry, comparisons, masking, bit manipulation, branching, etc.

Available, but not required in this system, are tape recorder, disc memory, more core memory, line printer, oscilloscope display, XY plotter and other devices. These may be easily added as required.

Complete System

The complete block diagram for the system is shown in Figure 8.

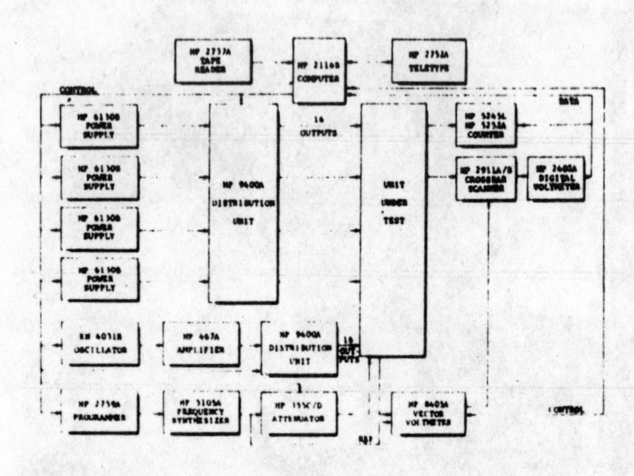

Figure 8.

Overall Block Diagram of Automatic Test System

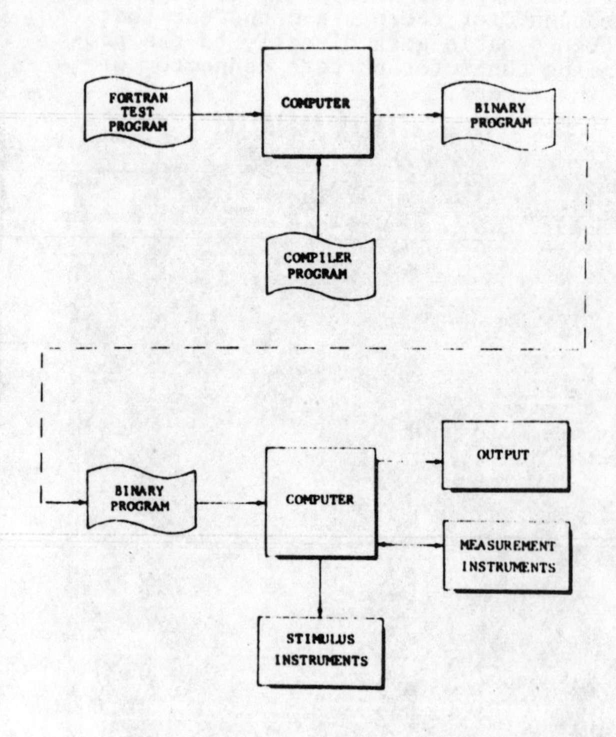

Figure 9.
Preparation of Operating Program Tape Using a Compiler.
(Not used with this system.)

Programming Language

Although this computer may be programmed in assembly language, FORTRAN or ALGOL, the language provided in which it is expected most test program will be written is BASIC. This language was developed at Dartmouth University, and is in extensive use on time share computer systems.

The BASIC language is simple to use and can be learned in a few hours. This makes it very valuable for inexperienced programmers.

Its operating mode is that of an interpreter rather than a compiler as is used with FORTRAN and ALGOL. With FORTRAN (see Figure 9), for example, a program is written with English-like statements. This, then, is totally converted to a binary machine language program on paper tape or other media. This binary tape is the one used when operating the system. Any changes or errors require recompiling, which for inexperienced programmers can be frequent, tedious and time-consuming.

An interpreter (see Figure 10) operates on each line of the English-like program as it comes to it, converting one line at a time to machine language. Errors are immediately presented to programmer, and may be corrected by retyping the line that contains the error. Similarly, program changes may be made at any time by retyping the English-like program statements.

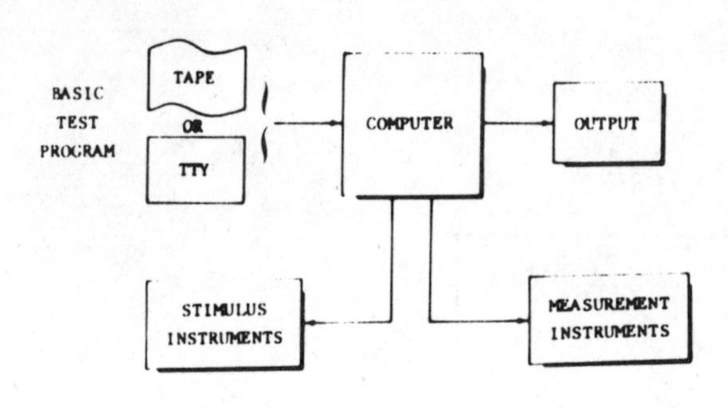

Figure 10
System Operation Using BASIC
Interpreter Language

The following example will illustrate a program written in BASIC:

```
100  PRINT "TYPE VALUE OF INPUT AND
     VOLTAGE OUTPUT"

110  INPUT I, V

120  LET G = V/I

130  IF G > 60 THEN 160

140  PRINT "AMPLIFIER GAIN TOO LOW,
     GAIN = " G

150  GO TO 170

160  PRINT "AMPLIFIER GAIN OK,
     GAIN = " G

170  END
```

Note each statement is numbered. They are performed in ascending number, unless directed to branch by a statement, like GO TO or IF statements.

PRINT statements cause a printout of exactly the words or symbols enclosed in quotation marks. The value of symbols outside quotation marks is printed.

The INPUT statement asks the operator to enter data; in the example above, two values, one for I, one for V.

The LET statement is the beginning of an algebraic calculation in which the symbols +, -, *, /, ↑ are used to represent addition, subtraction, multiplication, division, and exponentiation. Other symbols available are sine, cosine, logarithm, square root, and other commonly used functions.

The standard BASIC language, however, has no provision for working with instruments. Therefore, two statements have been added for this use.

The first is a CALL statement which is used to call a subroutine which controls an instrument. The CALL statement contains the program information to tell what range, function, frequency, voltage, or other parameter is desired. For example, the statement:

```
170  CALL  (3, 5000, 4.55)
```

sets the oscillator (instrument number 3) to 5000 Hz, with 4.55 volts output.

The statement:

```
180  CALL  (4, 2, 10, V)
```

tells the voltmeter (instrument number 4) to make an ac measurement (function number 2 is ACV) on the 10 volt range, and to label the answer as "V".

The second statement added to BASIC for instrumentation use is the WAIT statement, and it is used to insert a time delay up to 32,767 milliseconds to allow charging cable capacities, settling from switching transients, etc. The statement:

```
190  WAIT (35)
```

would cause a 35 millisecond delay before executing the next step of the program.

The block diagram of the ac testing section, annotated to show the CALL statement formats, is shown in Figure 11.

The program in Figure 12 is a program for testing an audio amplifier. This program includes setting all power supplies, checking the gain at ten frequencies from 1000 Hz to 10,000 Hz, and outputting appropriate messages including amplifier serial number.

After a program is typed in, the operator types RUN to start the program. If he wishes a copy of the program, he types LIST. If, before typing LIST, he turns on the teletype tape punch, a punched paper tape of the program will be generated. To enter this program in the future, the tape is placed in the photo reader and PTAPE is typed, causing the program to be read into the computer. As stated before, a program may be changed by

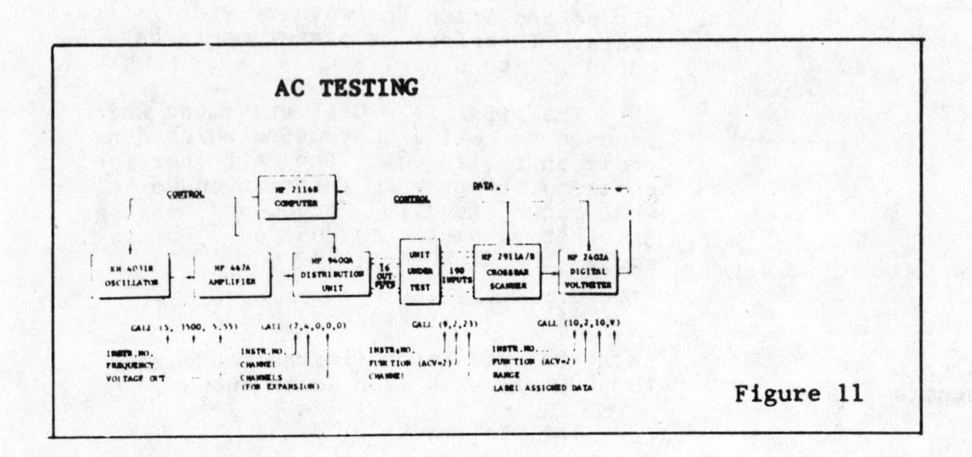

Figure 11

Block Diagram of AC Testing Section Showing CALL Statements

retyping a statement. New statements may be added between existing statement numbers, and statements may be deleted by typing only the statement number with nothing following.

The ease of programming will significantly reduce total test programming costs.

Self Test and Error Correction

A system such as this one allows a cross check of stimulus and measurement units. For example, by connecting one channel of the scanner to the output of the oscillator, a self test of both its frequency and amplitude may be made.

By providing a short on one channel of the scanner, the zero of the voltmeter may be checked, and the offset stored for subtraction from a measured value before calculations are made. Similar self tests utilizing a standard resistance or a zener diode voltage reference, external frequency references, etc., can be made.

Core Memory Used

The executive in BASIC uses about 5170 of the 8192 words of core provided in this system. The drivers for all of the instruments and digital interfaces use about 1200 words, leaving about 1820 words of core for the test program and data.

The program is stored in its compact English form, making best utilization of core space.

Since each word of core contains two characters of the test program, 1820 words of core can hold a program with about 3640 characters.

The example test program for the audio amplifier given earlier used about 540 characters, so 6 programs this long could be stored, or fewer, longer programs could be stored.

An additional 8192 words of core may be added simply, since the computer main frame is pre-wired for this. All of this memory could be used for storing test

PROGRAM	COMMENT
100 CALL (8, 1, 0, 1)	Sets Power Supply 1 to Zero
110 CALL (8, 2, 0, 1)	Sets Power Supply 2 to Zero
120 PRINT "PLUG IN AMPLIFIER"	Asks Operator for Information
125 PRINT "SERIAL NUMBER IS"	
130 INPUT S	Operator types in Serial No.
140 CALL (6, 4, 3, 0, 0)	Connects Supply #1 to Output 4 Supply #2 to Output 3
150 CALL (8, 1, -12, 100)	Sets Supply #1 to -12, 100 MA max.
160 CALL (8, 2, 12, 100)	Sets Supply #2 to +12, 100 MA max.
170 FOR F = 1000 TO 5000 STEP 1000	Establishes loop for changing frequency
180 CALL (5, F, .10)	Sets Oscillator to 1000 (then 2000, 3000, 4000, 5000 Hz), .1 V
190 CALL (7, 7, 0, 0, 0)	Connects Oscillator to Output 7
200 CALL (9, 5)	Connects Oscillator Output to DVM
210 WAIT (30)	Delays 30 Msec to allow settling
220 CALL (10, 2, .1, I)	Measures Input ACV on .1 range
230 CALL (9, 23)	Connects Oscillator Output to DVM
240 WAIT (30)	Delays 30 Msec to allow settling
250 CALL (10, 2, 10, V)	Measures Amp Output
260 LET G = V/I	Calculates Gain
270 IF G<5 THEN 320	Checks for Low Gain
280 IF G>10 THEN 340	Checks for High Gain
290 NEXT F	Return to 170 for next frequency
300 PRINT "AMPLIFIER SERIAL" S "GAIN OK, GAIN ="AT 5000 HZ"	
310 GO TO 100	
320 PRINT "AMPL SERIAL" S "GAIN LOW, GAIN =" G "AT" 100*F "HZ"	
330 GO TO 100	
340 PRINT "AMPL SERIAL" S "GAIN HIGH, GAIN =" G "AT" 100*F "HZ"	
350 GO TO 100	
360 END	

Figure 12

Test Program for an Audio Amplifier Written in BASIC

programs. With this additional core, about 18 programs this long could be stored in core memory.

Storage of Multiple Programs

When the operator types RUN, the program is executed starting at the lowest numbered instruction. If the first instruction is

1 GO TO 400

the computer will immediately start executing the test program beginning at 400, ignoring those starting at, say 100, 200, and 300. By then changing the program by typing

1 GO TO 200

and again typing RUN, the test program beginning at statement 200 will be executed.

Thus, a number of test programs may be resident in core at the same time, or they may be read into core via the tape reader, or from a disc or magnetic tape, if available.

Conclusion

What has been described is an example of an actual (see Figure 13) and typical automatic test system utilizing standard instruments and a general purpose instrumentation computer. It utilizes a simple programming language that may be learned in a few hours. It may be readily expanded or reconfigured to meet changing test needs. The combination of these factors provides an automatic test system that can be cost-justified in many applications.

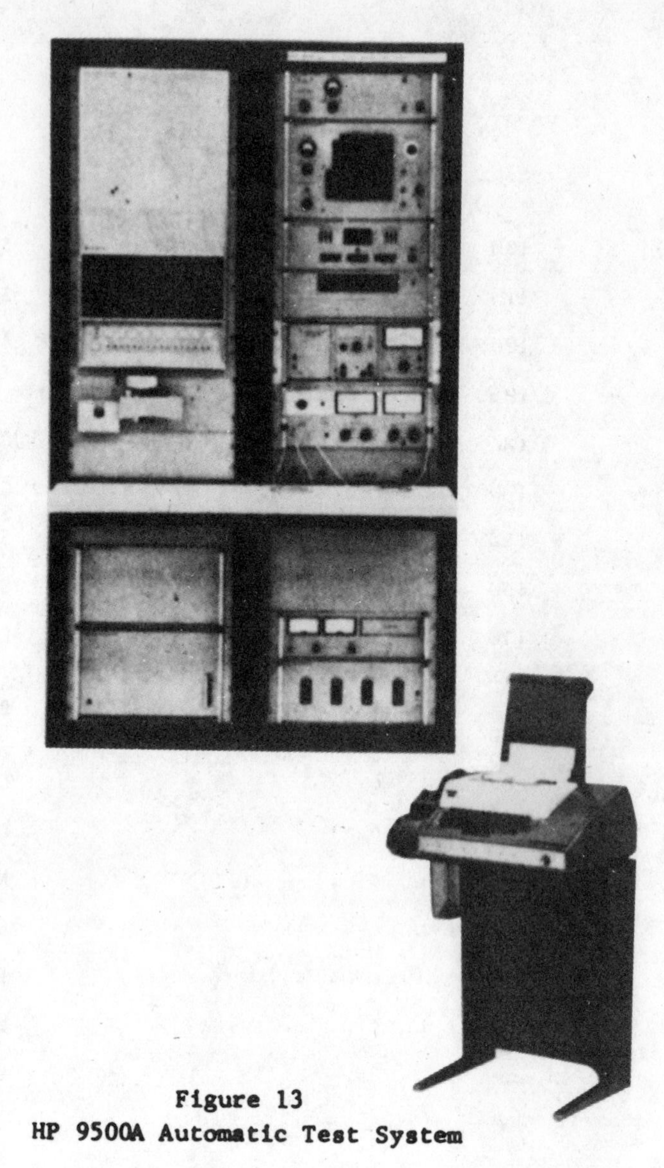

Figure 13
HP 9500A Automatic Test System

Automatic Test Systems Dedicated or Integrated

ARNOLD M. GREENSPAN, ASSOCIATE MEMBER, IEEE

Abstract—**An automatic test system is composed of parts that are basically refined versions of standard measurement and test instruments. The instruments used in the automatic test system must be automatically programmable to provide the stimulus or measurement range called for by the test program and they must be capable of receiving these programmed instructions from a computer or controller that serves as the test interpreter, director, and decision maker for the test system. Another characteristic of an automatic test system is that it requires its test instruments to be capable of being connected in any desirable electrical configuration by the test systems switching matrix. The final requirement of a test system is that it have some type of man-machine interface such as a printer or visual display to inform the operator of test results and/or provide him with instructions.**

The number of automatic test systems in use in Canada and the U. S. today has grown greatly over the last few years. These systems fall into two categories: they are either integrated systems, composed from building blocks of general-purpose programmable measurement and stimulus devices or they are dedicated systems, comprised of specially designed special-purpose stimulus and measurement devices aimed at a specific task.

This paper will examine the advantages and disadvantages of both approaches to test system design. It will highlight factors that must be considered before choosing either and will illustrate subtle and intangible aspects of this decision, which can prove crucial over the long run.

I. Introduction

THE WORDS automatic test system have different meanings for different people. This is not unusual because the tasks for which automatic test systems (ATS) are utilized are diverse, as are testing philosophies and techniques. This diversity has led to the proliferation of numerous ATS types throughout Canada and the U. S.

Manuscript received May 17, 1971.
The author is with the Aerospace System Division, RCA, Burlington, Mass.

This paper is intended to serve as an overview of the present status of the ATS field. It will discuss various applications of ATS in order to help identify those system characteristics that lend themselves to particular applications and thus act as a guide for the potential user/developer.

To represent the range of available test system types the paper will introduce the concept of the integrated and dedicated test system. The integrated system is one composed of separable, identifiable subassemblies (multimeters, pulse generator, etc.) that have been joined or integrated to configure an ATS. The dedicated system is one composed of special purpose elements not specifically designed for a test function (counters, half-adders, amplifiers, etc.). These elements lose their system functional identity when removed from the system environment.

The paper is not intended as an in-depth study of the many facets of ATS application, but rather an introduction to the ATS field for the novice and/or an introduction to segments of the ATS market with which the reader may not as yet have experience.

II. Test System Decisions

In examining his testing problem the potential user and/or developer of an ATS must make two major decisions. First, he must decide if he will perform the necessary testing manually or automatically. Second, he must decide upon a specific test approach or method within the constraints placed upon him by his first decision.

The first decision, manual or automatic, is usually easier to arrive at than the second and is related to characteristics of the tested item. There already exists a large body of data describing the advantages of automatic testing as compared to manual testing. These

Reprinted from *IEEE Transactions on Instrumentation and Measurement*, Nov. 1971.

include 1) greater speed in performing tests; 2) decrease in operator skill levels required to perform testing with commensurate reduction in training costs; 3) test results independent of individual operator skills; 4) test sequence and tolerances more consistently performed and increasing confidence in test results.

Assuming that the potential user is attracted enough by these advantages to select the automatic test, he must then determine the type of ATS he will use. This decision is no less crucial than the first, not only because this decision will play a significant part in determining the magnitude of the investment that is necessary, but more importantly because selection of the proper test system determines the ultimate utility and satisfaction the user will derive. However, the connection between the test approach or method and the type of test system selected is often not clear to the potential user/developer and is seldom emphasized by ATS vendors. This paper as it develops will attempt to make this relationship clear.

III. What Is Automatic Testing?

The simplest way to describe automatic testing is to compare it to manual testing with which most people are already familiar. Manual testing is usually performed by collecting individual pieces of test equipment, including measurement devices, special-purpose signal generators, power supplies, decade boxes, and a collection of clip leads. The test technician must plug in, set up, and connect all of this equipment to the unit under test (UUT) to make the tests he feels are pertinent. Other manual test equipment may interconnect the individual components in such a way that the operator performs these functions with simple switches. Normally, numerous tests are involved requiring the configurations and connections to be changed many times. Some sort of manual or set of instructions will normally be available to help the technician perform the tests. However, these instructions may not be complete or clear and usually will not be all inclusive. For example, the procedures or operations required to isolate a failure are too varied to allow specific instructions for every possibility, thereby lending uncertainty to the test and/or repair process.

In order to alleviate the multiple problems engendered by manual test procedures, the electronics industry's attention turned to automating the testing process. A typical ATS block diagram is shown in Fig. 1. It can be noted that the major ingredients required to perform manual testing are available within the ATS. There are power supplies, stimuli, measurement devices, and a switching system to allow the equipment to be arranged in desirable configurations. The instruction manual is replaced by a program tape that instructs the computer to carry out test instructions in the proper sequence, judge test results, and/or perform calculations. The requirement for writing tests results is performed by a printer. The keyboard and test results display have two functions: 1) to provide a means for man–machine communication within the testing process and 2) to provide the functional flexibility required in validating new test system programs. It is often necessary to use the test system operator to adjust a potentiometer, recognize if a lamp is on or off, or change the position of a switch. The printer is used to inform the operator of the action required, the display to indicate the parameter he must adjust to, and the keyboard to allow the operator to inform the test system that he has performed a necessary operation or to make a positive or negative response.

IV. Test System Operation

Generally, operation of an ATS starts with the operator informing the test system of the identity of the UUT either on the keyboard or by control switches. The test system will then seek the proper test program for that UUT and (usually) verify through keying or other means that the UUT has been correctly identified. This is necessary to avoid damage to the test system and/or UUT due to operator error.

Once the UUT identity is verified, testing will begin. The test system will read and interpret the instructions for the first test. These instructions will set up switching to connect stimulus and power to proper points on the UUT; they will also set up connections between a measurement point and measurement device. Control signals on the test tape automatically program the measurement devices to the proper scales and set the stimuli to the required voltage, frequency, slope, etc. A measurement command then will cause a measurement at the UUT test point. This measurement is transferred to the computer and compared against predetermined measurement criteria, which are also part of the test information. If the measurement is within the limits specified for that test the computer will instruct the system to seek the next test in the "go-chain" sequence. If the measurement is beyond the specified limits the computer will instruct the system to either halt and print that the UUT is defective, or direct the system to branch to a fault-isolation test sequence associated with that failure. In the case of a fault-isolation branch, testing will continue until some required fault-isolation level has been reached. As long as the results of each test are within limits, the test system will continue to rearrange switching, vary stimulus, change measurements, and evaluate the results of each test until every test in the go-chain sequence has been performed. At this time the testing will terminate and the test system will print out that the UUT is good. Thus the basic requirements for configuring an ATS are as follows.

1) A computer or controller to direct and control the testing process as well as interpret and evaluate test results.

2) Stimulus devices such as power supplies, signal generators, or pulse generators that can be automatically programmed to provide required amplitudes, pulsewidths, frequencies, and other inputs required to perform testing.

3) Measurement devices that can be programmed for the required ranges and scales needed to carry out testing.

4) A switching device to interconnect (under program

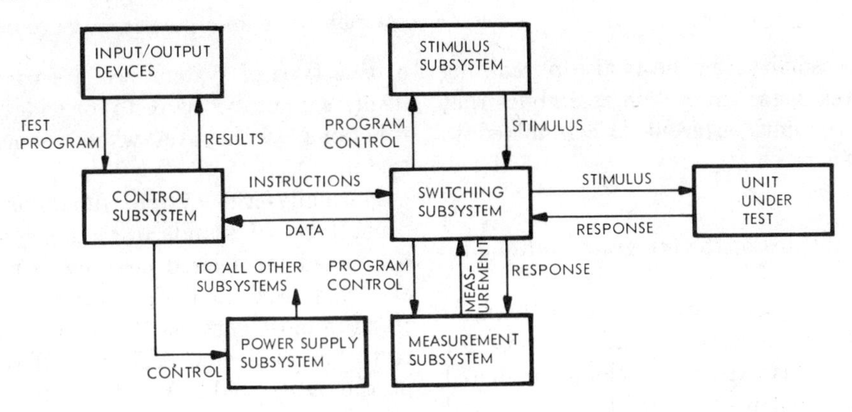

Fig. 1. Block diagram of a typical ATS.

control) the UUT to the proper stimulus and measurement devices.

5) Peripheral devices such as printers and displays to provide a man–machine interface to the degree required by the user. The sophistication of the peripheral devices will depend greatly upon the use for which the test system is planned.

6) Tape reader or equivalent device for loading the UUT program into memory. The requirements for and size of this unit are also dependent upon the test system's use. If the system is required to test a small number of UUTs it may be possible to keep all the UUT programs resident in memory. As the number and size of the UUTs grow larger it becomes necessary to store these programs on some medium and read them into the test system as they are required.

7) Computer programs to direct the testing operations.

8) Accessories and signal conditioners as required for specific test problems.

V. Integrated Test Systems

An integrated test system, for the purpose of this paper, will be considered as any system comprised of separable functional elements joined to operate as a system. For example, if the test system multimeter were removed from the system, the multimeter could still perform its function of making measurements under manual control; similarly the pulse generator could be removed and used in a manual test situation if necessary. Each element can operate independently of the remaining system elements as an entity.

This type of test system is quite common and is usually configured under two basic concepts. Either separate programmable test system entities are selected and modified to operate on a common data bus, or separate device controllers are used to act as an interface between the devices and data bus making them appear compatible to each other. In developing an integrated ATS it is desirable and sometimes possible to select devices from various suppliers that are already compatible and that meet the system test capability requirements. However, this fortuitous opportunity seldom presents itself; usually the compatibility and capability necessary for a dependable, accurate system must be engineered. An integrated system offers the following advantages to the user and/or developer.

Advantages

A. Wide Selection of Vendors from Whom to Select the Test System Building Block Components: The wide selection of vendors allows the user/developer configuring an integrated ATS to carefully select the elements of their system to optimally meet the requirements indicated by test requirement analysis and tradeoffs. In addition, the developer can often utilize the highly competitive atmosphere created by a wide number of vendors to obtain price and other advantages.

B. Expandable Capability Through Addition or Deletion of Building Block Components: The nature of an integrated ATS can often make modification or expansion of its parameter envelope less difficult than what would be required in attempting this change on a dedicated system. This is especially true when a software compatible building block can be used as a direct replacement. In this case the cost for expanded capability is merely the cost to purchase the replacement building block assembly. However, unless software-compatible units can be found, the problems and expense involved in modifying even the integrated ATS can be prohibitive.

C. The Initial Engineering Design and Development Costs May Be Lower: This advantage, as well as that stated in Section V-B refers to hardware rather than software expense. It is obvious that purchasing functional elements will eliminate design and development costs to the system developer. Of course, the price that is paid for the off-the-shelf components in the integrated ATS will contain some vendor development costs, but these costs will be distributed over many units and thus will not represent a large burden to the purchaser.

D. Expertise of the Building Block Vendors: A very significant advantage to the developer when choosing the integrated ATS is the opportunity to incorporate the expertise of the vendors whose assemblies he uses. Many years of design, development, skill, and specialized knowledge are purchased and accrue to the ATS developer when he opts for an integrated system. However, this advantage is mitigated by the possibility that he is also purchasing subtle and inherent weaknesses in system assemblies that will not become known until after he is committed to their use.

Disadvantages

Some of the disadvantages of the integrated automatic test system follow.

A. Hidden Development Cost: Development of the integrated ATS includes an initial engineering outlay for integration. The size of this expense is seldom predicted accurately because the problems that will be encountered when attempting to join the individual elements into a single operational system cannot be easily identified in advance. The types of problems that can be expected are grounding, EMI, EMC, timing, and loading problems. It is well known that these types of problems are the most difficult, frustrating, and consequently most expensive problems to solve.

B. Inefficient Utilization of Components: The building blocks selected for incorporation into an integrated ATS usually are designed for general purpose applications and thus will have capabilities beyond the requirements of the test system itself. The inefficiency of unused capability as well as reduced system reliability due to the presence of unnecessary failure mode candidates can prove troublesome.

C. System Size Is Greatly Increased: Because the packaging of the integrated ATS building blocks is not specifically designed for use in a system environment, a great deal of additional space is required. In addition, special cooling considerations often are necessary when individual test equipment components designed to operate in a free space environment are joined within a rack. The thermal analysis required to establish special cooling considerations may be complex and often must be resolved through a compromise not entirely satisfactory to any of the system's elements.

D. Increased Maintenance Requirements: Maintenance considerations of an integrated ATS is a drawback because the self-diagnostic capability of an ATS is somewhat circumvented when the system is composed of individual programmable components. The degree of diagnosis that is possible is limited because the programmable boxes are not usually designed to be tested automatically. Self-test is therefore reduced to determining which of these devices and its associated compatibility device, if the system is so configured, has become defective. Diagnosis and repair of these elements once they are identified must then be accomplished with standard manual test techniques requiring the requisite skill level and equipment to do the repair. It can then be seen that another of the major advantages of ATS (reduction of the required operator skill level) is also somewhat circumvented.

VI. Dedicated Test Systems

A dedicated test system is one that is specifically designed for a given task or class of tasks. The elements of this type of system, when separated from the system itself, cannot be used to provide characteristic system functions as they do when connected within the test system.

Normally, the system functions are configured from some type of standard modular devices arranged and connected for desired characteristics. Often the modular elements used to design the test system are the same (for the most part) as the modules that the system must test. The dedicated ATS has all the test characteristics of the integrated ATS. It is usually built about some selected computer that provides control and analysis, and it contains standard man–machine interfaces such as visual display, printer, and keyboard.

Advantages of This Type of System

A. Greatly Reduced Size When Compared to Integrated Systems: The dedicated test system will usually occupy less than half the space required by an integrated ATS. Not only because one does not have to accept the packaging idiosyncrasies of various and sundry test equipment manufacturers or allow redundant space for blowers, fans, and power supplies, and special rack mounting brackets, but also because space is no longer allocated to nonusable functions. When the space available for test and/or maintenance is limited, a dedicated ATS is often the only alternative.

B. Efficiency and Reliability Are Increased: The efficiency of a dedicated ATS is high because it contains only those functions required for the testing process; whereas, the integrated ATS will be configured from parts that usually possess capabilities and parameters well beyond the requirements of the test system itself. The dedicated ATS capabilities and parameters are custom designed for a specific test and/or support mission. The reliability as defined by its mean time between failures may also be increased merely through elimination of failure modes in superfluous circuitry. In addition, the reliability of a dedicated ATS is quite predictable because the components and design can be carefully selected to achieve the desired reliability. The reliability of the integrated ATS must be estimated on the basis of the sums of the claimed reliabilities of the various manufacturers whose components make up the system. The results obtained in this manner are often misleading because the claimed reliabilities are calculated in different ways under unknown conditions, which disallows direct comparison or addition of individual results.

C. Maintenance Is Simplified: By its nature the design of the dedicated test system lends itself to automated maintenance techniques. Development of stimulus measurement and switching subsystems through modularized elements designed for specific tasks allows recognition and isolation of failures with greater precision. This is because the modular design inherently has a large number of internal monitoring points available for self-diagnosis. The availability of these additional test points allows the designer of the self-test program to isolate a test system failure to a single or small number of modules.

D. Availability Is Higher: The mean time to repair for the dedicated system is shorter and hence, availability is greater than that of the integrated system due not only to the precision with which failure can be identified but also due to the ease with which the failure can be corrected. The modular nature of the dedicated ATS lends itself to rapid removal and replacement of the suspect element or elements. This rapid repair is not possible with the integrated ATS because removal tends to be difficult and it is too expensive to carry replacement spares on a building block basis. The cost for sparing the dedicated ATS modules is low because of the redundancy of module types within the typical dedicated system.

Repair of the defective element is also simplified in a dedicated ATS. The test programs required for repair of a module are easier to create and more practical than those that would be necessary for repair of an integrated ATS subsystem element. This is the result of limited test point availability and accessibility, larger volume of failure mode condidates, and failure analysis required for the integrated system building block.

E. Special Testing: Often when requirements are unique or when testing is necessary to check uncommon UUT parameters, a dedicated system is the only ATS solution. A good example of this is optical testing for which special equipment and circuits have been designed to accomplish the testing task.

Dedicated Test System Disadvantages

A. Cost: The development cost of a dedicated ATS tends to be greater than that of an integrated system. The difference between the two will be less than that implied by comparing the component acquisition costs of an integrated ATS with design and development costs of dedicated ATS. There are subtle and somewhat hidden engineering expenses associated with developing an integated ATS, such as the costs for compatibility devices and/or modification of the ATS building block elements, and the added expense to develop a more complex software package able to drive the diverse elements of the integrated system.

If the dedicated ATS is to be purchased rather than developed, the cost difference in acquiring the two types of systems could become smaller. The rate at which the developer wishes to recover these nonrecurring costs of design and development will determine the cost to the purchaser.

B. In-House Requirements: Developing a dedicated ATS requires a broader pool of expertise and knowledge than is required of the purchaser of an integrated ATS. That is, the skill required for the design of the test system must be available within the developer organization. The risk is high that troublesome or inadequate design approaches could become incorporated into the test system unless the design group is carefully chosen and supervised. Errors so incorporated must slowly be modified out again as experience with the test system is gained. This process can prove costly, painful, and is not always entirely successful.

C. Limited Test Set Capability: The purchaser/developer of a dedicated ATS must be very clear about what the system test capability must be; the typical dedicated ATS will be especially designed for optimum performance of a given task or class of tasks. This can be serious constraint if an error or miscalculation occurs when specifying system requirements. Modification of the dedicated system is of course possible, but the redesign tends to be expensive and sufficient space may not be available to incorporate necessary changes without mechanical redesign.

VII. Other Factors

This paper has so far addressed itself to the advantages and disadvantages inherent within integrated and dedicated test systems. It will now examine some of the peripheral factors affecting a decision between the two approaches.

Type of Testing

ATS are applied to many types of testing. Among these are: military and commercial test, production and support test, qualification and diagnostic test, component, printed circuit, assembly and system test. It will not be possible in this paper to deal with each of these in depth. However, a brief examination of each with a view to pointing out the more significant requirements and relationships between types of testing and types of ATS should prove valuable, since the dedicated and integrated test system approach is clearly not equally applicable to all test requirements.

Military and Commercial Testing

Military test systems must meet many criteria of reliability and maintainability that need not be considered when developing or purchasing a commercial system. Aside from the criticality of the military mission there are a number of unique logistics and training problems, which have no parallel in the commercial test world. The constraints imposed by the military test environment have led military procurement activities to depend heavily upon the dedicated ATS. This is due to the dedicated system's smaller size, ease of maintenance, and amenability to closely specified and controlled tolerances that can be oriented toward the military mission. The criticality of the support mission in a depot environment is not as severe as that in a tactical situation. However, even in the depot, the military user has opted to pay for the higher development costs of the dedicated system on the basis that it is cost effective when measured against their problem of training personnel to provide maintenance for an integrated ATS.

Commerical test systems cover a scope nearly as broad as military systems. These systems range from the very simple to the very complex and from direct comparison techniques utilizing the "known good unit" to comparison based upon detailed circuit analysis. Generalities regarding commercial test systems are difficult to formulate. However, it can be said that commercial tests systems tend to be more task oriented than general purpose.

Production Versus Support Testing

Production or factory testing is performed as part of the quality control function at interim stages or at the end of a production line. This type of testing becomes part of the production process and production efficiency depends in part upon the ease of use, speed, and reliability of the test system.

The factory test system is usually an integrated ATS, often configured in large part from available factory test equipment. The length of the production run and the test precision required are critical factors when determining the test system requirements. In many cases only slight modification, electrical or mechanical, is necessary to adapt a previously used test system to a new production run.

The repetitive nature of production line tests allows use of components that need not be as software flexible as those in a test environment where the testing requirements vary. The test process on the production line can use a relatively simple device designed to solve a single well-defined test problem. Often manual test equipment with minor modification and appropriate computer or controller interface can be used to configure this type of tester.

The reliability of the factory ATS must be high and and the mean time to repair relatively low to avoid shutdown of the production process due to an inability to test. Thus, this type of system must be designed to allow for rapid identification of failed components, easy replacement, i.e., repairability, and with an adequate number of spare building blocks so that production need not be delayed.

The requirement that factory ATS have high reliability and be easily maintainable sound very much like the requirements listed for the military ATS. The difference between the two are matters of approach and degree. The factory ATS is configured to less stringent environmental standards than the military ATS. This equipment carries an attendant reduced component cost that allows factory ATS to utilize redundancy to ensure availability. In addition, maintainability requirements for the factory ATS user is not as severe as that of the military user, due to the lowered expense of spares and because the lower turnover rate of personnel improves the average skill level.

Support testing usually takes place at remote or field locations. It is used to provide secondary level maintenance on prime hardware, reducing the repair time required and thereby lowering the spares level necessary for any given hardware item. The consequence of a failure is less immediate for a support ATS than it is for a factory ATS because the support ATS is not part of the on-line process it is supporting. That is, the factory ATS is often part of the production process and can stop this process in a relatively short period of time if it is not available. The reliability requirements will depend upon the specific support needs of the ATS user. The support ATS often must test many different types of equipment as well as be able to provide support for various versions of similar units. This requires broad test capability as well as a flexible software package. In addition the support ATS should be designed to store many programs either as resident memory or in some convenient protective device that can be used to load programs easily into the system as they are required.

ATS characteristics for support testing are not easily categorized into the dedicated/integrated classification. The field site constraints, the type of equipment supported, the capability of maintenance personnel, and the scope of the support mission all must be considered. Often hybrid systems are found where the concept of the integrated and dedicated system have been incorporated to meet optimally the specific support required.

Qualification and Diagnostic Testing

Qualification, go-no-go, or acceptance testing all refer to a test philosophy in which the UUT is examined only to determine if it is capable of meeting its specified performance parameters. No further testing is performed if the UUT fails to fall within the limits required by its performance criteria. It is either thrown away or moved to another test area where the cause of the malfunction is identified and corrected. Qualification testing is usually used for high-volume testing of low-value UUTs. The user/developer of an ATS for qualification test is usually primarily concerned with test speed and accuracy. This type of testing is often performed on large numbers of identical UUTs over long periods of time. Therefore, programming simplicity rather than flexibility is sought. Other desired qualities for this type of testing are a fairly sophisticated mechanical interface between the test system and the UUT. This interface should facilitate rapid and positive connect and disconnect between the test system and UUT. Finally, qualification testing allows utilization of a simple man–machine interface. Often nothing more than a bicolor light to indicate good and bad can suffice.

Diagnostic testing is required for more complex UUTs. The lowest level of these is the small printed-circuit board or a small assembly of only two or three components. The philosophy of this type of testing is to identify and replace a failed component to a level required in an established maintenance plan. For a printed-circuit board, this may be one or a number of component parts. For an assembly, it may be a chassis or a printed-circuit board. Often an ATS will be required to provide both qualification and diagnostic capability. This requirement imposes the most difficult challenge for the test system user/developer.

Test system requirements for diagnostic testing call for a sophisticated software/hardware combination allowing for a flexible interactive man–machine interface. Communication between man and machine by a printer and keyboard is usually a minimum. Many times photo displays are required and the industry is now considering computer-controlled video terminals that can be used in conjunction with light pencils and other operator-control devices to allow greater diagnostic capability and flexibility.

Diagnostic testing also requires that the test system come equipped with test probes that can be used to make measurements at UUT locations where a test point for automatic monitoring has not been provided. A degree of system complexity is added by this requirement because the system must be able to recognize the source from whence a measurement is made as well as differentiate between the lack of measured data and improper measured data. In addition, the test system must be able to communicate to tell the operator it is ready to make a probe measurement, then recognize a signal from the operator that a measurement has been made. Often a continuous type of monitoring requirement is encountered during diagnosis; thus this capability must also be provided by the test system.

The data ranges encountered in diagnostic testing are broader than those encountered in qualification testing. Diagnosis on one given unit could require dc to megacycle, microvolt to 100-V measurements in the process of isolating and identifying a single fault, whereas the measurement ranges used in qualification tend to be narrower and better defined.

Component, Printed Circuit, Assembly Test

Each of the UUTs heading this section possesses its own unique test problem. Many of these problems have been implied in the sections above dealing with the broader test categories. As a generality, the complexity of an ATS will vary directly as the complexity of the UUT for any given test category. This rather obvious statement was not put in merely as a gap filler, but rather to act as a warning to potential users to avoid over simplified and often misleading claims encountered when seeking a solution to a testing porblem.

VIII. Future Trends

Certain trends are becoming obvious for both the integrated and dedicated ATS. The system's size is decreasing for both, due to greater utilization of hybrid arrays, MSI and LSI. This decrease in size has been accompanied by an increase in system complexity, capability, and interactive characteristics of the test system elements.

The general purpose minicomputer is increasingly found at the heart of ATS. The availability of resident memory and utility of the software package appear to be the major factors in selecting the test system computer.

Programming for ATS is becoming simpler. This is especially true for component and digital printed-circuit board test where the programming task is becoming automated. Another advance in the programming process for automatic test systems is the increased use of on-line program validation and modification techniques.

New test techniques now in development may have a remarkable impact on the configuration of future test systems. Testing in order to recognize circuit characteristics by heat, noise, color, patterns, and other characteristics are being examined.

Multistation ATS in which minicomputers interface with and are controlled by large computers on a real time basis are presently being used very successfully in factory test of TV and computer circuitry. Other successful applications of this technique are found in monitoring of multiple remote phenomena at a centralized station.

Configuration of test systems and application of the test process is more and more becoming a system engineering problem in which entire subsystems rather than individual system elements are considered. In addition, the computer is becoming a smaller portion of the total system, due to the system elements becoming more complex and having built-in computational features.

IX. Summary

To achieve maximum utility and effectiveness, the user of ATS must be aware of the relative advantages and disadvantages of the various types of test systems that are available. Conversely, the developer of ATS should be keenly aware of the needs of the user in various test situations in order to be able to design his test system accordingly.

This paper has presented two basically different types of test systems: 1) the integrated system composed of separable subassemblies that could be removed from the ATS and still perform the same function as is performed within the test system; 2) the dedicated ATS that is configured of special purpose elements that have no system function identity when removed from the ATS environment. The relative advantages and disadvantages of the integrated and dedicated ATS were discussed in an attempt to point out and clarify some of the inherent as well as application problems that must be considered by both the user and developer in determining their ATS requirements. Finally, some ATS trends were discussed. Areas in which RCA and others have made some significant progress in applying and innovating ATS configurations were pointed out.

It is hoped that the broad range of subjects discussed in this paper will help the potential and present user and developer of ATS to better understand each other's needs and requirements for given test problems and environments.

ACKNOWLEDGMENT

The author wishes to thank J. F. Currier, O. T. Carver, and P. Bokros of the Program Management Office and D. B. Dobson of the Technical Publications Office for their helpful comments in the preparation of this paper.

AUTOMATIC TEST EQUIPMENT DESIGN

Philip W. LaClair, John H. Katsikas, and Wayne E. Hullett
Librascope Division
Singer-General Precision, Inc.

INTRODUCTION

This paper presents the specific principles and their implementation that are the basis for Librascope's approach to Automatic Test Equipment Design.

The best tool available for the control and analysis of test data is the digital computer. To apply this tool in the test system complex, Librascope has developed the following techniques and procedures:

A. English language communication with the computer and off-line program development.

B. Computer control of instruments generating input stimuli to the unit under test (UUT).

C. Accurate translation of UUT outputs from electrical signals to computer stored data.

D. Analysis of computer stored data to provide required performance parameters and measurement evaluations.

E. Automatic calibration and maintenance of the test system within specified error limits.

An automatic test equipment design that utilizes these principles results in a test system with increased accuracy, speed, and constancy. Integration of the remainder of the test system with the digital computer results in "hands off" testing, where the computer selects, controls, records, and analyzes test data. This ensures that the operator complements rather than compromises the test system because he directs only the computer.

The first principle - ELUCIDATE

The major problem in computer utilization by technicians of average skill and ability is computer programming that enables the technician to communicate with the computer. Librascope solved this problem by developing a simple and versatile programming language termed ELUCIDATE. To develop ELUCIDATE, Librascope used the systems devised for FORTRAN as a model. We have identified the basic English words engineers normally use to describe certain test functions and created a programming language that uses these words in a test system to accomplish direct testing. This language directly controls the input and output operation of the computer and rigidly constrains the internal design of the electronic test system; ELUCIDATE is not a language that can be used as a general-purpose programming tool. It is used with a Librascope-designed interface and test system only and, as such, provides programming and design efficiencies that result in an economical and effective test complex. The transformation from English commands and numeric locations to data words comprehensible by the test system is accomplished by an ELUCIDATE compiler. This compiler is written in FORTRAN and is stored in a large general-purpose computer incorporated in a commercial Time Share system accessible to authorized persons through normal Time Share procedures. Test programming is accomplished by a teletypewriter acoustically coupled to the time-shared computer.

The following economic advantages are derived from the off-line program compilation capability:

A. Only test control and function evaluation equipment required for a test system.

B. Full-time use of test system to perform its primary function-testing.

C. Programming requirement limited to an acoustic coupler and a Teletype.

These advantages allow a production facility to meet the competitive need for automation with a minimum capital investment.

The second and third principles - INSTRUMENTATION

The stimuli and measurement capabilities of a test system represent its major instrumentation requirements. Computer control of stimuli producing instruments (power supplies, oscillators, electrical loads, etc.) limits the required number of such instruments by increasing their utilization. The computer's accuracy and control permits less expensive elements to be used. Computer control also frees the test technician from repetitive control setting and checking functions, and decreases tremendously the probability of incorrect or out-of-tolerance stimulus by closing the control loop from stimulus generator to measuring instrument, through the computer.

Experience at Librascope has demonstrated that closed loop verification of stimuli signals is essential for reliable operation of an auto-

Reprinted with permission from *WESCON Conv. Rec.*, Aug. 1971.

matic test system. This method assures that each source instrument has been commanded to the correct signal output and that a second independent instrument has verified that the source instrument is, in fact, providing the desired output.

In a hypothetical test system, the setting for a signal generator carrier frequency is achieved by commanding the computer through the following steps:

A. Computer requests signal generator to 500 MHz.

B. Computer utilizes counter to establish existing output frequency of signal generator.

C. Computer uses the analog-to-digital (A/D) converter to measure the voltage controlled oscillator (VCO) control voltage.

D. From steps A through C, the computer calculates the required change in voltage to provide the requested frequency.

E. Computer inputs this voltage to the VCO control system and returns to step B.

This loop is continued until the oscillator output frequency is within the value specified by the original program command.

The method of adjusting a test set stimulus is illustrated in Figure A.

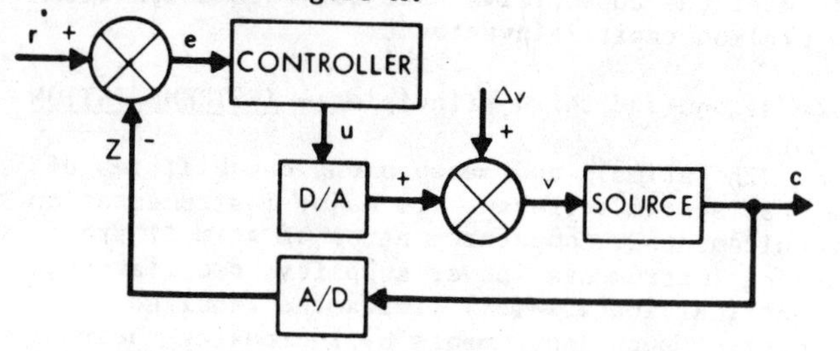

Figure A - Stimulus Adjustment System

The signal generator (source) frequency is controlled; therefore, r is the desired (reference) frequency, and c is the frequency produced by the signal generator. The error between the required frequency and the actual frequency is denoted by e; the controller is an algorithm that drives the error to zero.

The source is usually represented as a constant gain, or proportionality constant. Thus, for the signal generator, which is a VCO, the source is represented as a block with an input voltage v with a transfer function K (Hz/volt).

In general, the value of K is not known exactly because of non-linearities in the system and day to day drift, etc. Consequently, the problem is one of system identification, as well as one of control. Furthermore, the output of the digital-to-analog (D/A) converter is usually subject to some error. By using a closed loop system, this error is cancelled.

The Librascope design assumes that the control and identification problem can be separated. With an assumed value of K, the system can be described by the equations

$$c_n = Kv_n \tag{1}$$

$$u_n = u_{n-1} + \frac{1}{K}e_n \tag{2}$$

$$e_n = r_n - y_n \tag{3}$$

$$y_n = c_{n-1} \tag{4}$$

In the Z domain, this system has the block diagram representation of Figure B.

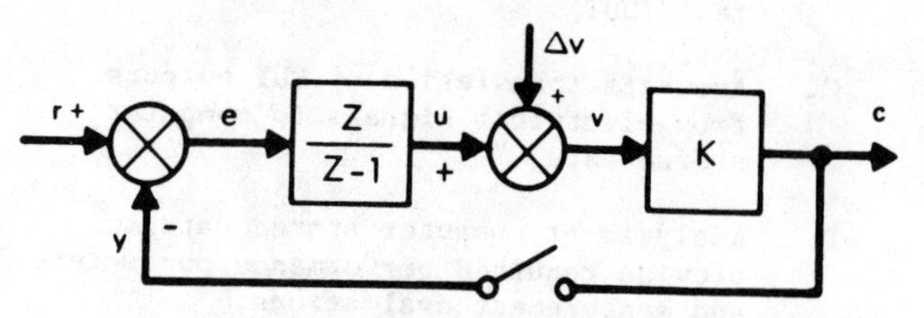

Figure B - System Block Diagram

The position error constant is

$$K_p = \lim_{Z\to 1} G(Z) = \infty, \tag{5}$$

while the velocity error constant is

$$K_v = \frac{1}{T}\lim_{Z\to 1} (Z-1)\, G(Z) = \frac{K}{T} \tag{6}$$

where T is the sampling period.

This is a Type 1 System, with a zero steady-state error for step inputs.

Let K_o be an initial assumed value of K, with accompanying error ΔK. The system identification algorithm is

$$K_n = \frac{(c_{n-1} - c_{n-2})}{(u_{n-1} - u_{n-2})} \tag{7}$$

and equation (2) becomes

$$u_n = u_{n-1} + \frac{1}{K_n} e_n \tag{8}$$

The complete system was simulated on a digital computer in the case of signal generator frequency adjustment. For example, a voltage of 100 volts was assumed to produce a frequency of 500 MHz. Thus, the true value of K is 5×10^6 Hz/volt. An initial guess of K was chosen to be 6×10^6 Hz/volt. A voltage error $\Delta v = 1$ volt was used. The normalized system error is shown in Figure C. For a sampling period of 3 ms, the output error is zero after 9 ms.

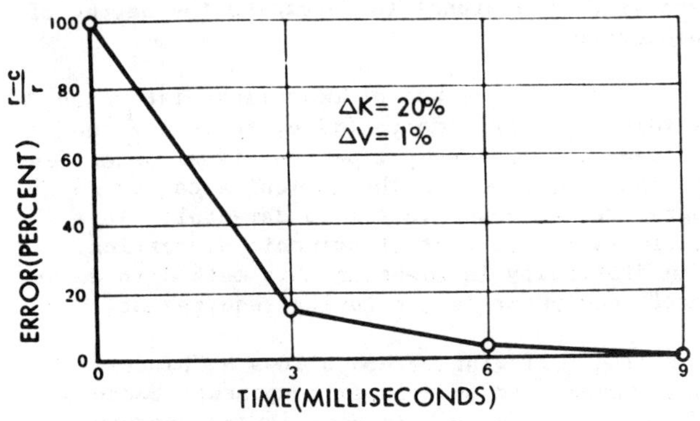

Figure C - Output Error

Integration of the computer into the measurement instrumentation of a test system provides evaluation capabilities for the testing complex that was previously unachievable because of equipment mass or cost. By sampling signal amplitudes in the time domains, the Librascope design permits the computer with a minimum of instrumentation to evaluate sophistocated electronic information. For example, analog-to-digital conversion under time and sequence control is used to transform analog and digital waveforms to computer stored data. This transformation is accomplished within the constraints of sampling theory to ensure accurate, relatable digital data.

Use of digital instead of analog techniques is possible by representing a continuous waveform by its values at a discrete number of time instants. The continuous signal shown in Figure D is represented by the set of points given in Figure E.

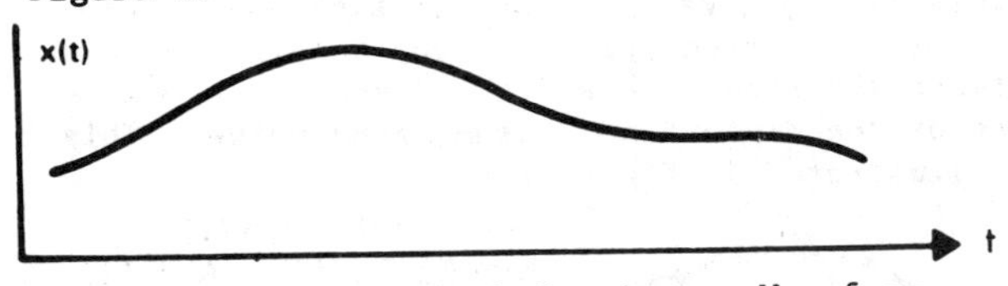

Figure D - Typical Continuous Waveform

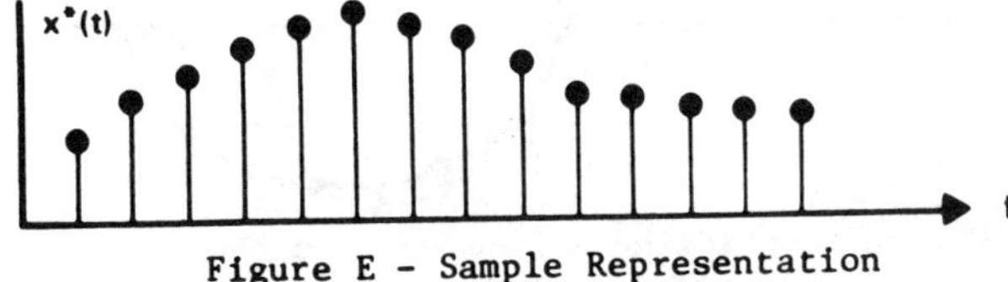

Figure E - Sample Representation of a Continuous Waveform

For certain applications, such as distortion measurements, the question arises as to how much information is lost by this representation. It will be shown that no information is lost for certain types of signals, i.e., the continuous waveform could theoretically be reconstructed exactly from its values at the sampling instants.

The sampling process can be viewed as an amplitude modulation of the unit impulse train by a continuous waveform. The unit impulse train is depicted in Figure F.

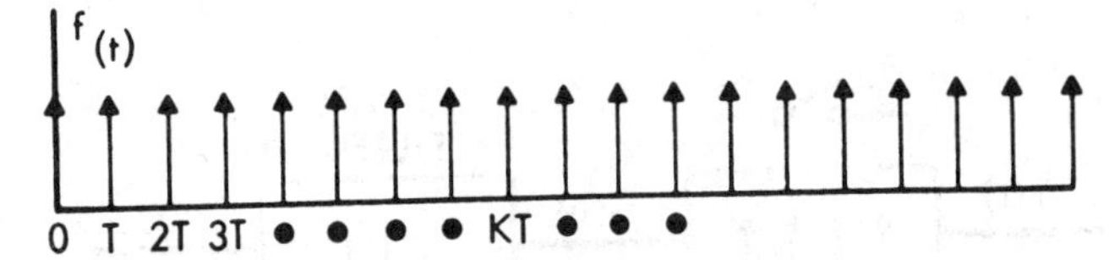

Figure F - Unit Impulse Train

The frequency spectrum (magnitude of the Fourier Transform) of the unit impulse train is shown in Figure G, with $\omega_s = 2\pi/T$.

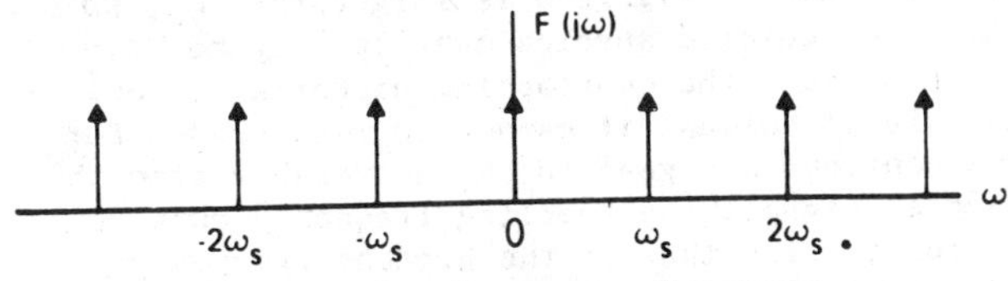

Figure G - Frequency Spectrum of Unit Impulse Train

Assume the spectrum of the continuous signal has a finite bandwidth, as illustrated in Figure H. The spectrum of the sampled signal $x^x(t) = X(t)f(t)$ is simply the spectrum of x(t) repeated at intervals of ω_s (sampling frequency) along the frequency axis, as shown in Figure J.

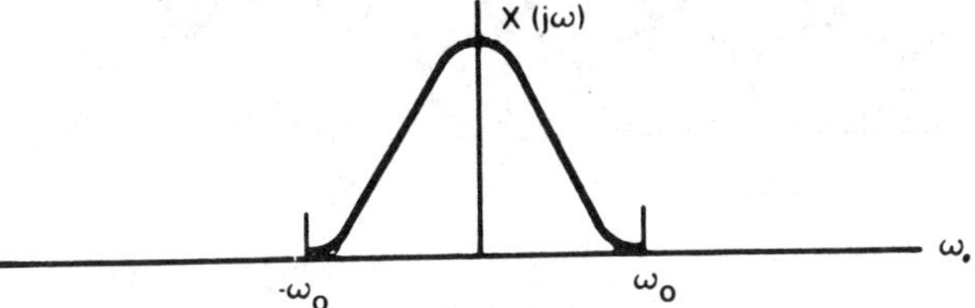

Figure H - Spectrum of Signal with Finite Bandwidth

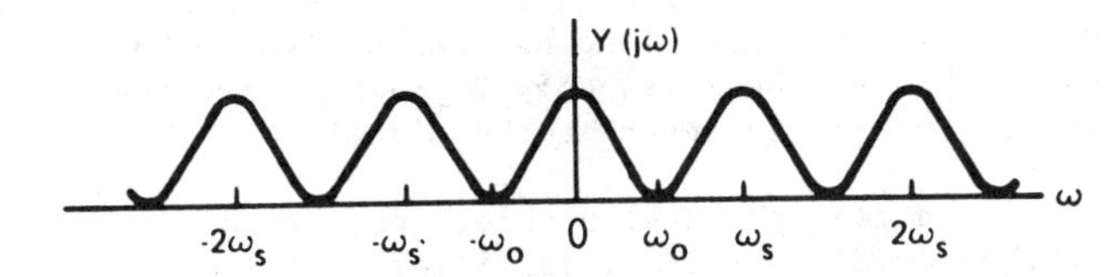

Figure J - Spectrum of Sampled Signal

The spectrum of x(t) can be recovered exactly from the spectrum of the sampled signal (Figure J) if $x^x(t)$ is passed through the ideal lowpass filter shown in Figure K.

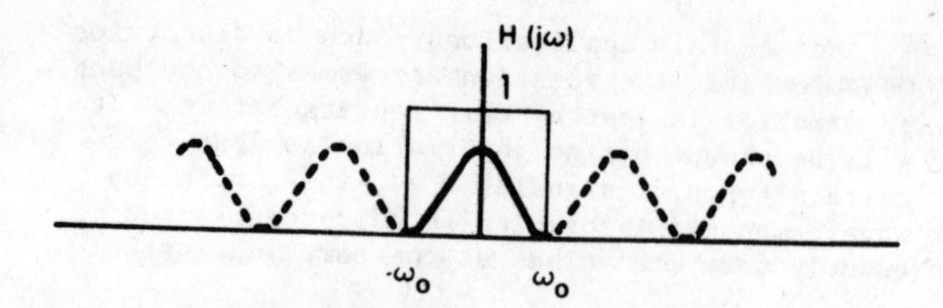

Figure K - Ideal Low Pass Filter

The sampling and recovery is illustrated in Figure L.

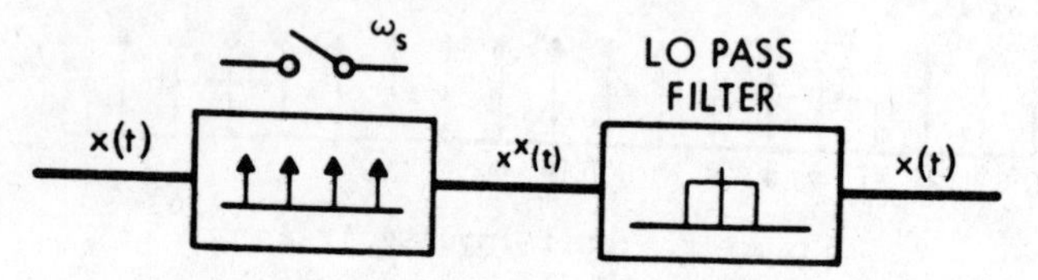

Figure L - Sampling and Recovery

Assume that the bandwidth of the continuous signal shown in Figure J is larger than ω_o, so that the repeated spectra overlap (Figure M). In this case, the overlapping distorts the original signal so that it cannot be recovered. For the continuous signal to be recoverable from the sampled signal, the sampling frequency must be at least twice that of the highest frequency component of the continuous signal. That is, $\omega_s \geq 2\omega_o$.

This is the content of Shannon's sampling theorem.

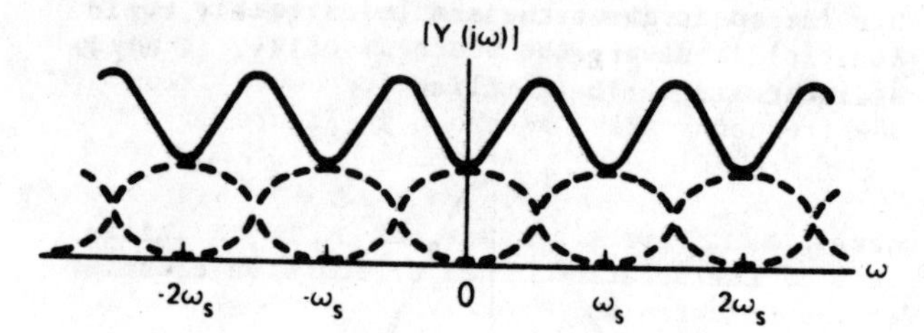

Figure M - Overlapping Sidebands Caused by a Low Sampling Frequency

For example, if an 8 kHz signal is sampled at 200 kHz, the 8 kHz signal can be recovered exactly if it contains no harmonics higher than the 12th. Furthermore, only 25 samples are required to exactly describe the signal.

In a real physical problem, neither a finite bandwidth nor an ideal filter exist; however, good approximations are available.

The fourth principle - ANALYSIS

Once transformed, the test data is in suitable form for analysis by any technique that can be reduced to a mathematical model, i.e., amplitude, frequency, noise, distortion, time and phase relation; etc. The advantages of combination of principles two, three, and four lies primarily in the ability of the test system to perform any type of analysis without the expense of buying, maintaining, and operating special analysis equipment. Additional benefits accrue from the accuracy and speed of analysis and the ability of the computer to output the results in convenient and easily comprehensible form.

The following is a sample of a mathematical analysis of a signal to determine the degree of distortion.

The usual method of measuring distortion involves tuning a notch filter to remove the fundamental frequency from the signal under test, so that the power in the harmonics can be compared to the power in the fundamental. This ratio is the percent of harmonic distortion. The difficulty in adapting this method to automatic equipment is the tuning requirement.

The following method allows a "hands-off" measurement and uses fewer components because the bulk of the work is done in the computer.

Basically, the concept is as follows: It is desired to determine the percent of distortion in a signal x(t). The signal is sampled at a 200 kHz rate, as illustrated in Figure N. Because the maximum audio frequency to be tested is 6 kHz, then sampling at 200 kHz will detect harmonic components up to 100 kHz.

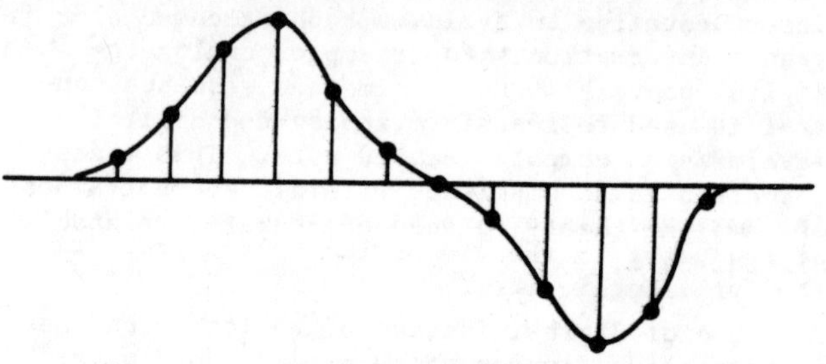

Figure N - Sampled Signal Under Test

The sampled signal is converted to digital form in an A/D converter, yielding a set of points $[x(t_k)]$, k=1,2,...N. If a sine wave can be fitted to these points, then the difference between the sine wave and the signal is a measure of the distortion, assuming no noise. This is illustrated in Figure P.

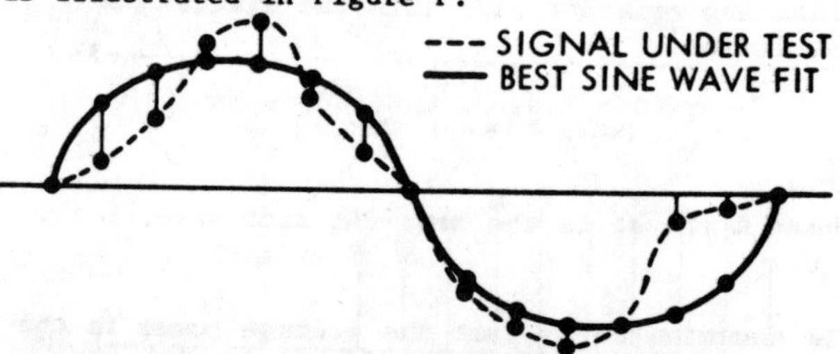

Figure P - Fitting a Sine Wave to the Sampled Signal Under Test

The sine wave is fitted so that the sum of the squares of the differences between the sine wave and the distorted signal is minimized. Mathematically, the problem is to determine the values of A and ϕ that minimize the function

$$J(A,\phi) = \sum_{i=1}^{N} [x(t_i) - A \sin(\omega t_i + \phi)]^2 \tag{9}$$

Let the minimizing of values be denoted by $\hat{A}$ and $\hat{\phi}$. Then, the best-fit sine wave is given by:

$$y(t) = \hat{A} \sin(\omega t + \hat{\phi}) \tag{10}$$

The measure of the distortion is given by:

$$D = \frac{2J}{NA^2} \tag{11}$$

which is derived as follows:
The distorted signal under test is written

$$x(t) = s(t) + d(t) \tag{12}$$

where s(t) is the pure sine wave (undistorted) portion of x(t) and d(t) is the distortion. Conceptually, if the signal x(t) were expanded in a Fourier sine series

$$x(t) = \sum_{k=1}^{\infty} a_k \sin k\omega t \tag{13}$$

then equation (12) represents the decomposition

$$x(t) = a_1 \sin \omega t + \sum_{k=2}^{\infty} a_k \sin k\omega t \tag{14}$$

into the fundamental plus the harmonics (distortion).

The harmonic distortion is defined as the ratio of the power in the distortion to the power in the fundamental, i.e.,

$$D = \frac{\frac{1}{T}\int_0^T d^2(t)\,dt}{\frac{1}{T}\int_0^T s^2(t)\,dt} \tag{15}$$

where s(t) and d(t) have period T.

The numerator of equation (15) is simply $\Delta t/T$ times the least square value, J, as seen by substituting equation (12) into the discretized version of

$$J = \int_0^T [x(t) - \hat{A} \sin \omega t]^2\,dt \tag{16}$$

where A sin st is the best-fit sine wave, i.e., s(t).

The denominator is just the average power in the best-fit sine wave

$$s(t) = A \sin \omega t \tag{17}$$

which is

$$\frac{1}{T}\int_0^T S^2(t)\,dt = \frac{A^2}{2} \tag{18}$$

Substituting (16) and (18) into (15), we have

$$D = \frac{\frac{\Delta t}{T} J}{\frac{A^2}{2}} \tag{19}$$

or

$$D = \frac{2J}{NA^2} \tag{20}$$

where N comes from the discrete form of (16):

$$J = \sum_{i=1}^{N} [x(i\Delta t) - \hat{A} \sin \omega i\Delta t]^2 \Delta t \tag{21}$$

thus,

$$T = N\Delta t \tag{22}$$

Minimization and evaluation of the distortion factor D is performed numerically in the computer. Because a good initial guess of amplitude A and phase ϕ can be readily made, the minimization is performed by solving the simultaneous equations by Newton's method:

$$\frac{\partial J}{\partial A} = 0 \qquad \frac{\partial J}{\partial \phi} = 0 \tag{23}$$

This method is characterized by extremely rapid (quadratic) convergence. Specifically, if it is desired to solve the equation

$$f(x) = 0 \tag{24}$$

where f and x are n vectors, an x_0 is an initial guess of the solution; then a recursive equation for the solution is:

$$x_{i+1} = x_i - \frac{\partial f}{\partial x_i}^{-1} f(x_i). \tag{25}$$

where $\frac{\partial f}{\partial x_i}$ is the n x n matrix

$$\frac{\partial f}{\partial x_i} = \begin{bmatrix} \frac{\partial f_1}{\partial x_1} & \cdots & \frac{\partial f_1}{\partial x_n} \\ \vdots & & \\ \frac{\partial f_n}{\partial x_1} & \cdots & \frac{\partial f_n}{\partial x_n} \end{bmatrix} \tag{26}$$

Solutions correct to several decimal places within three or four interations are typical with a good initial value of x_0.

The fifth principle - AUTOMATIC CALIBRATION

Automatic test system down time during scheduled production testing cannot be tolerated. To prevent unscheduled down time and verify system operation readiness and instrument calibration, Librascope employs an automatic calibration technique. This technique begins with system transfer standards (i.e., crystal clock, standard cell, etc.) and checks each instrument's operating characteristic against its computer stored calibration data. Using this data, a current calibration correction factor for each instrument is generated, stored, and applied to that instrument's measurements. In addition, a stored record of previous calibrations is updated and interrogated to determine any trend toward out-of-tolerance operation. Should this trend occur, the faulty instrument can be scheduled for recertification or replacement. During the calibration procedure, all test system switches, relays, and control devices are individually exercised. Should any device malfunction, the operator is informed of the nature and location of the fault.

The Librascope Automatic Calibration Technique (ACT) performs three functions:

A. Increases overall system accuracy.

B. Increased productivity by reducing the frequency of periodic calibrations.

C. Provides a system self-test.

This is accomplished by an adaptive procedure that measures the test equipment errors and corrects for them.

Basically, the procedure works as follows: Upon a command from the operator, the system performs a sequential self-test of the instruments comprising the test system. Because the theoretical output of the instrument at each operating point is known, the instrument error is found by comparing the theoretical output with the measured output. The resultant errors form a calibration curve, as shown in Figure Q, and are stored internally in the computer.

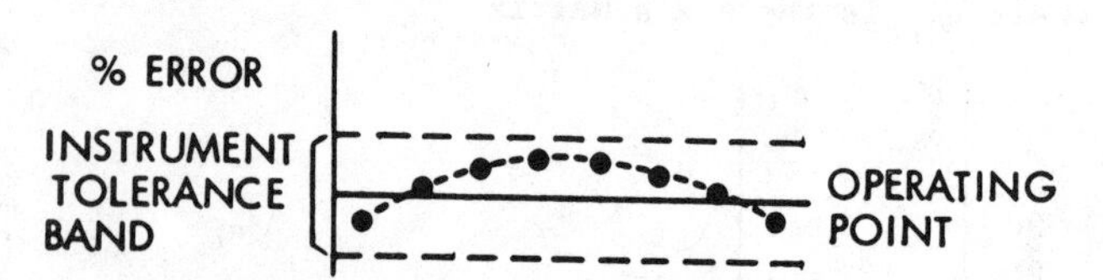

Figure Q - Calibration Curve

The test system accuracy is defined to five discrete levels:

A. Accurate

B. Acceptable

C. Degenerating

D. Calibration Required

E. System Failure

A system printout defining a level of A and B is an acceptable operating condition. A level C printout requires verification of the individual equipment tolerance level to determine the degenerating unit.

The testing sequence is designed so that the instruments are verified in descending order of tolerance. An oscillator is used to verify the counter and sampling system, a voltage source is used to verify the A/D converter, then the signal stimuli are verified, and so on until all stimuli and measurement instruments are verified. Furthermore, during each measurement of an instrument, the computer corrects instrument errors by referencing to a previously stored calibration curve.

At the end of the self-test a calibration curve has been stored for each instrument in the test system. These calibration curves are used during the normal (production) operating mode of the test system, so that each measurement of the UUT is corrected for the error of the testing instrument. Thus, the accuracy of the test system has been extended to beyond that of the tolerances of the individual instruments.

To provide a figure of merit that would indicate when a manual calibration of the unit is due, the normalized difference between instrument error and instrument tolerance is computed at each operating point for each instrument. Thus, if the tolerance is T percent and the measured error is M percent, the normalized difference is

$$\frac{T \cdot M}{T} \times 100 \text{ percent} \qquad (27)$$

This is normalized in the sense that the figure for a perfect (zero error) instrument would be unity. In a sense, this number is the percent of the percent of error. The result is a set of numbers with one number for each operating point of each instrument. These numbers are positive if all the instruments are within tolerance. The smallest of these numbers is the figure of merit, C.

The graph of C vs. Time, shown in Figure R, provides an indication of a trend toward an out-of-tolerance condition, due, for instance, to drift.

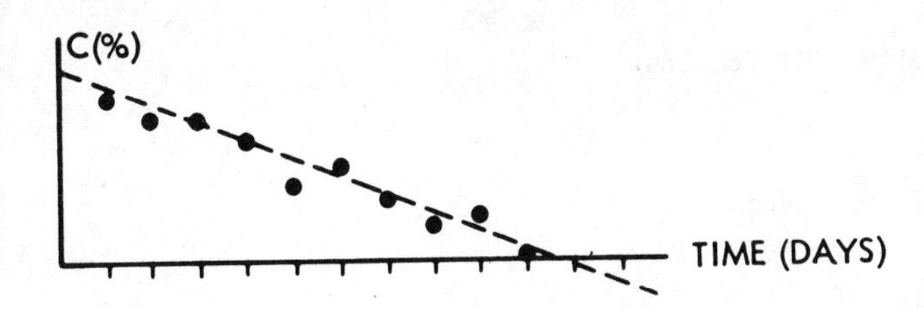

Figure R - Example of C vs Time

In this manner, the test system is recalibrated before an out-of-tolerance condition exists. Furthermore, even if an out-of-tolerance condition develops, the preceding production test results are not suspect because the errors are controlled by use of the calibration curves. Using this technique, the test system needs to be calibrated only when an out-of-tolerance condition exists, rather than periodically.

System accuracy is now on the same order as the standard within the test system; in this case, a crystal oscillator and a precision voltage source.

EQUATE - NEW CONCEPTS IN AUTOMATIC TESTING

Mr. J. Kelly and Mr. P. M. Toscano

ECOM Philadelphia, Pa.
and
RCA Burlington, Mass.

The EQUATE is a computer controlled automatic test system for use in the performance of acceptance testing on a wide variety of circuits, assemblies and systems and the on-line generation, editing and validation of test programs. The system is designed for use by the engineer or technician inexperienced in computer programming. The computer is used to control the generation of *UUT waveforms and analyze UUT response signals. In this manner both standard and non standard waveforms may be generated by one general purpose device. All waveforms and normal parameters are measured and analyzed. Special design features of EQUATE will be presented, such as the optical dial and meter reader and the UUT control driver whose functions are to reduce operator participation in the acceptance tests. Conclusions will include a summary contracting EQUATE with classic ATE designs.

*Unit Under Test

INTRODUCTION

Background

Until now the computer in automatic test equipment performed essentially administrative functions such as directing test sequencing, making go/no go decisions and recording results. Generally the testing goals were achieved by connecting stimulus and measurement "black boxes" to given points on a UUT, however, if the computer is a part of the operational subsystems, the hardware content will be reduced. This inclusion is effectively performed by two procedures:

1. Computation of desired parameters from simple standard tests such as the measurement of modulation by taking a set of 64 sample measurements and using the computer to perform a Fast Fourier Transform.
2. Elimination of special equipment computer generated waveforms closely resembling the UUT's actual operating stimulus (i.e. pulse on a pedestal).

Since actual signals in operational systems have a wide variety, an effective test station must have the capability of generating any waveform. Based on a mathematical formulation and using digital techniques, the needed variety can be synthesized with a minimum of hardware.

Additional advancement in test systems has been realised in the use of high order test languages (such as ATLAS) to write application programs. A highly pure subset of ATLAS (called EQUATE/ATLAS) is compiled on line to yield an interpretive code, which is executed at run-time on the test system.

Applications

EQUATE was initially conceived as a quality acceptance tool, hence the acronym EQUATE (Electronic Quality Assurance Test Equipment), to provide capability for performing quality acceptance of electronic devices and systems at a manufacturer's plant without resorting to sampling plans. Actually ingredients needed for that requirement creates an all-purpose automatic test base. The proper source program accomplishes the particular type of test that is desired. The capability to do diagnostic and calibration testing (as well as performance testing) is intrinsic in the basic EQUATE system.

The current development model consists of a disc operating system (including computer and peripherals) and a stimulus/measurement system with test capability from DC to 18.0 GHz. Application studies indicate that specific deployments might be compartmented as follows:

1. Low Frequency and Digital
2. RF
3. Microwave and Radar
4. Non Electronics (Optical, Hydraulic, Pneumatic, Automotive, etc.)

These systems can be configured by utilizing elements of the total EQUATE system.

SYSTEM DESCRIPTION

General

The system is described in four (4) parts:

1. Control
2. Stimulus - generation of low frequency, digital, RF and microwave.
3. Measurements - by use of low and high speed sampling units basic properties of all signals are determined.
4. Programming Language - EQUATE/ATLAS - executed by the computer (NOVA800) and its peripheral devices, under the direction of the resident executive and the application program in core at any given time and is

Reprinted from *IEEE Auto. Support Systems Symp. Rec.*, Nov. 1972.

described in order to highlight improvements over ARNIC ATLAS.

Switching straddles both the stimulus and measurements systems making connections as programmed.

A functional block diagram of EQUATE is shown in Figure 1.

Control

The control group is shown in the extreme left hand portion of Figure 1. The paper tape punch, paper tape reader and line printer are normally used only in system software development, hence would be part of the programming center and not deployed. Control of the peripheral devices is accomplished using a combination of logic cards located within the computer's main frame.

The computer also contains hardware multiply/divide, power monitor and auto-restart, and automatic program load (bootstrap). A user oriented operator-control panel provides all the control functions required for normal operation during UUT testing and while debugging test programs. Details of the control panel are shown in the photograph of Figure 2.

Stimulus

The Low Frequency Stimulus provides for the synthesis of all low frequency signals, both for the stimulation of UUT's, and for use by the remainder of the test system (e.g. for modulating a high-frequency carrier). It consists of circuitry for converting computer-generated binary-coded signals into required waveforms.

To the maximum extent possible, waveform characteristics are synthesized directly by the computer itself, thereby providing a means of handling varying stimulus requirements, i.e. in the analog and servo case, this is accomplished through the use of D/A converters driven by a digital buffer unit; for pulse and TTY outputs the digital buffer will be loaded with computer-derived values necessary for the direct synthesis of width, repetition rate and waveform.

Low Frequency Stimulus output will be routed to the Low Frequency Switching Subsystem which contain all switches necessary to interface between UUT's, the stimulus and measurement functions. In this way, the number of wires running from the test system I/O to the stimulus and measurement subsystems is kept to a minimum.

The Low Frequency Stimulus is illustrated in Figure 3.

The Low Frequency Stimulus will be capable of providing main and delayed pulse trains, either continuous or burst, at frequencies up to 20 MHz, both for UUT stimulation, and for driving the Digital Buffer Unit (DBU) and RF Synthesizer.

The DBU itself will consist of 64 uncommitted shift registers of 64 bits, whose configuration is programmable to provide variable-length digital and low-frequency waveform outputs. As a 10 MHz digital message generator/receiver, the DBU will stimulate up to 64 separate UUT input pins, with as many as 64 test patterns, while simultaneously reading test results from up to 64 UUT output pins. Alternatively the various DBU shift registers may be programmably linked to provide longer serial outputs, burst or recirculating, of up to 4096 bits, thereby providing a TTY and pulse-width modulation capability. The DBU is also organized such that one-half of the circuitry may be clocked by the main pulse, and the remaining half by the delayed pulse, so that variable time relationships may be provided.

As a waveform generator, the DBU, together with a D/A converter, filters, attenuators and amplifiers will produce any arbitrary waveform at frequencies up to 1 MHz utilizing as many as 256 separate computer-generated samples. To produce a sine wave, for example, the DBU will be loaded with a set of samples representing the desired waveform and passed through a D/A converter, active or passive Chebycheff filters, attenuators or amplifiers and routed to the UUT. Output levels will be adjusted by measuring the signal internally, and reprogramming the attenuator, thereby compensating for losses in the filters and amplifiers. Desired resolution is obtained at low frequencies by means of the pulse generator, and at higher frequencies by the RF Synthesizer driving the External Time Base.

Finally, the Low Frequency Stimulus will be capable of simultaneously outputting (or inputting) digital signals while outputting low frequency waveforms. This capability will prove useful in testing A/D and time-division multiplex UUT's, as well as devices in which analog and digital circuitry is intermixed.

The UUT Power Subsystem supplies both AC and DC power as required to all units under test. All of the DC supplies contain remote sensing and short-circuit protection, are directly programmable from the computer and are stackable either in serial or parallel (except for the 0 - 1000 volt supply which is restricted by voltage breakdown limits). The subsystem also provides variable AC power (0 - 140 V), either 60 or 400 Hz, up to 2 KVA.

Two identical RF synthesizers are used in EQUATE and cover the band from 10 KHz to 100 MHz. The output frequencies are derived from a precise standard frequency mathematically by using combinations of mixers, programmable dividers and multipliers to accomplish the operations. Each synthesizer consists of six programmable decade units, mixers, multipliers, filters and amplifiers (each contains a programmable phase locked oscillator). The functional block diagram of the basic EQUATE RF Synthesizer is shown in Figure 4. The synthesizer is programmable in 100 Hz steps from 10 KHz to 99.9999 MHz.

The AM and Pulse Modulator provides the means to modulate the synthesized signal (with the low frequency waveform generator output signal). The modulated signal is then amplified in the output amplifier and routed to the programmable attenuator for program controller levelling. Modulations possible in the first synthesizer are CW, AM, FM, SSB, FSK, pulse and FDM to 18 GHz. Modulation in the second synthesizer will be AM and FM, with the carrier extended to 500 MHz.

The microwave stimulus provides signals for the 100 MHz to 18 GHz spectral region and is housed with the RF synthesizer in a single rack. The microwave stimulus receives its initial input from RF stimulus "A" (10 KHz to 100 MHz) which is multiplied and filtered a number of times to yield the following groups of outputs:

1. 100 MHz to 4 GHz
2. 1 GHz to 4 GHz

3. 4 GHz to 12 GHz
4. 12 GHz to 18 GHz

All of these signals may be AM or Pulse Modulated. FM modulation is possible from 80 MHz to 18 GHz. Outputs are switchable (after final attenuation) to the UUT lines, to the self test line or to act as the Local Oscillator of the Spectrum Analyzer and Impedance Measurement Units. The functional block diagram of the microwave stimulus is shown in Figure 5.

Measurements

A simplified block diagram of the EQUATE Low Frequency Measurement Subsystem is shown in Figure 6. The subsystem consists of three basic measurement units as shown.

The Low Frequency Measurement Subsystem preprocesses UUT response waveforms for subsequent computer analysis and evaluation. The major circuit within the subsystem is the Low Speed Voltage Sampling Unit. The Low Speed Voltage Sampling Unit contains a 15 bit analog to digital converter, a 100 ns sample and hold circuit, a programmable amplifier/attenuator and a programmable resistor which may be driven by a DC or AC signal for use in measuring resistance and complex impedance. This floating measurement unit is fully isolated from system ground as digital inputs and outputs are photon coupled through light emitting diodes and photo transistors while pulse and AC signals are transformer coupled.

The unit has its own isolated power supply and will be used to make the following measurements:

1. Voltages - DC and AC
2. Impedance - both resistive and complex together with phase
3. Synchro/Resolver and Servo Circuits

Prior processing (mixing, detecting, etc.) extends the overall utility of the device. The digital output of the sampler is processed mathematically by the computer (e.g., Fast Fourier Transforms) to actually derive the required value. In effect, then, the computer becomes an integral part of the Measurement Subsystem. The block diagram for this unit is shown in Figure 7.

The High Speed Voltage Sampling Unit with an effective window of 4 ns consists of two programmable attenuators, a mixer, video amplifier and high speed sampling unit. The analog output of the sampler is routed to the Low Speed Voltage Sampling Unit where it is converted to digital date. The block diagram of the High Speed Voltage Sampling Unit is shown in Figure 8.

The Frequency Time Interval Measurement Unit consists of a modified *Hewlett Packard 5360A counter with a 5379A Time Interval Plug In. Pulse rise and fall times, frequency, period, TTY, and other time intervals are measured by this frequency/time interval counter operated in conjunction with the computer. The block diagram for this unit is shown in Figure 9.

The RF Measurements Subsystem, as a single unit, serves as a preprocessor of RF measured data. Inputs and outputs of this subsystem will be switched at the unit itself, thereby reducing line losses. Different types of high frequency measurements that are made directly by the subsystem include peak power, average power, VSWR phase and frequency. In conjunction with the computer and Low Frequency Measurements Subsystem, it is also possible to measure other RF parameters, such as AM and FM modulation, recovery time, pulse width, PRF, etc. A separate detector is used for measurements involving pulse and AM signals. Pulsed carrier frequencies are measured by computer-aided automatic spectrum analysis techniques.

The microwave measurement subsystem is illustrated in a block diagram in Figure 10. Signals to be measured are attenuated and switched for either power or frequency measurement and may be routed for up or down mixing, filtering, and discriminating as required to do spectrum analysis. VSWR and complex impedance are measured via the dual directional coupler and the harmonic frequency converter working in conjunction with the network analyzer.

In addition, the measurements subsystem utilizes a DC and AC standard to augment calibration and UUT certification, Setting accuracies range from one-fourth millivolt at DC to one millivolt at 20 KHz.

Programming Language

The EQUATE programming language is an extended version of ATLAS, a test oriented language originally developed for Airline Avionic Equipment Testing, that allows engineers with no software training to write programs by use of a simple reference manual. The programmable characteristics of each stimulus and measurement device in the system are defined in this manual.

Standard vocabulary and syntax are used almost entirely with minor differences to align the language to the stimulus and measurement hardware. Noun modifier, in most cases an appropriate modifier from standard ATLAS, for each programmable hardware characteristic was selected. A few new modifiers had to be designed for the unusual capabilities of EQUATE. Some extensions to the ARINC ATLAS language have been incorporated in EQUATE by addition of a small number of significant improvements to make the language easier to use. Resource allocation is handled by device identifiers which are attached as a suffix to the signal type identifiers (e.g. AC-SIGNAL SINE, DC-SIGNAL DC1, DC-SIGNAL DC3). The measured characteristic subfield has been extended to include an optional label for specifying the memory location where the measured value is to be stored and the dimensional units of the measured value must also be specified (e.g. MEASURE, (FREQ 'X' KHz)....). In the example the frequency of the input signal would be measured, converted to KHz and then stored in location 'X'. Any of the standard multiplier prefixes could have been used, such as MHz, GHz, Hz.

EQUATE may be programmed to perform computations using FORTRAN mathematical expressions on measured values and to compute limit values or stimulus values at run-time. Since these expressions may be used in any statement where a variable is allowed, the CALCULATE verb (included for compatibility with ARINC ATLAS) is not really needed.

A modified procedure definition format has been adopted. EQUATE ATLAS requires the user to

* Modification consists of EQUATE packaging with an interface permitting NOVA program control.

specify dimensional units in the procedure definition while standard ATLAS permits the dimensional units to be specified in the procedure call which requires association of dimensional units with each variable storage location. This latter procedure complicates the compilation of procedures. EQUATE ATLAS requires the user to supply dimensional units in his procedure definitions avoiding the need for code to handle this information in the compiler. Since the information must be supplied any way, it makes little difference to the user whether it is supplied in the procedure call or in its definition.

SYSTEM OPERATION

General

The three basic ways in which EQUATE is used are:

1. Software Development - performed at programming center.
2. Program Validation - performed "on line" at center.
3. UUT Test - performed at deployed locations - limited validation.

In all cases implementation is with the "on line" computer which is integral to the EQUATE system.

Software Development

EQUATE is provided with the following programs used for systems software development:

1. ALGOL Compiler - A compiler for the ALGOL-60 language with extension for definition of non-standard operators, and for bit, byte and string data manipulation. (The ATLAS compiler is written in ALGOL).
2. FORTRAN Compiler - A compiler program for FORTRAN IV extended for real time multitask programming.
3. Relocatable Assembler and Linking Loader - An extended assembler is provided that differs from more basic assemblers in four respects:
   a. Relocatability - programs are assembled so that they can be loaded anywhere in core memory by the relocating loader.
   b. Interprogram Linkages - programs can be assembled which reference data, instructions and addresses defined in other programs. All such linkages are resolved by the loader.
   c. Number Definition - simple methods for defining double precision, decimal and floating point constants are provided.
   d. Conditional Assembly - whole programs or portions of programs can be assembled or bypassed on the basis of a conditional expression.
4. Text Editor - A text editor program developed by RCA provides an extremely effective means of editing source programs. The video terminal provides hardware for character and line insertion/deletion and also provides extra storage for text exceeding the limits of the visible screen. The editor programs moves text to and from the video terminal using the following commands:
   a. SCAN - scan forward, page by page over the text stored in a file.
   b. PAGE M - locate and display the Mth page of text.
   c. LOCATE "word" - locate and display the page containing the designated character string.
   d. CURRENT - recover the current page. Used when the user has made some mistake such as erasing the whole page.
   e. RETURN - returns control to the operating system.
5. Utility Programs - EQUATE includes a full range of utility programs for debugging, binary file edit, library file edit and arithmetic functions.

All of the above programs are controlled from the video terminal keyboard via an interactive command line interpreter. At a software development center all of the above programs would be permanent files on the disc memory which can be brought into execution simply by typing the program name. At deployed installations these programs would be available on a tape which could be loaded onto the disc as required. After the loading process (which takes only a few minutes) from cassette tape, all of the programs become instantly available, as before, by typing the name of the desired program.

Program Validation

In using EQUATE for program validation the operator would go through the following steps:

1. Apply power to system.
2. Set operating mode to "manual".
3. Type "test" command and identify UUT.
4. Connect UUT.
5. Run selected tests by using appropriate control switches.
6. If necessary to make changes load the UUT source program, type "text edit" command and make changes.
7. Type "ATLAS" command to run the compiler.
8. Repeat the change cycle until all tests compile and run satisfactorily.

Testing (UUT)

In using EQUATE to test a UUT, the operator would go through the following steps:

1. Apply power to system.
2. Check that mode select switch is in "automatic".
3. Use the video terminal keyboard to type in the "test" command and the UUT name.
4. Check hook-up of UUT(s).
5. Press "START TEST" pushbutton.

At this point the system would proceed through the test of UUT 1 and UUT 2. During the time that UUT 2 is being tested another UUT can be connected at UUT station 1 etc, etc. In the manufacturing (quality assurance) test application the system merely accepts or rejects the UUT. In a depot or field maintenance application the program would have been written to automatically commence diagnostic testing after rejection (no/go) of the UUT. It is evident then, that the different applications of EQUATE (Quality Assurance, Diagnostic, or Calibration), are merely a matter of program selection.

SOFTWARE DESCRIPTION

EQUATE is supplied with an extremely powerful and capable state-of-the-art software system for test program development, debugging and validation. The EQUATE software system shown in Figure 11 is a hybrid system incorporating the best features of conventional compilers and interpretive systems. The marriage facilitates a powerful programming language, excellent program error checking, compact object code, fast execution and efficient program debugging and validation.

EQUATE software is designed for the experienced user as well as for those who have never written a program. Consequently as a new user gains experience he will not find himself limited by artificial restrictions built into the software. Modern compiler design techniques have been applied to produce an efficient on-line compiler program with capabilities previously obtainable only with very large programs that could only run in powerful and expensive off-line support computers. On-line operation facilitates program changes in a matter of minutes in contrast to the many hours or even days at a supporting computer center.

The ATLAS compiler program produces interpretive code that is stored in the disc memory. Compiled programs are brought automatically into core memory for execution in blocks as required. Test system designs that do not use disc memories are severly limited in this respect since large blocks of core memory must be allocated to resident programs to achieve satisfactory performance. This imposes very severe restraints on the design of the language, the core resident language translator, and test execution programs. These restraints generally result in primitive (and deceptively simple) programming languages that do not provide run-time variables, formatted I/O, nor an arithmetic capability equivalent to FORTRAN. With disc memory it has been possible to provide a compiler with all of these important capabilities while also providing very comprehensive error checking. In many cases the compiler will correct trivial errors (such as missing commas) but it will always detect significant errors which are spelled out on the display (and in the program listing, if one was requested) with an arrow pointing to the location of the error. The test program need not be segmented to fit the core memory. Only a very small part of it is devoted to resident programs and needed programs are swapped in and out as required.

Real Time Disc Operating System (RDOS)

The EQUATE operating system is a modified version of the system provided by the computer vendor. The modifications include driver programs for the video display, the operator's control panel, the test results printer and the cassette tape units.

RDOS provides a comprehensive file system that gives the user a simple command language to edit, compile, execute, debug, assemble, save and delete files. File protection is provided using a number of system defined file attributes. File directories are maintained. All peripheral devices are named and treated as files, providing device independence by device name. RDOS provides an I/O facility with buffered and spooled operations. Unused core storage is allocated by the operating system for dynamic system buffers and overlays. RDOS supports both single and multitask environments. Multitask environments may include a hierarchy of up to 256 classes of priority with any number of tasks at each priority. Tasks may be synchronized by communicating with each other or they may compete for CPU control asynchronously in a real-time environment.

With this powerful operating system EQUATE can be adapted to virtually all conceivable automatic test system applications.

ATLAS Compiler

The EQUATE compiler (Figure 12) is a two-phase system. The first phase analyzes ATLAS source statements and generates an intermediate code file. The second phase converts test numbers to byte addresses and resolves all test number references. The output of the second phase is the UUT program file ready for execution. The operation of the compiler is controlled by a Monitor routine that processes each statement and calls the appropriate processor for each verb. A Lexical Analyzer is used by the Monitor and statement analyzer modules to scan the ATLAS source statements and obtain information concerning each language element in the statement.

All of the statement analyzers perform syntax analysis and generate messages when errors are found. The I/O Statement Analyzer processes ATLAS input/output statements (DISPLAY, INDICATE, etc). The Test Statement Analyzer handles device control statements (APPLY, MEASURE, etc). The Preamble Statement Analyzer performs analysis of the DECLARE and DEFINE statements and creates a symbol table. The symbol table, along with an equipment table and dimensions table are used by the analyzer modules to generate the intermediate code.

The UUT program file generated by phase two consists of machine independent interpretive code. This code is structured in the following basic form: Operation code, argument 1, argument 2, ..., argument N. This code format eliminates the need for an assembler pass and linking loader pass. Also the user does not need to segment the UUT program to fit into available memory. This is handled automatically by the test-execution system which inputs segments of the program as required.

Sample listings generated by the compiler are illustrated in Figures 13 and 14. Figure 13 illustrates the compiler's diagnostic features with errors identified by a row of asterisks and a specific error type. The location of the error is identified by an up arrow (∧). Figure 14 illustrates a corrected listing of the same program.

Test Execution System

The EQUATE ATLAS run time system shown in Figure 15 operates as a task under the disc operating system. This system controls test program execution and provides a user-oriented environment for high volume testing as well as for test program development and validation. The execution environment includes an interpreter program for the interpretation of the object code file generated by the compiler program: Test Equipment Control routines which drive stimulus, measurement and switching devices; Analytic Routines which process raw measurement data to generate the measured values requested by the ATLAS program (Fourier analysis is one of these routines); and an Executive routine to interface with the test system operator.

SELF TEST CONSIDERATION

A system self check capability is provided to insure proper operation of EQUATE. It provides a diagnostic capability to detect faulty operation and locates the fault to a subassembly or circuit board. These checks and tests are made up of the following types and levels of tests:

1. A short operational check of instruments that will be used for a particular test. This type of test is fully automatic and is sometimes referred to an a "survey" test. It is software implemented and is available for inclusion in UUT programs as a confidence check at the program writer's option.
2. A short check of all the functional capabilities of the system. This test will last from five to ten minutes and is somewhat more comprehensive than the "survey" test.
3. A full detailed self check and diagnostic test of all functional elements of the system. This test is fully automatic and covers all functions to their specified tolerances. This test is also used to determine or certify the need to perform calibration at given intervals. Calibration instructions are included in the printouts as required.
4. A fault isolation procedure is provided to detect and locate faults to the printed circuit card(s) or module level. This test includes operator participation such as probing, moving a cable, connecting external test equipment etc., etc. This procedure is used when one of the previously described programs indicates a malfunction.

In all cases the operator monitors system status by use of these programs in conjunction with the peripheral media (printers and video terminal).

Other Considerations

EQUATE incorporates a unique capability which will be evaluated for cost effectiveness after deployment. This capability eliminates the need for the operator to perform tuning and loading adjustments on the UUT. A fixture containing miniature TV cameras monitors the meter pointer position of three respective meters on the UUT and at a particular point in the program measures the video signal to determine the pointer deviation. The program then computes and orders the number of pulses required to drive a stepper motor to a specific position for proper tuning. This procedure is used for "antenna tuning" and "antenna loading". A third multi-meter, internally normalized to read any one of seven parameters is controlled by a seven position switch. This switch is driven by a stepper motor controlled by the program while the camera monitors for go/no go position of the meter pointer.

Acknowledgements

Many thanks are due the following people for their contribution to the EQUATE project and to this paper:

A. Amato	Overall Mechanical Design
R. Beigel	Power & Optical Meter Reader Design
B. Bendel	Control Switching & L. F. Stimulus Design
J. Brodie	Mechanical Design
L. Dickman	Compiler & Interpreter Design
J. Fay	Software System Design
H. Hale	Mechanical Design
A. Krisciunas	R. F. Stimulus Design
J. McGrann	Mechanical Design
E. Richter	Microwave Design
F. Schwedner	Measurements System Design
E. Sutphin	Measurements Design
E. O'Brien	Measurements Design
A. Vallance	Software Design
B. Wamsley	Overall System Design

The work is being accomplished as part of the US Army Manufacturing and Technology program under contract DAAB05-71-C-2461 issued by US Army Electronics Command, Industrial Management Division, Philadelphia, Pennsylvania 19103. These programs have as their objectives the timely establishment of manufacturing processes, techniques or equipment to insure the efficient production of current or future defense programs.

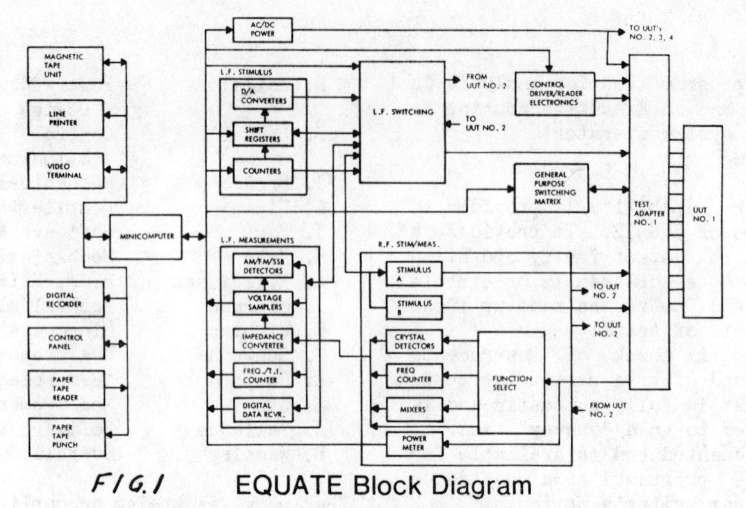

FIG. 1 EQUATE Block Diagram

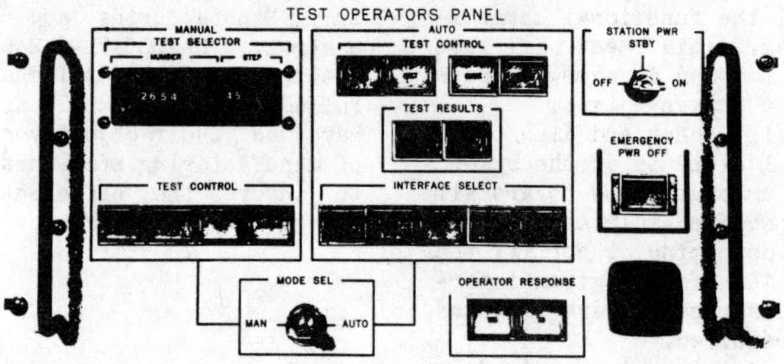

TEST OPERATORS PANEL

MANUAL TEST SELECTOR NUMBER STEP
AUTO TEST CONTROL
STATION PWR STBY OFF ON
TEST RESULTS
EMERGENCY PWR OFF
TEST CONTROL
INTERFACE SELECT
MODE SEL MAN AUTO
OPERATOR RESPONSE

FIG. 2 EQUATE Control Panel

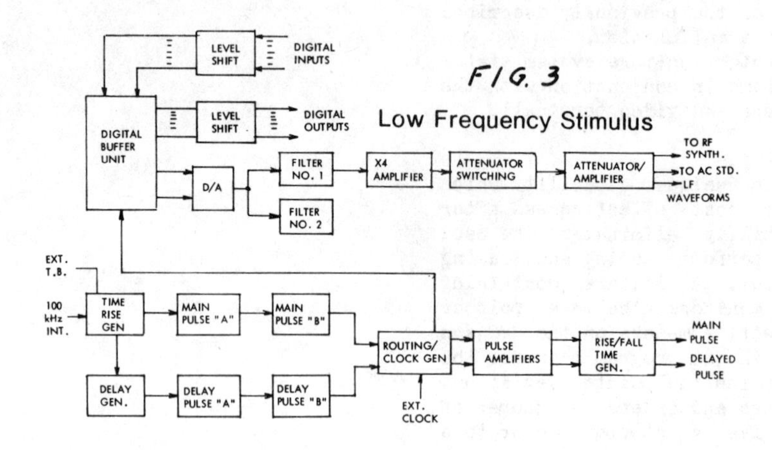

FIG. 3 Low Frequency Stimulus

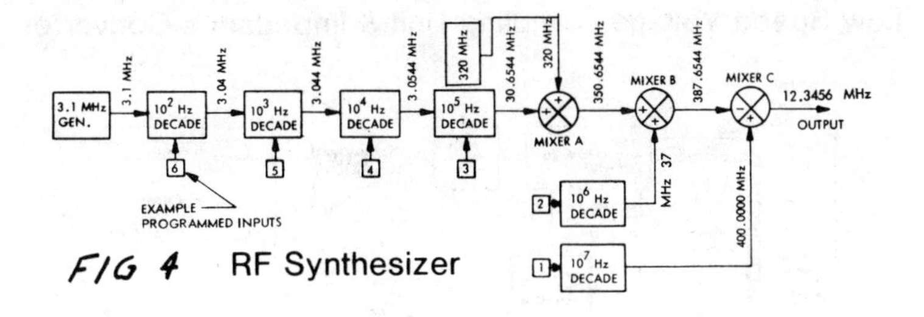

FIG 4 RF Synthesizer

FIG. 5
EQUATE Microwave Stimulus

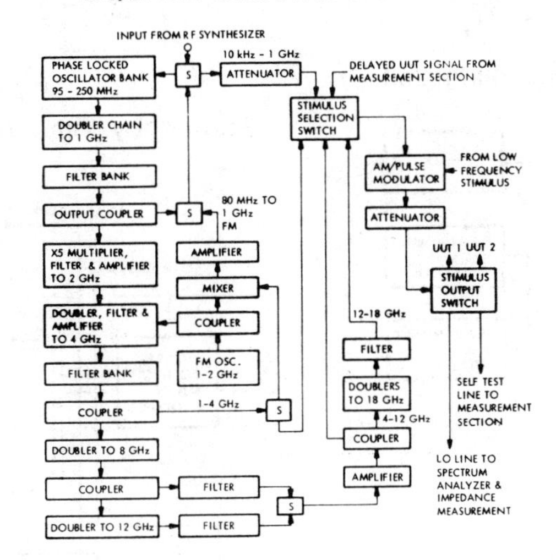

FIG. 6
EQUATE Low Frequency Measurement System
Block Diagram

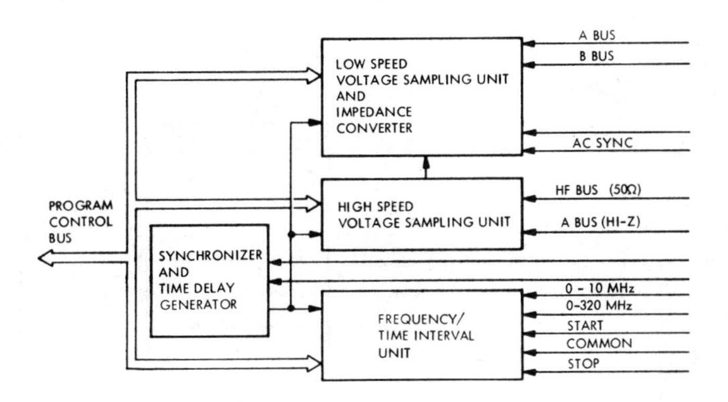

FIG. 7
Low Speed Voltage Sampling Unit & Impedance Converter

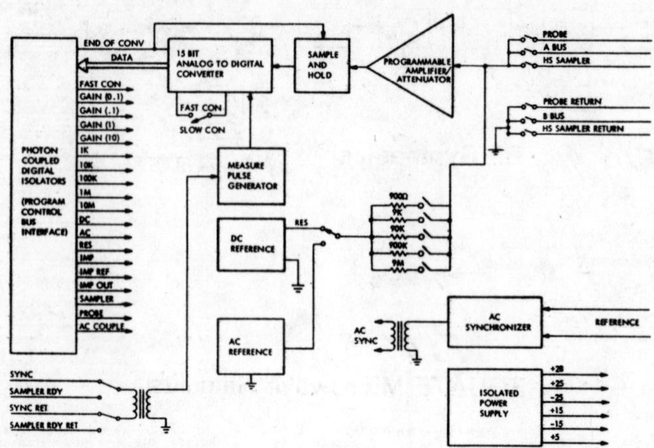

FIG. 8 High Speed Voltage Sampling Unit

VIDEO AMPLIFIER
MIXER
TO LOW SPEED VOLTAGE SAMPLING UNIT
PROGRAMMABLE RF ATTENUATOR
HIGH SPEED SAMPLE & HOLD
PROGRAMMABLE HI-Z ATTENUATOR
LF SOURCE TRIGGER
EXT MOD
HF SYNC
TIME DELAY GENERATOR
RANGE & DELAY REGISTER
COMPUTER I/O

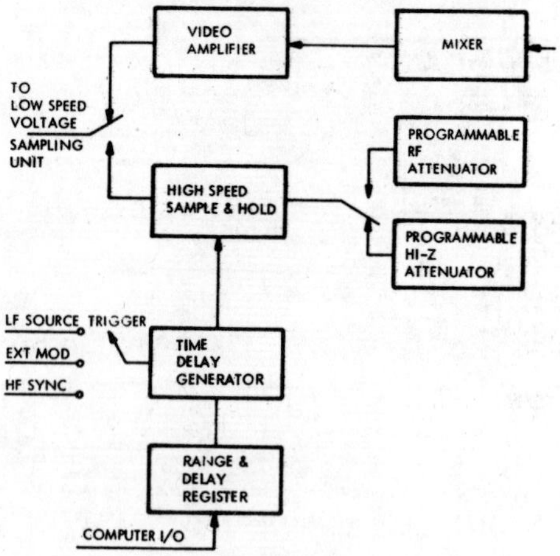

FIG. 9
Frequency/Time Interval Measurement Unit

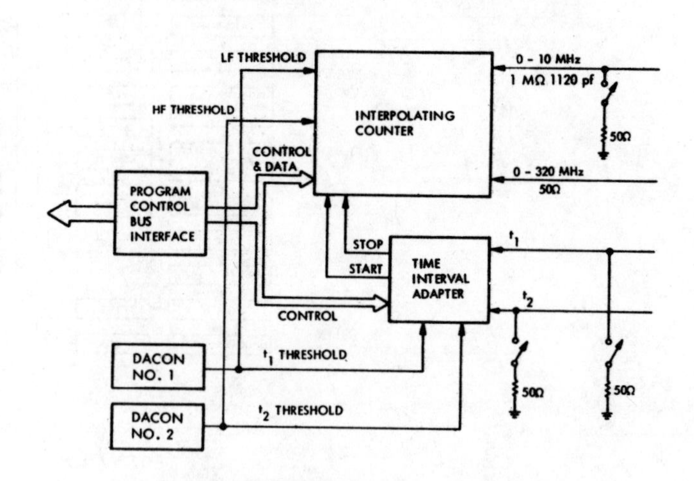

FIG. 10
EQUATE Microwave Measurements

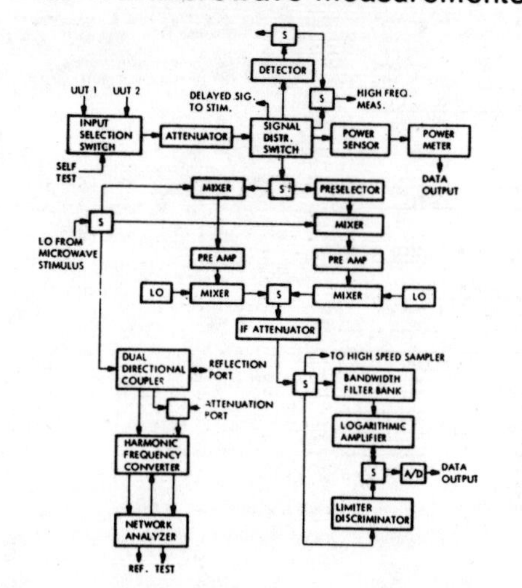

FIG. 11
EQUATE Software System

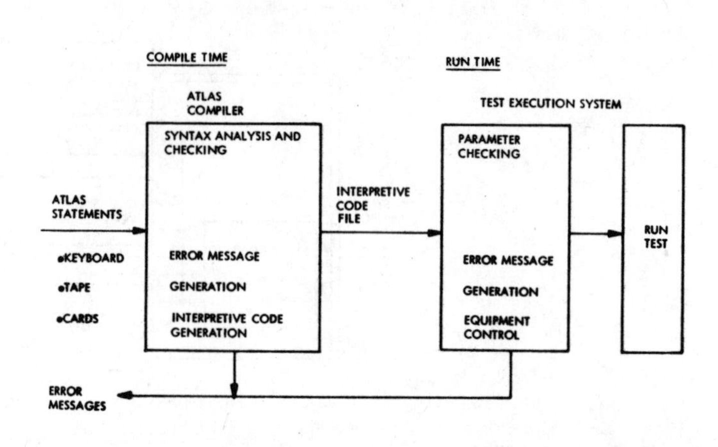

FIG. 12 ATLAS Compiler

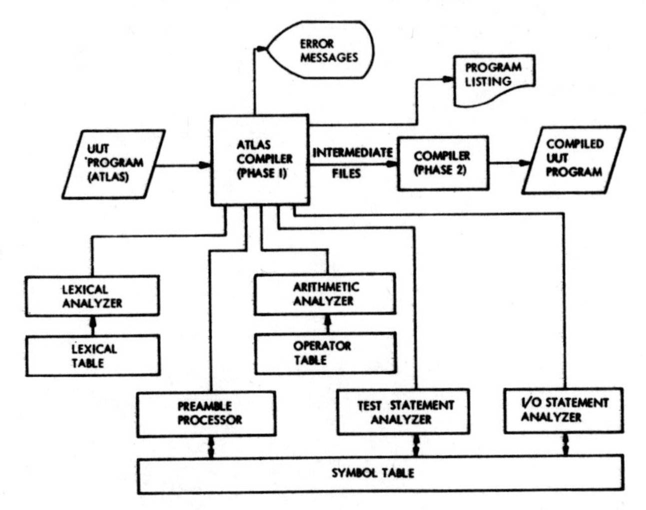

FIG. 15 EQUATE Test Execution System

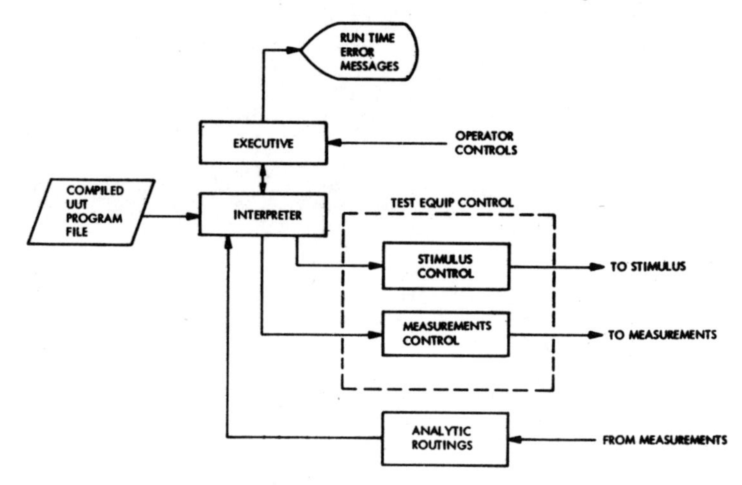

FIG. 13 Program (With Errors)

```
EQUATE-ATLAS  -  VERSION 0.1 (DEVELOPMENTAL)  -  7 AUGUST 1972

  1:         BEGIN, ATLAS DEMONSTRATION PROGRAM $
  2:         DECLARE, DECIMAL, 'FREQUENCY', 'I' $
  3:         DECLARE, DECIMAL, LIST, 'RESPONSE', 5 ELEMENTS $
  4:         DEFINE, 'RF-SOURCE', SOURCE, AC-SIGNAL RFA$
  5:         DEFINE, 'RFM', SENSOR, AC-SIGNAL RFM$
  6:         DEFINE, 'DC-SUPPLY', SOURCE, DC-SIGNAL DCA, VOLTAGE ( ) $
  7:
  8: C       RF AMPLIFIER BAND-PASS TEST
  9:
 10:         1)  SET UP POWER SUPPLY
 11:         2)  MEASURE OUTPUT VOLTAGE FOR LOGARITHMIC
 12:             INPUT FREQUENCIES FROM 10 HZ TO 100 KHZ
 13:         3)  CHECK BAND-PASS CHARACTERISTICS
 14:         4)  PRINT RESULTS $
 15:
 16: E001000 APPLY, 'DC-SUPPLY', 10 V $
 17:         CALCULATE, 'FREQUENCY':.01, 'I'=1 $
                                   ^
****** EQUAL SIGN MISSING
 18:
 19: B       MEASURE OUTPUT VOLTAGE AT APPROPRIATE FREQUENCY $
 20:
 21:      10 APPLY, 'RF-SOURCE', VOLTAGE 1 MV, FREQ 'FREQUENCY' KHZ,
 22:         FREQ MAX 100 KHZ, THRU-IMP 600 OHM, CNX HI 3 LO 6 $
 23:         DELAY, MIN 1/2 SEC $
 24:         MEASURE, (VOLTAGE 'RESPONSE'('I') MV), 'RFM',
 25:         VOLTAGE MAX 10 MV,FREQ MAX 'FREQUENCY' KHZ, CNX HI 5 LO 2 $
 26:         CALCULATE, 'FREQUENCY'='FREQUENCY'*10, 'I'='I'+1 $
 27:         COMPARE, 'I', GE 5 $
 28:         GOTO, STEP 001010 IF LO $
 29:
 30: C       CHECK BAND-PASS CHARACTERISTICS $
 31:
 32:         COMPARE, 20*LOG('RESPONSE'(3)/'RESPONSE'(2)), LT 3 DB $
 33:         GOTO, STEP 001020 IF HI $
 34:         COMPARE, 20*LOG('RESPONSE'(3)/'RESPONSE'(4)), LT 3 DB $
 35:         GOTO, STEP 001020 IF HI $
 36:         COMPARE, 20*LOG('RESPONSE'(3)/'RESPONSE'(1)), GT 20 DB $
 37:         GOTO, STEP 001020 IF LO $
 38:         COMPARE, 20*LOG('RESPONSE'(3)/'RESPONSE'(5)), GT 20 DB $
 39:         GOTO, STEP 001020 IF LO $
 40:
 41:         RECORD, "RF AMPLIFIER BAND-PASS OK" $
 42:         GOTO, STEP 001040 IF GO $
 43:
 44: B       FAILED BAND-PASS TEST $
 45:
 46:      10 PRINT, "FAILED RF AMPLIFIER BAND-PASS TEST" $
          ^
****** STATEMENT NUMBER PREVIOUSLY DEFINED
 47:         PRINT, "FREQUENCY(HZ)    OUTPUT VOLTAGE(MV)" $
 48:         CALCULATE, 'I'=1, 'FREQUENCY'=.01 $
 49:      30 PRINT, 'FREQUENCY"*1000,"    ******   ",'RESPONSE'('I'),
                              ^
****** IMPROPER LABEL
 50:                "   ***.**" $
 51:         CALCULATE, 'FREQUENCY ='FREQUENCY'*10, 'I'='I'+1 $
 52:         COMPARE, 'I', GE 5 $
 53:         GOTO, STEP 001030 IF LO $
 54:
 55: B       END OF RF AMPLIFIER BAND-PASS TEST $
 56:
 57:      [illegible] TERMINATE, ATLAS PROGRAM $
```

FIG. 14 Program (Corrected)

```
EQUATE-ATLAS  -  VERSION 0.1 (DEVELOPMENTAL)  -  7 AUGUST 1972

  1:         BEGIN, ATLAS DEMONSTRATION PROGRAM $
  2:         DECLARE, DECIMAL, 'FREQUENCY', 'I' $
  3:         DECLARE, DECIMAL, LIST, 'RESPONSE', 5 ELEMENTS $
  4:         DEFINE, 'RF-SOURCE', SOURCE, AC-SIGNAL RFA$
  5:         DEFINE, 'RFM', SENSOR, AC-SIGNAL RFM$
  6:         DEFINE, 'DC-SUPPLY', SOURCE, DC-SIGNAL DCA, VOLTAGE ( ) $
  7:
  8: C       RF AMPLIFIER BAND-PASS TEST
  9:
 10:         1)  SET UP POWER SUPPLY
 11:         2)  MEASURE OUTPUT VOLTAGE FOR LOGARITHMIC
 12:             INPUT FREQUENCIES FROM 10 HZ TO 100 KHZ
 13:         3)  CHECK BAND-PASS CHARACTERISTICS
 14:         4)  PRINT RESULTS $
 15:
 16: E001000 APPLY, 'DC-SUPPLY', 10 V $
 17:         CALCULATE, 'FREQUENCY'=.01, 'I'=1 $
 18:
 19: B       MEASURE OUTPUT VOLTAGE AT APPROPRIATE FREQUENCY $
 20:
 21:      10 APPLY, 'RF-SOURCE', VOLTAGE 1 MV, FREQ 'FREQUENCY' KHZ,
 22:         FREQ MAX 100 KHZ, THRU-IMP 600 OHM, CNX HI 3 LO 6 $
 23:         DELAY, MIN 1/2 SEC $
 24:         MEASURE, (VOLTAGE 'RESPONSE'('I') MV), 'RFM',
 25:         VOLTAGE MAX 10 MV,FREQ MAX 'FREQUENCY' KHZ, CNX HI 5 LO 2 $
 26:         CALCULATE, 'FREQUENCY'='FREQUENCY'*10, 'I'='I'+1 $
 27:         COMPARE, 'I', GE 5 $
 28:         GOTO, STEP 001010 IF LO $
 29:
 30: C       CHECK BAND-PASS CHARACTERISTICS $
 31:
 32:         COMPARE, 20*LOG('RESPONSE'(3)/'RESPONSE'(2)), LT 3 DB $
 33:         GOTO, STEP 001020 IF HI $
 34:         COMPARE, 20*LOG('RESPONSE'(3)/'RESPONSE'(4)), LT 3 DB $
 35:         GOTO, STEP 001020 IF HI $
 36:         COMPARE, 20*LOG('RESPONSE'(3)/'RESPONSE'(1)), GT 20 DB $
 37:         GOTO, STEP 001020 IF LO $
 38:         COMPARE, 20*LOG('RESPONSE'(3)/'RESPONSE'(5)), GT 20 DB $
 39:         GOTO, STEP 001020 IF LO $
 40:
 41:         RECORD, "RF AMPLIFIER BAND-PASS OK" $
 42:         GOTO, STEP 001040 IF GO $
 43:
 44: B       FAILED BAND-PASS TEST $
 45:
 46:      20 PRINT, "FAILED RF AMPLIFIER BAND-PASS TEST" $
 47:         PRINT, "FREQUENCY(HZ)    OUTPUT VOLTAGE(MV)" $
 48:         CALCULATE, 'I'=1, 'FREQUENCY'=.01 $
 49:      30 PRINT, 'FREQUENCY'*1000,"    ******   ",'RESPONSE'('I'),
 50:                "   ***.**" $
 51:         CALCULATE, 'FREQUENCY'='FREQUENCY'*10, 'I'='I'+1 $
 52:         COMPARE, 'I', GE 5 $
 53:         GOTO, STEP 001030 IF LO $
 54:
 55: B       END OF RF AMPLIFIER BAND-PASS TEST $
 56:
 57:      40 TERMINATE, ATLAS PROGRAM $
```

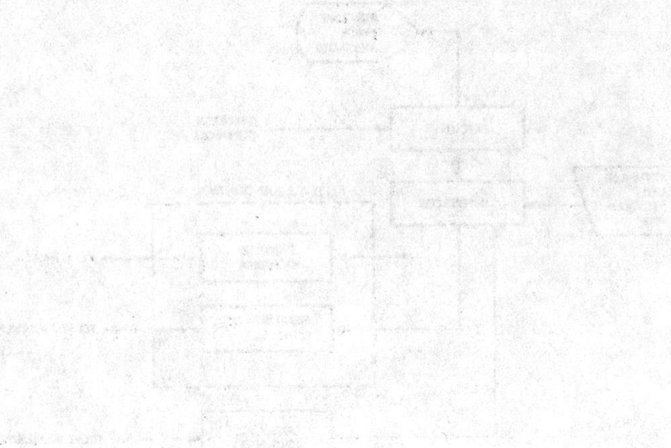

PART III: Test Design Philosophy and Techniques

The most fundamental and critical element of any ATE system is the test design embodied in the application programs. The preparation of control programs for ATE has purposely been called test design to emphasize that test preparation is a design process more so than a programming process. It has been estimated that 40 to 50 percent of the cost of a test program is attributable to the test design function. Yet, despite the criticality of test design, very little attention has been given to techniques for effective test design for ATE. Digital test design is getting substantial attention, but the bulk of today's testing problems involve at least some analog tests. One can argue that a good test is based on stimulus and measurement techniques and has little to do with whether a test is performed manually or automatically. In analog testing that is largely true. But ATE requires a more disciplined approach to testing if quality is to be high and prices reasonable. Without documentation to guide test engineers in good test design practice, ATE does not produce the promised consistent quality of testing. It remains just as subjective as manual testing with all of its improvisations except that this design subjectivity is perpetuated in the test program.

Why test design has been slighted is not clear. Perhaps it is thought of as too mundane. Somehow the better engineers in the ATE industry seem to have been preoccupied with the sophisticated problems of test language and compiler development rather than attempting to develop better testing techniques. RCA has done a fair amount of work in the area of developing testing ground rules. The best reference available, however, is a document prepared for the Navy which is not generally available to the industry (see Part VII-B, [8]). The papers included in this section represent the best that could be found in commercial publications.

The usual approach to test design is to begin with available factory or maintenance test procedures and develop a series of performance tests. These are referred to as "go-chain" tests because if the UUT is performing all right, all test results are "go". In maintenance applications, most requirements call for some degree of fault isolation in the event that a UUT fails its go tests. In the factory environment, it might be more practical to simply indicate which test failed and leave troubleshooting to other procedures. Even in factory testing, however, it would be advantageous to have the ability to add automatic fault isolation as common failure modes become evident. Hence it is important to address both performance and fault diagnostic testing in any comprehensive coverage of test design. It is up to the user to decide if the troubleshooting elements are to be omitted in the program design. In most applications to date, fault isolation has been incorporated into the test programs. The preferable approach is to design these tests after performance tests have been designed.

Since little is known about failure modes for the UUT at the time of test design, the fault isolation tests tend to be theoretical. They are based on a logical analysis of all possible failures assuming only one failure at a time. The single-failure assumption, of course, cannot be made in the factory environment. This is another reason why automatic fault isolation in factory testing tends to be less practical than for units that have been fielded. To develop fault isolation logic allowing all possible multiple failures is virtually impossible.

At best, fault isolation test design is a "guessy" process without adequate failure mode information available at the time of original test design. The first paper, "New Techniques for Implementing UUT Testing," advocates more effective use of information gained from go-chain tests to deduce most likely failures before designing special tests for fault isolation. It also recommends using "most probable failure" information for test design when available. Both of these recommendations deserve serious consideration as a means for designing more effective and more efficient test programs that minimize the expensive ingredient of engineering test design.

Digital circuitry is rapidly replacing analog circuitry. With this transition, some radical changes in test design philosophy are required. A number of new sophisticated techniques are being developed to help generate the enormous numbers of tests required to test a digital UUT. Some of these techniques will prove useful, but some could lead the test designer into making invalid assumptions that render his test program useless. The paper, "Diagnostic Test Generation for Digital Logic," provides a review of the basics of digital testing and some of the techniques available for test design. Since much of this is new to test design engineering, it remains to be seen which techniques are most effective. One thing is clear, however, digital testing cannot rely on visual evaluation of circuit performance as has been the process for analog test design. Some computer-aided test design appears mandatory. This aspect of test design is covered in Part V.

The paper entitled "Diagnostics for Logic Networks" introduces the concept of fault models as applicable to digital testing. The "stuck-at" concept of digital component failure prevails in most fault analyses. It assumes that digital logic elements fail only at logical 0 or logical 1 and not in between. This seems reasonable because of the design approach taken in logic circuits. However, as microminiaturization overtakes logic circuit manufacture, the safe limits of element and lead separation may be violated and intermittent failures may become significant. This will destroy most of the validity to existing concepts in digital testing whether the model is manually generated or computerized. One thing seems clear, testing is rapidly becoming the limiting factor of production in modern electronics. It is therefore a top priority requirement that practical techniques be developed for test design. For the first time in history, testing has the opportunity to become the lead function in electronics design. Hit or miss approaches cannot be tolerated.

The fourth paper in this section, "Transfer Function Testing," offers a "big picture" approach to test design by considering the UUT as input/output device with an associated transfer function. If the transfer functions can be determined, this approach would work nicely with the third-generation concept of sample-measurement analysis. The nice thing about the approach described is that it is a comprehensive technique which, once mastered, becomes a general solution to the test design problem. It is probably wise, however, to approach this concept cautiously because, like the digital modeling concept, it can lead to gross errors if invalid assumptions are made during the equation or algorithm definition stage. There is such a dire need for innovative approaches to test design that all ideas should be pursued.

NEW TECHNIQUES IN IMPLEMENTING UUT TESTING

CY VEDOMSKE

Raytheon Company, Oxnard, California

ABSTRACT

Two new considerations in computerized testing that can greatly speed the testing process are Most Prominent Failure items and test analysis arrays. Average fault localization time for a batch of units under test can be reduced using MPF items derived from failure trends. Test flow also can be optimized by structuring test and fault isolation logic using test analysis arrays to minimize test time and rapidly localize faults.

INTRODUCTION

Various techniques can be successfully applied as part of the application software for testing units under test (UUTs). This paper will expand on two techniques that, when used properly, can reduce test time and provide adequate flexibility for incorporating changes and adapting to different UUTs. These techniques are the use of Most Prominent Failure (MPF) items and the use of three test analysis arrays (no-go, no-test, data).

DEFINITIONS

Most Prominent Failure (MPF) items are those components that rank statistically high in the list of failed components for a UUT. Their conception can easily be seen: As we gain experience in testing a UUT, we learn to associate certain types of failure with a particular component . . . the MPF item associated with that failure, e.g., nine times out of ten, the failure of a certain amplifier was due to transistor Q1, which is the MPF item for that amplifier. An MPF item can be a component, a PC board, or subassembly depending on the level of testing. MPF items can be selected by listing failed components for a batch of UUTs in descending order of occurrence and choosing the top items in the list as MPF items, or by taking those items that reliability studies have indicated as problem areas.

Test analysis arrays are three common core arrays used in a bookkeeping manner to efficiently use a test library to direct testing flow and communicate data between tests. Basically, each array can be defined as follows:

"No-go" array - a list of failed tests or tests that must be performed due to a failure, e.g., if functional level receiver test failed due to insufficient output level, that functional test plus its corresponding amplifier gains and amplifier MPF item test would be listed in the "no-go" array.

"No-test" array - a list of those dependent-type tests not performed due to failure of the test they are dependent on (if the power supply for an amplifier has failed, the gain test for that amplifier would be meaningless and thus a "no-test").

"Data" array - an array made up of the measured values, computed values, etc., determined in one test and required for use in other tests; the "data" array is the means by which tests communicate with each other.

IMPLEMENTATION OF MPF ITEM TEST TECHNIQUES

Just by reading the definition one can probably immediately envision how to take advantage of or make use of MPF items in UUT application testing:

Establish list of MPF items for UUT

Examine the tests for that UUT to ascertain which tests would fail for each MPF item

Generate an MPF test for each MPF item: a test that will positively ascertain whether its MPF item (component, PC board, subassembly) failed

Insert the MPF tests into the appropriate places in the testing sequence: in the beginning of the mainstream of testing or at appropriate no-go branchings. (Note that the MPF tests are not intended to be only an extension of a troubleshooting tree, but a technique to speed up fault finding, and they should be executed at the first sign that they are suspect. In fact, if an MPF fails often enough, perhaps it should be performed prior to any other tests.)

IMPLEMENTATION OF TEST ANALYSIS ARRAYS

Implementation of test analysis arrays is intimately related to testing flow; therefore, array implementation will be illustrated by an example of how they might be incorporated into a testing flow. Following this example, a brief list of the actual steps we might follow in implementing test analysis arrays is given.

Figure 1 illustrates an approach to a basic overall test and repair flow. Note that one of the assumptions in figure 1, although not necessary, is that as many routine tests (those tests that must be satisfactorily performed to declare the UUT "GO") as possible will be done prior to making repairs (it usually saves time if all online/offline repairs and replacements are done at one time).

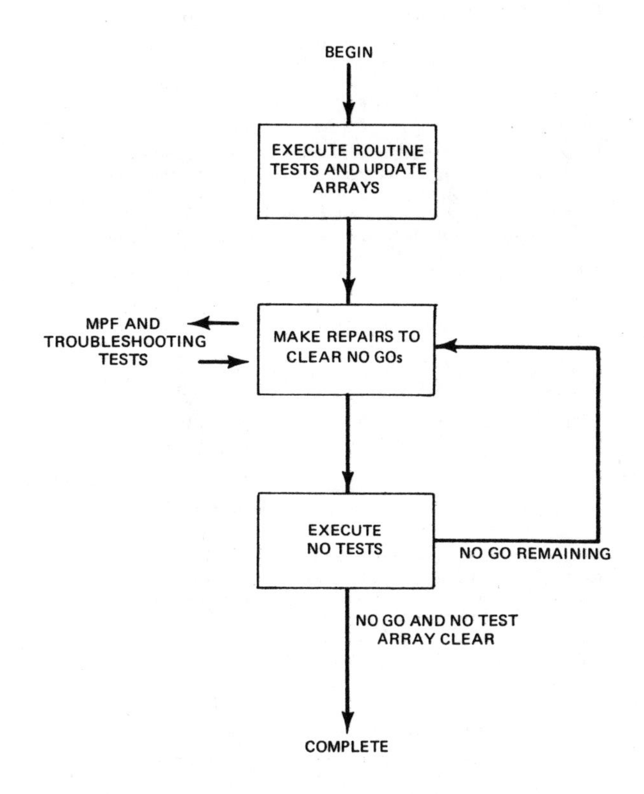

FIGURE 1. BASIC OVERALL TEST AND REPAIR FLOW

Reprinted from *IEEE Auto. Support Systems Symp. Rec.*, Nov. 1973.

Assuming that we have determined the sequence of tests (based on figure 1), have knowledge of the interdependence of these tests, and can treat each test as independent of the others (by making use of the "data" array for test intercommunications and subroutines for minimizing redundant code), we might proceed as follows:

Referring to figure 2, the tests are structured so we enter and execute the first test, put any information (measured values, calculations, etc) required for other tests in "data" array, and if a failure occurs, enter appropriate test number(s) (current test number, MPF item test number, or troubleshooting test number) in "no-go" array, and enter test number for any subsequent "no-tests" (tests that it would not make sense to perform due to failure of this test) in "no-test" array. As one can see in figure 2, prior to executing the next test, the "no-test" array is examined to see if that next test number is in it, and if it is, that test is bypassed; otherwise, that test is executed and the test analysis arrays are updated. This process is repeated until all routine tests have been executed.

The second phase of testing is the repair phase, in which faults are localized and necessary online/offline replacements are made. As shown in figure 2, the repair phase of testing involves examining and executing the "no-go" array test numbers, which are a list of appropriate MPF or troubleshooting tests that localize failures and the routine tests that initially failed and need to be repeated after a repair to ascertain that they are now "go" tests.

The third phase of testing involves executing those tests in "no-test" array; as these tests are being executed, the "data," "no-go," and "no-test" arrays are again updated as necessary, and if more failures are found, the repair and "no-test" execution phase are again repeated and this reiteration process is performed until the "no-go" and "no-test" arrays are cleared.

From figure 2, perhaps we would list the following steps to implement test analysis arrays:

Order the routine tests for minimum operator intervention and in descending order of independence (power supply tests first, amplifier that runs off of power supply second, etc). Note that this is usually already done when given the tests.

Determine "no-test" matrix for the routine tests; that is, for each routine test, determine which succeeding tests would be "no-tests" if that test were to fail.

Write execute portion of each routine test essentially as if it could be performed independently of other tests (not independent of subroutines accessed by several tests) and that, when necessary for inter-test communication, makes use of "data" array.

Determine fault localization technique for each routine test failure and write the execute portion for it; note that this may involve examination of "no-go" and "no-test" array if several tests can be used to decide where a problem is; also note that this fault localization technique may involve several test numbers when there is more than one reason a test could fail, and note that some of these test numbers could call for offline repair.

If MPF test technique is to be used, generate execute portion of MPF tests.

At the beginning of each routine test, add the necessary logic to direct testing flow dependent on that phase of testing:

a) If in the first phase of executing the routine tests (figure 2), add logic to cause each test to be skipped if it is in the "no-test" array.

b) If in a random search mode, add logic to prevent tests from being skipped even if they are in the "no-test" array.

Generate the necessary control logic for directing test flow into the trouble localizing or "no-test" execution phase and to the first test in one of those phases. At the end of each routine, fault localizing, or MPF test, add the necessary logic to update the "no-test," "no-go," and "data" arrays per the results of the execute portion of that test; also add logic to direct testing flow dependent on that phase of testing.

a) If in the first phase of executing routine tests, add logic to cause each routine test to be performed in sequence.

b) If in the trouble localizing and repair phase of testing, add logic that will direct testing flow according to "no-go" array entries

c) If in a random search mode, pause

d) If in execute "no-test" phase, add logic to direct testing according to "no-test" array entries

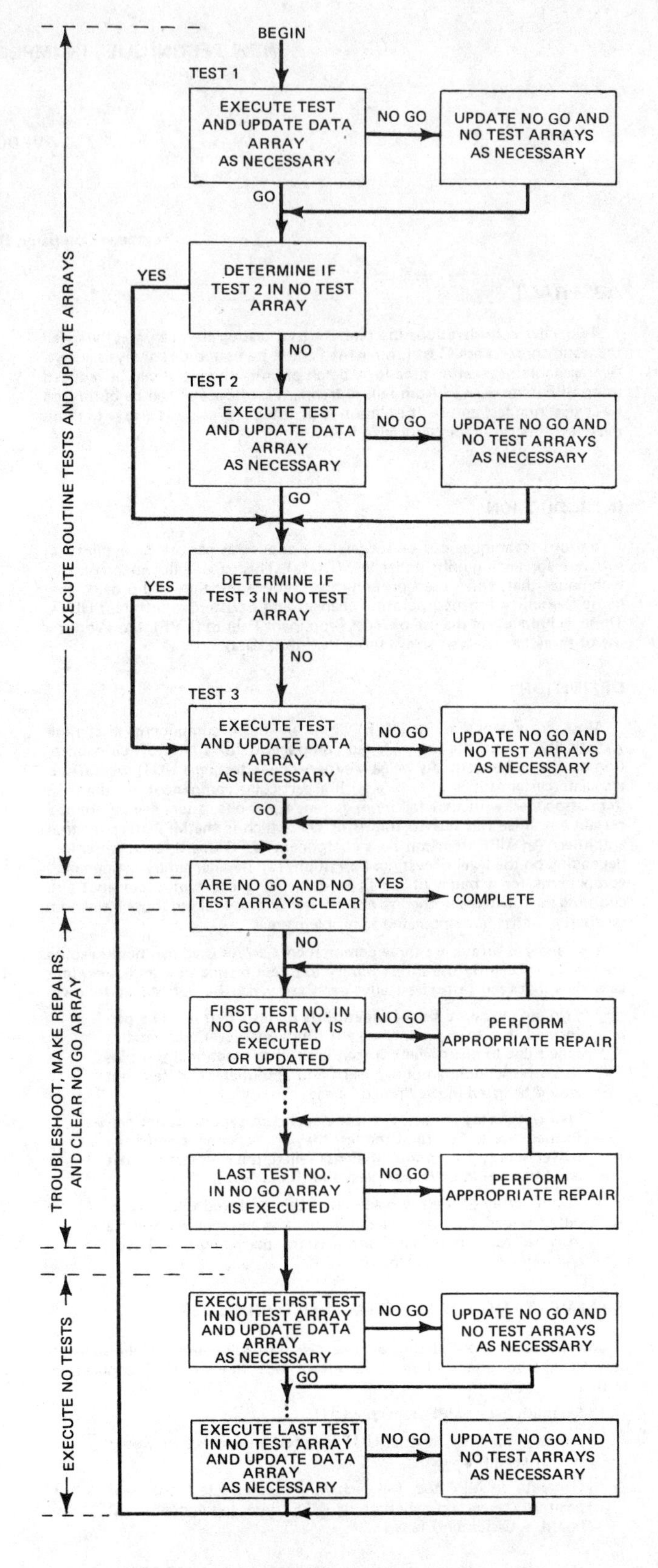

FIGURE 2. OVERALL TEST AND REPAIR FLOW

CHARACTERISTICS OF THESE TECHNIQUES

Some of the basic considerations for using MPF test techniques or test analysis arrays will now be mentioned.

The most important consideration of the two techniques under discussion is that of minimizing test and repair time:

1) Application of the MPF test technique, i.e., make use of current statistical failure data to aid in rapid localization of faults, by definition reduces the average troubleshooting time for a UUT. As one example, if test 10 checks Q1 in an amplifier, and Q1 fails in 1 out of every 10 UUTs, then one could save the time it takes to execute 9 tests in 10 UUTs by making Q1 an MPF item and checking it first in each UUT (figure 3). As another example, for the case where fault localization is done "offline," incorporation of MPF tests into "online" application tests can in the long run save time by eliminating the offline troubleshooting efforts for that item.

CONVENTIONAL TEST APPROACH

Time to Localize Q1 in UUT 10 =

$$\frac{10 \text{ Tests}}{\text{UUT}} \quad \frac{1 \text{ Min}}{\text{Test}} \quad 10 \text{ UUT} = 100 \text{ Min}$$

MPF TEST TECHNIQUE

Time to Localize Q1 in UUT 10 =

$$\frac{10 \text{ Tests}}{\text{UUT}} \quad \frac{1 \text{ Min}}{\text{Test}} \quad 9 \text{ UUT} +$$

$$\frac{1 \text{ Test}}{\text{UUT}} \quad \frac{1 \text{ Min}}{\text{Test}} \quad \text{X } 1 \text{ UUT} = 91 \text{ Min}$$

FIGURE 3. MINIMIZING FAULT LOCALIZATION TIME CAN BE ACCOMPLISHED

2) Utilization of the test analysis arrays, i.e., "no-go," "no-test," "data," can reduce test time by optimizing the test flow to eliminate execution of "no-tests," by making use of MPF tests, by acquiring as much data as possible for a UUT without doing redundant testing, and by having an efficient approach to resolving "no-go" tests.

Flexibility, as another consideration of the techniques under discussion, is an advantageous characteristic in that it meets the requirement for a general-purpose test structure. Being able to apply the MPF test technique to various levels (system, detail, etc) of testing and various types of components (discrete components, PC boards, subassemblies, etc) and being able to easily change the MPF item list as UUT batches change or reliability improvements are incorporated implies that the MPF test technique is flexible.

Flexibility, as a characteristic of test analysis arrays, shows up in that when test analysis arrays are used to structure test flow, it is relatively easy to change the execution portion of tests, add new tests, or change test sequence without severely impacting the application software (to make these kinds of changes, only the particular tests involved must be reworked).

Another characteristic of the MPF test technique and use of test analysis arrays is that they are compatible with existing test techniques. For example, special techniques such as brute force testing (applying stimuli and measuring response at various test points), comparison testing (stimulating known good device and unknown UUT and comparing results at various test points), transfer function testing (using impulse or noise stimuli) etc, can be factored into an application software structure that makes use of MPF item test technique and test analysis arrays.

One more consideration of the MPF item test technique and use of test analysis arrays is their advantage when many tests or subassemblies are involved. For one thing, a test structure based on these two techniques is relatively easy to understand; therefore, several people can learn to quickly apply it and thus can jointly work to generate a large quantity of application software and the end user can easily learn how to make changes to it. Another aspect is that of time saving; as pointed out earlier, use of the test analysis arrays attempts to maximize the efficiency of a lengthy testing process by eliminating redundant testing and minimizing online/offline trips.

LOOKING BACK

In summary, two general-purpose techniques that can be used to help minimize test time and provide an efficient and flexible approach to computerized UUT application software have been presented. One technique, the Most Prominent Failure item test technique, takes advantage of statistical failure rate data to minimize fault localization time. The other technique, test analysis arrays, makes use of three common core-storage arrays ("no-go," "no-test," and "data") to efficiently direct testing flow.

DIAGNOSTIC TEST GENERATION FOR

DIGITAL LOGIC - A REVIEW

John L. Fike and Robert J. Smith, II, Institute of Technology, Southern Methodist University
Dallas, Texas 75275

Stephen A. Szygenda, Electrical Engineering Department, University of Texas
Austin, Texas 78712

Abstract

The purpose of testing a logic network may be to simply determine if the network functions correctly, or additionally, to identify the location of defects. The testing process requires a source of input patterns, and a means of determining the correct output response. A number of combinations of techniques are in use, with varying costs, accuracies, and flexibility. These are discussed and compared, with suggestions for best utilizing whatever methods are available in a given testing situation.

I. Introduction

The purpose of this paper is to provide a tutorial review of test generation techniques for digital logic assemblies.

The scope of the paper is those methods which are either currently in use, or potentially useful, for detection and location of defects in actual digital circuits. The intent is to view these methods from the point of the manager in charge of production test and associated repair. Such a person is assumed to be more concerned with costs (acquisition, application, and maintenance), feasibility (background required, ease of use, applicability to problem), and flexibility (extension to more complex circuits, increase in resolution) than with theoretical elegance. References are included for those who wish to pursue the latter aspect.

An overview of the problem may be simply stated: given a digital sub-assembly (UUT), determine a) is it "good" or "bad", and/or, b) if it is bad, why? The sub-assembly may consist of one or more MSI/LSI arrays, circuit boards, etc. The measure of "goodness", or definition of required function may be a logic diagram, truth table, or functional description. Some sort of tester (ATE) is available, through which we may apply input pattern sequences and observe the resulting outputs. Secondary questions which may occur include "Is this design testable?", and "How should this design be produced so as to provide maximum diagnostibility?"

II. Terminology

An author's viewpoint is often implied by his definition of common terms; the basis for our development is computer-assisted fault simulation. The following definitions, while not universally agreed upon, are intended to be reasonably understandable and yet theoretically defensible.

A logic diagram is the intended logic function represented by interconnected Boolean functions, which are either primitive, or defined elsewhere (as in the case of MSI functions). The intended function may also be represented by Boolean equations and/or truth tables, or by a word description (as in the case of an adder or shift register). Whatever the form, the intended function is called the good circuit or good machine, which terms carry no connotation of correctness, completeness, or fitness for the end use. In other words, testers cannot test the "quality" or "correctness" of the design.

A defect is a physical occurrance which alters the intended design. It may be a shorted or open signal lead, a burned-out transistor, or something similar. Many defects will never manifest themselves as erroneous (different from expected) operation; those which do cause failures. The occurrance of certain restricted sets of defects may be modeled in hardware, by a computer program, or on paper. These modeled defects are called faults, and the process of determining the operation of the modeled good and faulted designs (or fault machines) is called fault simulation.

The presence of one or more defects in the physical circuit, or of the defect or fault is propagated to the output terminals, and results in a measureable difference between the good and the defective or faulted circuits. Input points and measured output terminals may be modified to

Reprinted from *IEEE Auto. Support Systems Symp. Rec.*, Nov. 1973.

provide opportunity for such detection. The "normal" terminals of the circuit are called primary inputs and primary outputs, while those used only for testing are called test points. We shall refer primarily to the modeled circuit, and consequently will ordinarily use "fault" in our discussion.

An input test pattern is a simultaneous or parallel definition of all circuit inputs, and the corresponding output patterns for the good and the faulted circuit are the good output and the fault output, respectively. A test sequence is a set of such patterns arranged in a particular order. A go/no-go or detection test sequence is a pattern sequence which purports to detect the presence of any defects or faults of some class, should they be present. A diagnostic test sequence purports to not only detect the presence of defects or faults of some class, but to enable the user to determine or distinguish, to some degree, which defect or fault occurred. This degree is a measure of the diagnostic resolution of the test sequence, and is determined by the characteristics of the logic, the defects or faults considered, and the test sequence. Average diagnostic resolution to within two faults means that the average number of faults per distinguishable group is no greater than two; a number this low is seldom achieved in practice. Methods for resolving faults include the fault dictionary approach, where the output patterns of a fault machine are matched or "looked up" in a suitably-organized list; the diagnostic tree approach, where the set of possible faults is partitioned into ever-smaller subsets until no further division is useful; and the walk-back approach, where a probe is guided through the logic, beginning at the incorrect output pin, until the defective element or interconnection is isolated.

No mention has yet been made of how the test pattern sequences are derived or created. This is because the foregoing definitions apply to test sequences obtained from any source. Test generation is the process of creating such patterns; one example might be the use of random binary input sequences, another might be the use of an algorithm in conjunction with the logic diagram. Test acquisition is not a common term; we use it to identify the process where suitable pattern sequences are available as a by-product of another procedure. Test grading is the process of determining which modeled faults will be detected by a given pattern or sequence; the percentage (and possibly the relative location) of such faults detected is a measure of the thoroughness of a test sequence. An exhaustive test for a combinational circuit (without memory) consists of all possible input patterns is any order; for a sequential circuit an exahustive test is, in general, all possible sequences of all possible input patterns.

III. Summary of Approaches

The diagnostic testing problem may be divided into three tasks: 1) obtain a set of input test patterns which are "sufficient" or "good enough" according to some criterion; 2) obtain a corresponding set of output patterns for the good circuit, and perhaps for the circuits containing the faults of interest; 3) apply these test inputs to the UUT, monitor the resulting outputs, and utilize the diagnostic information thus obtained. Tasks 1 and 2 are combined in entries of Table 1.

Ways to Obtain Input Patterns ↓ / Ways to Obtain Output Patterns →	Reference Board	State Table or Equations	Simulator (Good Ckt only)	Fault Simulator	Algorithm or Heuristic
State Table	1.	2.	3.	4.	N/A
Random Number	5.	N/A	6.	7.	N/A
Exhaustion	8.	N/A	9.	10.	N/A
Heuristic	N/A	N/A	N/A	11.	12.
Algorithm	N/A	13.	14.	15.	16.
Diagnostic Programming	17.	18.	19.	20.	N/A

(Numbers Refer to Paragraphs in Section IV; N/A means not applicable)

Table 1

Input/Output Pattern Generation Techniques

The following general comments apply to the "Ways to Obtain Input Patterns" Column:

State table approaches are the basis for testing simple circuits, such as ordinary IC's. Unless the logic is highly regular (as in the case of registers or counters, for example), increasing complexity makes these approaches unwieldy. For random sequential circuits, such as control logic, the state table is often not known, and deriving it may be prohibitively expensive.

Random Numbers are used in many test methods, and are quite useful for testing combinational and certain types of sequential logic. The required randomness may pose a problem of repeatability; this is overcome through use of pseudo-random techniques that always yield the same sequence. Methods which create regular patterns of ones and zeros, such as alternation, or shifting a one through a field of zeros, are also included here. The advantage of random number and regular pattern tests is the speed with which they can be generated; the disadvantage is that they bear no relation to the logic under test.

Exhaustion is the generation of all possible pattern or pattern sequences. This method is extremely useful for circuits having small numbers of input pins and few internal states, but since the number of possible patterns for a sequential circuit is 2^{m+n}, (where m is the number of states and n the number of input pins) the applications of this method are limited.

Heuristic methods refer to the class of procedures which succeed most of the time, but which do not guarantee to generate a test, even though one may exist. Logic-diagram-based procedures for general sequential circuits fall into this category, and are discuesed in more detail later.

Algorithmic methods refer to the class of procedures which analyze the logic in conjunction with faults of a given class, and guarantee to generate a test if one exists. Unfortunately, these are generally restricted to combinational or highly-regular logic because test generation costs become impractical for more difficult problems.

The Diagnostic Programmer is most effective as a source of patterns for logic which is highly regular, or which he has designed himself. He is less effective for extremely complex logic. The following general comments apply to the "Ways to Obtain Output Patterns" row:

A Reference Board is a copy of the actual logic which is known to operate properly. If available, it is an excellent source of good machine outputs, but is subject to the following problems: a) It may be difficult to determine that the reference is, in fact, good; b) It may be difficult to initialize the reference and the UUT to the same internal state; c) It is impractical to introduce known faults into the reference; d) timing variations may introduce untentional differences in behavior of the two circuits.

State Tables or Equations are a useful way to predict outputs for the good circuit, if available, and techniques exist for using these for analysis of faulty circuits. In practice tables and equations are used primarily for simple networks, since these descriptions are usually not readily available for large circuits.

A Simulator for the good circuit is, if available, more useful than a reference board, state tables, or equations, simply because it has wide applicability. A simulator does not require fabrication or manual testing of the logic, nor must the state table or equations be determined or analyzed. A well-designed simulator may also indicate such design problems as non-initialized logic; a general purpose logic simulator is a design aid with many applications besides test generation.

On the other hand, a simulator may have a large initial cost, and computer time may be expensive, especially for inefficient programs and/or very accurate timing models. Since output is worthless if the wrong logic is simulated, thorough checking of the simulated logic description is mandatory. Finally, a simulation of only the good machine yields no information concerning the behavior in the presence of faults.

Fault Simulation has all of the advantages and disadvantages of good machine simulation; operation in the presence of faults of certain classes is also simulated. This process may limit the size of the logic that may be handled, and will certainly increase the computer run costs. On the other hand, simulation of a proposed design in the presence of faults is a "must" to measure its diagnosability and/or fault tolerance. Finally, many test generation methods are based on the use of a fault simulator to evaluate the proposed pattern sequences.

Some algorithms and heuristics yield, as part of the circuit analysis process, a predicted set of good machine and fault machine outputs, so these may be considered as still another way of predicting output patterns. This method of determining output patterns is rarely used for actual circuits, due to restrictions on logic handled and faults modeled; in practice, fault simulation is used instead.

The following numbered paragraphs correspond to entries in Table 1, and summarize more detailed experiences derived from application of specific combinations of input pattern generation methods and techniques for obtaining resulting outputs:

1. State-Table approaches using a reference board are primarily useful for testing IC's and small networks. The state table must be found (if not give), and no diagnostic information is provided. The method can be relatively inespensive, since the tester can be very simple.

2. State-table inputs, and the use of state-tables or equations for output prediction are typical of incoming inspection tests for IC's. When the complete state-table is available and no diagnosis is necessary, this is the method of choice.

3 & 4. State-table inputs to simulators are used primarily to obtain dianostic information through fault simulation. Unfortunately, if the logic is sufficiently complex to warrant simulation, then the state table is unlikely to be available.

5. Random-number testing using a reference board is probably the most widely-used of all test methods. Where the logic is combinational or easily reset and a correct re erence is available, this method is rapidly set up (special purpose test boxes are often built for one board design) and economically applied. The method is ordinarily limited to relatively simple logic.

6. Random-number input to good-machine simulators is not often used, simply because there is no way to measure the thoroughness of the test sequence.

7. On the other hand, random number input to fault simulators is often used, especially as a way to quickly find patterns that will detect the first 30%-70% of the modeled faults. It is rare that all faults in a sequential circuit of even moderate complexity will be detected by even a very large number of randomly-generated patterns.

8. Exhaustive testing using a reference board is less often used than state table or random methods. This technique would have value only for small networks.

9 & 10. Exhaustive input to a simulator would be of value only to determine fault characteristics for very small networks. Simulation costs would quickly become prohibitive.

11. Heuristic pattern generation is only used in connection with a fault simulator, because detection of faults is not guaranteed, and thus a means of scoring proposed patterns is needed. Such techniques are quite effective, especially large, complex networks. They are expensive to use, but compare favorably with other methods.

12. For simple circuits, some heuristics may also predict output without the necessity of simulation. These are of little practical value.

13. Algorithms exist for operating on state-table or equation representation of circuits. These can be quite expensive for circuits of even moderate size, and are primarily of theoretical interest.

14 & 15. Algorithms are, however, of great usefulness in connection with simulators, especially when faults are modeled. This is primarily because an algorithmically-generated test for a fault will usually detect many others. Unfortunately, no algorithms exist for general sequential circuits; these techniques are actually heuristics.

16. Algorithms for combinational circuits can predict outputs of both good and faulted networks without simulation. These are little used for actual circuits.

17. Diagnostic programming in conjuction with a reference board is widely used, typically in situations where the tester has a FORTRAN-like input language. Principal disadvantages of this method are a lack of a measure of thoroughness, and difficulty of the task for complex logic.

18. The diagnostic programmer can also determine the correct output pattern, which can be stored in table form on tape or in the tester. This is quite difficult, and rarely used.

19 & 20. The diagnostic programmer is often used as a source of patterns for a simulator, especially a fault simulator. In most such applications, the programmer designs patterns to detect faults which remain undetected after other pattern generation methods have been used.

V. Comparison of Approaches

With the variety of possible (and practical) methods described above, it is obvious that no single appraoch is ideal for all testing situations. The following comparisons refer to the more common techniques.

Network complexity affects the choice of method in several ways. Not only do the simpler pattern generation methods become less effective as complexity increases, but more importantly, the investment represented a given board increases, thereby making rework and repair of faulty boards more justifiable. However, the problem of diagnosis becomes more difficult as complexity increases. For these reasons it is possible to roughly order test generation methods by complexity of the logic to be tested, as follows:

State-table or exhaustive inputs using a reference board (nos. 1, 2, & 8) are useful for simple networks, such as incoming IC test. These are suitable for high volume, and can use homemade testers. Such methods are low in cost but offer no diagnosis.

Random numbers using reference board (no. 5) are useful for general logic and can use homemade tester. No measure of thoroughness; not appropriate for highly sequential logic. Low in cost with diagnostic resolution via probing.

As logic becomes more complex, the user will probably utilize a more sophisticated tester (possibly computer-controlled) which allows a diagnostic programmer to prepare input sequences. Outputs may be determined by a reference board (no. 17), by the programmer himself (no. 18), or by on-line or off-line simulation (no. 19). Problems with this approach are the labor required (much of it repetitive), and the fact that the circuit test is no better than the programmer who prepared the test pattern sequences.

Perhaps the "ultimate" is use of a computer-controlled tester in conjunction with a fault simulator (some testers include such a simulator as part of a software package). Such a facility will at least make the diagnostic programmer more productive (no. 20), and may also provide semi-automatic pattern generation (nos. 4, 7, 11, 15, and 20 in varying combinations). Such capability may represent an investment of approximately $40,000 and more for the simulator alone, plus maintenance and operating expenses. Such systems may also be leased, or test generation services purchased on a quotation basis for each design. Fault simulation is no panacea; patterns must still be generated, applied and evaluated. Costs increase with accuracy of the model, and certain common defects are modeled only with difficulty, if at all. But fault simulation is the method of choice for complex logic and large design volume. It is probably less costly in the long run, and is being widely adopted.

References:

Available publications in the area of testing tend to fall in two areas:

1) Theoretical/historical, which are readily available but which answer few immediate or practical questions;

2) Applied, which appear frequently in the trade press but too often are attempts to justify a particular manufacturer's approach to building testers. Accordingly, we present a variety of sources, each of which will provide many more references than can be included here:

Books:

Breuer, Melvin A. (Ed.), Design Automation of Digital Systems, Volume I, Prentice-Hall, 1972).

Chang, Herbert, Eric Manning and Gernot Metze, Fault Diagnosis of Digital Systems, (Wiley, 1970).

Friedman, Arthur and P. R. Menoa, Fault Detection in Digital Circuits, (Prentice-Hall, 1971).

Papers:

McClure, Robert M., "Fault Simulation of Digital Logic Utilizing a Small Host Machine," Proceedings of the ACM-IEE Design Automation Workshop, (June 1972).

Szygenda, S.A., John L. Fike, E.W. Thompson, J. Hemphill, C. Hemming, A.A. Lekkos, "Implementation of Synthesized Techniques for a Comprehensive Digital Design, Verification, and Diagnosis System, " Proceedings of the 5th Hawaii International Conference of Systems Sciences, (1972).

Proceedings:

The annual proceedings of the ACM-IEEE "Design Automation Workshop," and of the IEEE "International Symposium on Fault-Tolerant Computing" are both recommended.

Articles:

Cromer, Earl G., "Fault Isolation and Repair Techniques: Current Strategies and Equipment," Electronic Packaging and Production, Vol. 13, No. 6, (June 1973).

Bibliographies:

Carroll, B.D. and E.W. Smith, "A Bibliography of Fault Tolerant Computing," 46 pp., NTIS #AD - 739 522.

Bennetts, R.G. and D.W. Lewin, "Fault Diagnosis of Digital Systems-A Review," Computer, Vol. 4, No. 4, July/August 1971, pp 12-21.

Diagnostics for logic networks

Single or multiple faults may be located using a choice of four techniques applicable to both combinational and sequential logic

Despite substantial success in improving circuit reliability in recent years, there is growing interest in techniques for detecting and locating failures in complex digital networks. The reasons for the increased emphasis on diagnostics center around the lack of test points in LSI circuits, the need for efficient test procedures in the face of increasing network complexity, and the increased need for complete testing of electronic equipment.

The techniques covered here are economically feasible only for networks of modest size—typically, hundreds of gates. In contrast, a 1973 digital cabinet is likely to contain thousands of gates; therefore present work in the field of diagnostics is devoted to seeking better ways of dealing with large network arrays. One way to expand this limit is to partition large networks so that they may be treated as a series of smaller networks. However, such partitioning is not always practical. Therefore, present work in this field is devoted to expanding these techniques for large network arrays.

There are two basic approaches to testing digital systems: functional and structural. In functional testing, the aim is to verify that the unit under test behaves as required, by determining whether it executes its tasks properly. Thus, a functional test verifies that a counter increments (or decrements) correctly; that memory locations can be read from and written into; that transfer of control occurs under the proper conditions; etc. Tests based on structural considerations, on the other hand, are designed to assure that the individual hardware constituents of a unit are operating correctly. Thus, a "structure test" assures that every AND gate gives an output of 1 if and only if all of its inputs have a logic value 1; that every flip-flop can be properly set, reset, or toggled (as appropriate); that clock signals occur at the proper time; that parity checks are correct; and so forth.

Tests that verify function, "function tests" for short, are most appropriate for major assemblies such as central processing units or memory systems, and these tests find their widest applications in final assembly and field maintenance. Typically, function tests start by verifying that a small portion of the system (called the hard-core) is operating correctly and then progress outward through the circuits that perform the remaining repertoire of operations. The difficulty with these tests is that they generally are written without close examination of the hardware details because the programmer usually cannot be burdened with circuit details. Therefore, it is likely that function tests are somewhat incomplete. In fact, it is not unusual for function tests to be limited in fault-detecting capabilities to only half of the possible faults in a digital unit.

But tests derived solely on the basis of functional considerations need not always be so limited in coverage; microprogrammed machine organizations, in particular, are especially amenable to fairly complete, functionally derived tests.

Function tests are based on heuristics and are derived manually on the basis of familiarity with the system, but without regard for fine structural details. When tests are derived on the basis of structural considerations, common practice is to work at the level of individual gates. Thus, structure tests are particularly appropriate for testing in the manufacturing process, for acceptance testing prior to assembly of major units, and for off-line repair. In addition, tests based on structural details are used for periodic checkout of real-time systems where particularly stringent dependability requirements exist.

Examination of the fine network structure can lead to efficient tests, particularly when appropriate assumptions are allowed regarding the possible consequences of failures. When logic networks are treated from the point of view of structure, it becomes possible to develop diagnostics with the aid of algorithms developed for this purpose, as will be described later. In fact, there now exists a body of knowledge about structure tests that is broad and general enough to justify the label "theory of diagnostics." Further, the currently rigorous research into this branch of switching theory has already produced powerful tools for developing tests.

Of course, as with all theories, the applicability of diagnostic theory is circumscribed by the models on which it is based. In particular, the assumptions made regarding the results of various failures are of paramount importance. If no assumptions are made, then the test has to contain all possible input combinations or input sequences.

The key features of the two classes of tests can be summarized as follows: To the extent that the fault model is truly representative, structure tests can be made complete, although they may be excessively long for large systems. However, they can be optimized with regard to test length and can be developed by means of algorithms (implemented as computer programs). Function tests are feasible for large systems, but are usually not complete. They are based on heuristics that are derived manually on the basis of familiarity with the system, but without regard for fine structural details.

Because of difficulties in developing functional

A. K. Susskind Lehigh University

Reprinted from *IEEE Spectrum*, Oct. 1973.

techniques offering complete test procedures and due to the lack of universally optimal methods, the balance of this article deals with structural techniques.

The "stuck-at" model

The most universally-accepted fault model today is the "stuck-at" (SA) model. It supposes that *failure mechanisms in a gate result in its inputs or outputs being either stuck at 1 or at 0.* Thus, an AND gate with one or more inputs or the output stuck-at-0, supplies a logic signal of 0 to all of the gates to which it is connected. A j-input AND circuit with k inputs stuck-at-1 becomes a j-k input AND circuit, while the output stuck-at-1 causes the gate to supply a permanent 1 to all of its loads. Similar statements can be made about OR gates: one or more inputs stuck-at-1 cause the OR to act as though its output were stuck-at-1, and OR inputs stuck-at-0 can no longer affect gate operation. An inverter input stuck-at-0 is equivalent to its output permanently at logical 1, and an inverter output stuck-at-0 is equivalent to its input permanently at 1.

Note that under the SA model, failures cause fixed signals to appear at leads—i.e., leads become "clamped." Thus, tests based on the SA model deal with *static* faults. Parameters that affect dynamic behavior, such as switching speed, are verified by other tests. Function tests combined with SA tests can be quite effective in these situations.

The stuck-at (SA) fault model was first proposed for dealing with the early logic-circuit families (DTL and RTL) when discrete components were used. Just how well it fits LSI is not clear and perhaps the current popularity of the stuck-at model is not justified.

Prominent among faults that are not adequately covered by the SA model is shorting between adjacent conducting lines. In some technologies (RTL, DTL, ECL), this failure can be modeled as the insertion of an AND or OR function between the shorted leads. Even when this model is adequate, consideration of shorted lines introduces great complexity into the testing process because of the large number of pairs of lines on a chip. So it is necessary to eliminate all lead pairs that are sufficiently distant.

The SA model also falls down in dealing with another type of failure, intermittent faults, so, as in the past, the success in finding intermittent faults is based on the resourcefulness of the troubleshooter.

Tests under the SA model can be conveniently described by the conventional analytical tools for logic circuits, such as Boolean algebra, or simple extensions of it. Thus, the SA model offers analytical convenience as well as good representation of many, but not all, of the failure mechanisms.

Fault distribution and location

No matter how the effect of a failure is modeled, one must decide whether or not the assumption can be made that faults occur only one at a time, or in combinations. Tests based on single faults are simpler and shorter than those that consider all possible failure combinations. However, the adequacy of single-fault tests needs careful assessment. In production testing, the single-fault assumption may not be valid, especially for high-density IC technology. But for equipment that has been operational, the single-fault assumption is frequently justified. In any case, it is true that a complete test for all single faults will also detect the bulk of simultaneous faults. However, the results must be carefully studied since multiple faults can produce misleading results.

Redundancy in tests can be a great aid in fault location. To illustrate, assume a fault exists in one of four units, A, B, C, and D. Suppose test 1 in a sequence can reveal only that A or B is faulty; test 2 detects faults in C or D only; test 3 detects faults in A or C. Then tests 1 and 2 together are sufficient for fault detection, but all three tests are required for fault location. The conditions for complete fault location are simple:

If a fault exists, it can be located to a single unit, if for every pair of units U_i and U_j there exists a test for which the response in the presence of a fault in U_i is different from the response in the presence of a fault in U_j.

The role of simulation

When simulation is available, there is a temptation to relax care in deriving test procedures. In fact, attempts have been made to use trial-and-error test-generation techniques. The first trial may consist of rather arbitrarily derived test signals. In all but very simple cases, however, the percentage of faults detected by the initial test sequence is usually too small. So, additional trials usually are called for. However, since simulation is rather expensive, it is unlikely that this kind of approach is profitable where thorough tests are desired.

Whether tests are generated solely by consideration of function (in disregard of network structure), by analysis of the logic network (in disregard of the functional use), or on the basis of mixed strategies, it is necessary to appraise the effectiveness of the test procedure. For that purpose, there are two alternatives: one can insert faults in a copy of the equipment to be tested and score the effectiveness of the test procedure in detecting (locating) the faults, or one can simulate the equipment on a computer and run the test procedures on the simulation model, appropriately modified for faults. The former will be called fault *insertion,* the latter *simulation.*

Fault insertion has the advantage of complete fidelity. Just what a set of tests will reveal about a given error is fully shown by running the proposed test sequence on the actual equipment in the presence of the fault. If the insertion of faults can be economically implemented, running the actual equipment in the presence of the fault is likely to be cheaper than simulating it. But, of course, evaluation of tests by fault insertion can take place only after a model of the equipment has been built, and so this method of test evaluation cannot be used in the design process. Design of MSI and LSI circuits demands consideration of testing concurrent with the design process. Therefore, fault insertion in these devices is useful only in a final verification of test procedures after a network or machine has been committed to production in substantial quantities.

Although test simulation may consume many hours of high-priced, host-computer running time, it can be very helpful. It fits in well with some of the procedures used in other design-automation tasks, such as logic simulation, to verify the validity of the proposed design. Because of its hardware independence, simulation offers not only lead time, but it can also handle faults that are difficult, if not impossible, to insert physically and so it overcomes a significant limitation of fault insertion where LSI is used. But no simulation is better than its model; hence, simulation is not necessarily a final arbiter of test procedures.

The most prominent ways of using network structure in developing tests are based on algorithms and are surveyed below. More details can be obtained from textbooks, articles, and bibliographies in the reading list on p. 44.

Path sensitization

Consider the network of Fig. 1 and assume that the SA model is to be used. A test to determine whether lead 1 (the top input to the AND gate) is stuck-at-1 (abbreviated s-a-1) would require: (a) that the nominal logic value of the signal on lead 1 be set to 0; (b) that the output of the gate becomes 1 if and only if lead 1 is s-a-1; and (c) that the change in the output of the AND gate due to the fault be reflected in a change in the network output. Condition (a) is met by setting a = 0; condition (b) by setting to 1 the signal values on both lead 7 and lead 9; and condition (c) by setting to 0 the signal on lead 12. Another way to state this situation is that (a) *provokes* the fault; (b) *sensitizes* the gate under test; and (c) sets up a *sensitive path* from the gate to the network output. Together (b) and (c) *propagate* the fault to the output. In the example, the fault under test is provoked by making a = 0; the gate is sensitized by setting b = 0 and setting either c or d (or both) to 1.

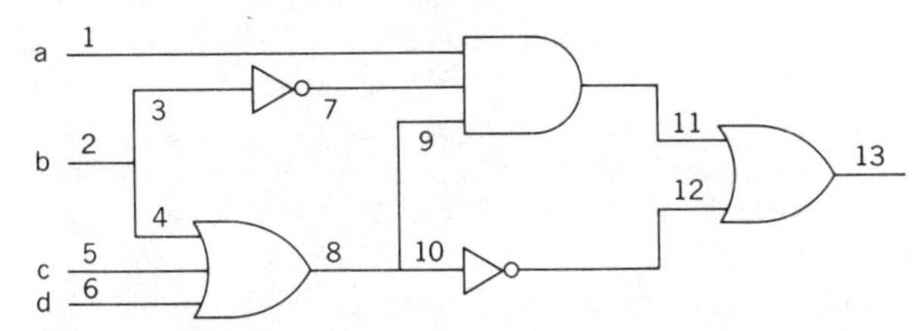

[1] Path sensitization is based on the assumption that failure mechanisms in a gate result in its inputs or outputs being stuck-at-one or stuck-at-zero (SA model). This fault is then traced along a path to the network output and its effects noted.

This example illustrates a procedure always applicable to finding tests for combinational logic, (i.e., logic in which the output at any time t is a function only of the inputs at that time, and independent of all past inputs): one determines how the fault can be provoked and how it can be propagated to the output by means of a sensitized path (or paths) from the fault through the associated gate to the output(s) of the network.

Consider now a test for lead 9 s-a-1 in Fig. 1. To provoke this fault, we must set b = c = d = 0 and to sensitize the gate, we must set a = 1. Because b = 0, lead 7 has logic value 1, as is needed to sensitize the AND gate. But leads 8, 9, and 10 have logic values 0, so that lead 12 has logic value 1. The latter, alas, blocks the sensitive path from the AND gate to the output, and so the conditions for provoking the fault and sensitizing a path to the output are incompatible. No test exists for lead 9 s-a-1. Of course since this is a nondetectable fault, it has no effect on operation of the circuit.

The network used here is simple, and for each fault a test can be readily found by searching for inputs that simultaneously provoke that fault and let its effect be observed at the output. For more complex networks, however, an orderly procedure for simultaneously provoking the fault and sensitizing a path from it to the output is needed. Furthermore, this procedure should be capable of indicating in a straightforward manner when a failure is not detectable, and it should also reveal the various choices available when a fault can be detected by more than one test, i.e., by more than one combination of inputs to a network.

Boolean difference

An important analytical technique for finding a sensitive path is based on the Boolean difference. Consider a Boolean expression z which relates the network output to the signals on a set of leads, $x_1, x_2, \ldots x_z$. Let x_t be the lead for which a test is to be found, so that $z = Z(x_1, x_2, \ldots x_t, \ldots, x_z)$. Then there is a sensitive path from x_t to the output, if logic values are chosen for the other leads x_i, in such a manner that the value of z changes as the logic value on x_t changes. This condition may be stated as

$$Z(x_1, x_2, \ldots, 1, \ldots, x_z) \oplus Z(x_1, x_2, \ldots, 0, \ldots, x_z) = 1$$

I. D-Algorithm test example

	Lead numbers (signals)												
	1(a)	2(b)	3	4	5(c)	6(d)	7	8	9	10	11	12	13
1. Provoke	0												
2. Propagate through AND							1		1		D		
3. Propagate to output												0	D
4. Consistency I								1		1			
5. Consistency II			0										
6. Consistency III		0		0									
7. Consistency IV					1	0							
					0	1							
					1	1							

where the indicated values replace the variable x_t in Z. The operator $\oplus$ is the ring-sum, also called EXCLUSIVE-OR, and the expression has the value 1 when one of the operands has value 1, but when not both are 1s. The above ring-sum is called the Boolean difference (BD) of z with respect to x_t and is denoted by dz/dx_t.

In Fig. 1, $z = a\bar{b} + \overline{bcd}$, so that the conditions for a sensitive path from lead 1 through the AND gate to the output are all the independent inputs that satisfy the expression

$$\frac{dz}{da} = (\bar{b} + \overline{bcd}) \oplus \overline{bcd} = \bar{b}c + \bar{b}d = 1$$

To provoke a stuck-at fault, the independent input conditions must be selected so that the lead to be tested for s-a-1 (or s-a-0) is due to the independent inputs in the absence of the fault, at logic value 0 (or 1). In the example, to provoke the fault lead 1 s-a-1, it is required that $a = 0$, hence $\bar{a} = 1$. This condition together with the one that assures a sensitive path are met by the solutions to

$$\bar{a}(\bar{b}c + \bar{b}d) = 1$$

which comprise the set of all tests for lead 1 s-a-1.

Similarly, for the fault lead 9 s-a-1 previously discussed, we have $z = a\bar{b}x_9 + \overline{bcd}$, $dz/dx_9 = (a\bar{b} + \overline{bcd}) \oplus \overline{bcd}$, and so the tests for x_9 s-a-1 are given by the solutions to

$$\bar{x}_9[(a\bar{b} + \overline{bcd}) \oplus \overline{bcd}] = 1$$

But $\bar{x}_9 = \overline{bcd}$, which when substituted makes the left-hand side identically zero. Thus, no solution exists for lead 9 stuck-at-1, which is in agreement with our previous finding. (However, solutions for 9 s-a-0 do exist. They are $a = 1$, $b = 0$, and c or d or both set to 1. These are also the conditions for testing lead 1 s-a-0.)

The following generalization can now be made. Given a fault on lead i that is provoked if and only if the Boolean function P_i is true, where P_i is expressed in terms of the independent inputs, then all the tests for this fault are given by the solutions to

$$P_i \frac{dz}{dx_i} = 1$$

Note that there is no need to restrict consideration to the SA fault model; the latter is still convenient, however, because it makes the function P_i so readily apparent. To illustrate the power of the general formulation, suppose in the previous example one wants to test for an assembly error in which the three-input OR gate is replaced by an exclusive-OR. Here we would use $z = a\bar{b}x_8 + \bar{x}_8$, so that

$$\frac{dz}{dx_8} = a\bar{b} \oplus 1 = \bar{a} + b;\ P_i = bc\bar{d} + b\bar{c}d + \bar{b}cd$$

The tests are $b = 1$ and c or d (but not both) set to 1; $a = 0$, $b = 0$, and $c = d = 1$.

D algorithm method

An alternative method for finding tests in combinational logic is given by the popular D algorithm. Its execution involves signal tracing through the network, whereas the Boolean-difference approach is entirely algebraic. Another difference is that the latter reveals directly all the tests for a given fault, while the D algorithm provides only one test at a time. When all possible tests for a given fault are desired, the D algorithm may involve repeated (but finite) trial and error. This feature can lead to a waste of effort in a case where a given fault is not detectable, as in the example when lead 9 s-a-1 was considered. On the other hand, for a given test the D algorithm reveals all of the faults.

To illustrate this point, let us consider again the problem of finding a test for lead 1 s-a-1. To provoke the fault, set $a = 0$ on lead 1. This is indicated on the first line of Table I. Then propagate the fault through the AND circuit by demanding that leads 7 and 9 have value 1, as shown on the second line, where we indicate the propagation by placing a D under the lead number for the output of the AND gate in column 11. The D, which denotes that the signal will vary as the fault is or is not present, is propagated to the output by demanding 0 on lead 12, as indicated on the third line. Signals on the remaining lines are then specified to be consistent with the values already specified: the entry in column 12 demands 1s in columns 10 and 8 which, fortunately, is consistent with the existing specification for line 9. For lead 7 to have value 1, lead 3 must have value 0. This is the second consistency requirement. Because leads 2, 3, and 4 are tied together, they must carry identical entries, and so in columns 2 and 4 entries of 0 are made on line 6. Finally, to achieve the 1 on lead 8, one or more of leads 4, 5, and 6 must be set to 1, and this introduces the final consistency specification. Fortunately, no inconsistencies were encountered. When they are, one must seek a different path for propagating the fault to the output. (In the simple example on hand, however, only one path exists.) Because of the variety of propagation paths that may have to be considered and because of the attendant consistency operations, the D algorithm can require substantial trial and error. But as our example also shows, success can yield a test for more than one fault, because wherever a D occurs, the corresponding lead also is checked. Thus, in Table I, the test for lead 1 s-a-1 is also a test for lead 11 s-a-1 and lead 13 s-a-1. (Actually, it is also a test for lead 10 s-a-0 and lead 12 s-a-1. A more complete presentation of the D algorithm would show how these additional tests are uncovered.)

Subscripted algebraic expressions

For the purpose of test generation, the approach based on the Boolean difference offers the advantage of completeness. On the other hand, the D algorithm has appeal because of the insight that it provides. Neither, however, can efficiently handle multiple faults. For that purpose algebraic tools have become available which combine the best features of the Boolean difference and D algorithm methods of fault location.

Topological information can be carried in a Boolean expression by subscripting the variables according to the network leads on which they appear. In Fig. 1, the input to the AND circuit on lead 9 is given by $b + c + d$, and to indicate the paths which the various independent inputs take to reach lead 9, one can write this expression in the subscripted form $b_{2,4,8,9}$ +

$c_{5,8,9} + d_{6,8,9}$ (the subscripts refer to lead designations). Similarly, the expression for the signal on lead 12 can be written in subscripted form as $\bar{b}_{\bar{2},\bar{4},\bar{8},\overline{10},12}$ $\bar{c}_{\bar{5},\bar{8},\overline{10},12}$ $\bar{d}_{\bar{6},\bar{8},\overline{10},12}$, where subscripts are complemented to show the inversion from lead 10 to lead 12. These expressions are examples of a *modified* Boolean formulation that places into evidence details of the network structure as well as its function. The form used here is called SPOOF-an acronym for Structure and Parity Observing Output Function. The SPOOF for lead 11 is particularly interesting and is given by

$$
\begin{aligned}
S_{11} &= a_{1,11}\bar{b}_{\bar{2},\bar{3},7,11}(b_{2,4,8,9} + c_{5,8,9} + d_{6,8,9,11}) \\
&= a_{1,11}\bar{b}_{\bar{2},\bar{3},7,11}b_{2,4,8,9,11} + a_{1,11}\bar{b}_{\bar{2},\bar{3},7,11}c_{5,8,9,11} + \\
&\qquad a_{1,11}\bar{b}_{\bar{2},\bar{3},7,11}d_{6,8,9,11}
\end{aligned}
$$

Note that the first product in the final sum is of the form $ab\bar{b}$ and that for the purpose of finding tests, it is *not* set to 0. In fact, terms (i.e., subscripted literals) having different path lists are considered *distinct algebraic variables* in forming SPOOFS. With this restriction, SPOOF terms may be manipulated according to the rules of Boolean algebra.

To illustrate how SPOOFs are used to find tests, consider the SPOOF for the output of the network in Fig. 1:

$$
\begin{aligned}
S_{13} = S_{11,13} + S_{12,13} &= a_{1,11,13}\bar{b}_{\bar{2},\bar{3},7,11,13}b_{2,4,8,9,11,13} + \\
& a_{1,11,13}\bar{b}_{\bar{2},\bar{3},7,11,13}c_{5,8,9,11,13} + a_{1,11,13}\bar{b}_{\bar{2},\bar{3},7,11,13}d_{6,8,9,11,13} + \\
& \bar{b}_{\bar{2},\bar{4},\bar{8},\overline{10},12,13}\bar{c}_{\bar{5},\bar{8},\overline{10},12,13}\bar{d}_{\bar{6},\bar{8},\overline{10},12,13}
\end{aligned}
$$

Through this expression, the effect of a fault is readily determined. For example, lead 1 s-a-0 causes every term with subscript 1 to be set to 0, so that the first, second and third products are 0, and there remains only (when subscripts are omitted)

$$S_{13}^{\,1\text{-at-}0} = \bar{b}\bar{c}\bar{d}$$

Similarly, lead 9 s-a-1 causes every term in which 9 appears to be set to 1, and every term in which $\bar{9}$ appears to be set to 0. Thus,

$$
\begin{aligned}
S_{13}^{\,9\text{-at-}1} &= a\bar{b} + a\bar{b} + a\bar{b} + \bar{b}\bar{c}\bar{d} \\
&= a\bar{b} + \bar{b}\bar{c}\bar{d}
\end{aligned}
$$

Because the fault-free behavior of the network is also given by $z = a\bar{b} + \bar{b}\bar{c}\bar{d}$, the fault 9 s-a-1 is not detectable, as was discovered before.

To find tests for any one fault f in Fig. 1, it suffices to find solutions to

$$z \oplus S_{13}^{f} = 1$$

where S_{13}^{f} denotes the output SPOOF in the presence of the fault. In particular, with lead 1 stuck-at-0, we have

$$(a\bar{b} + \bar{b}\bar{c}\bar{d}) \oplus \bar{b}\bar{c}\bar{d} = 1$$

which has the three previously-mentioned solutions: $a = 1$, $b = 0$, and at least one of the inputs c and d set to 1.

There are more powerful ways of using SPOOFs in developing tests for SA faults, as discussed in Flomenhoft, M. J., Si, S. C., and Susskind, A. K., "Algebraic Techniques for Finding Tests for several fault types, 1973 International Symposium on Fault Tolerant Computing, June 1973. There it is shown how SPOOFs can also be used to derive fault-location tests and to find tests for detecting other kinds of faults, such as shorts between leads that are in physical proximity and thereby cause behavior which can be modeled as a wired AND or OR.

When the subscripts are dropped in the SPOOF

Where to find more information

Almost every issue of the *IEEE Transactions on Computers* in recent years has contained at least one article pertinent to diagnostics, with major emphasis on theory. Particularly relevant are the special issues on fault-tolerant computing (Nov. 1971 and Mar. 1973), which contain revised versions of some of the papers given in the 1971 and 1972 conference digests.

There are two textbooks in the field:

Chang, H. Y., Manning, E., and Metze, G., *Fault Diagnosis of Digital Systems.* New York: Wiley-Interscience, 1970.

Friedman, A. D., and Menon, P. R., *Fault Detection in Digital Circuits.* Englewood Cliffs, N.J.: Prentice-Hall, 1971.

Practicing engineers may find good use for Chapters 3 and 7 in *Design Automation of Digital Systems, vol. 1—Theory and Techniques,* edited by M. A. Breuer. Englewood Cliffs, N.J.: Prentice-Hall, 1972.

The following survey papers each contain detailed bibliographies:

Bennets, R. G., and Lewin, D. W., "Fault diagnosis of digital systems—a review," *Computer,* pp. 12-21, July/Aug. 1971.

Breuer, M. A., "Recent developments in the automated design and analysis of digital systems," *Proc. IEEE,* vol. 60, 1, pp. 12-27, Jan. 1972.

McCluskey, E. J., "Test and diagnosis procedure for digital networks," *Computer,* pp. 17-20, Jan./Feb. 1971.

The most comprehensive bibliography is P. Scola, "An annotated bibliography of test and diagnostics," *Honeywell Computer Journal,* vol. 6, 1972. (On enclosed microfiche only.)

Examples of functional testing can be found in "Digest of papers, testing to integrate semiconductor memories into computer mainframes," Oct. 4, 1972, IEEE publication 72CHP810-2C.

The following publications are particularly good sources of detailed information:

Digest of Papers, 1971, 1972, and 1973 International Symposia on Fault-Tolerant Computing, IEEE publications 71C6-C, 72CH0623-9C, and 73CH0772-4C, respectively. In the 1971 and 1972 programs, nearly half the papers were on diagnostics. The papers from Session 3 of the 1973 symposium are specifically on diagnostics, but several other papers are also relevant.

Other good sources are:

Bossen, D. C., and Hong, S. J., "Cause-effect analysis for multiple fault detection in combinational networks," *IEEE Trans. Computers,* vol. 20, pp. 1252-1257, Nov. 1971.

Thayse, A., "Testing of asynchronous sequential switching circuits," *Philips Research Reports No. 27,* pp. 99-106, 1972.

Farmer, D. E., "Algorithms for designing fault detection experiments for sequential machines," *IEEE Trans. Computers,* vol. 22, pp. 159-167, Feb. 1973.

Breuer, M. A., "Testing for intermittent faults in digital circuits," *IEEE Trans. Computers,* vol. 22, pp. 241-246, Mar. 1973.

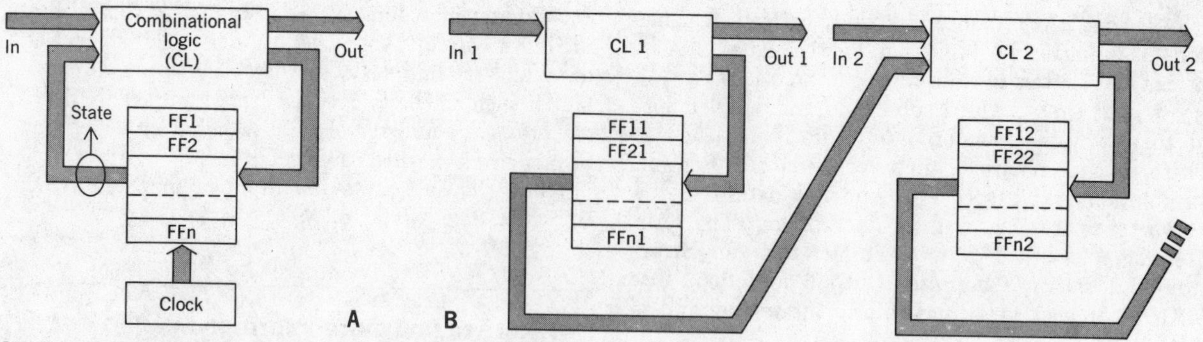

[2] Sequential network transformation. The input/output time sequences of the circuit in A (a sequential network) are analogous to the spatial sequences of the circuit in B (a combinational network).

and no algebraic simplifications are made thereafter, it is called an "E-expression," which is an expression that allows one to find, algorithmically and without trial and error, tests that detect any combination of stuck-at faults in combinational logic. No assumption whatever need be made about the number and distribution of simultaneous faults.

Multiple SA faults. The algorithms that lead to multiple fault detection tests are very simple and include two major steps: (1) scanning the E-expression for enumeration of possible tests and (2) selecting the input combinations that constitute a complete and irredundant test set. It is noteworthy that complete scanning, which requires only one pass, assures that all detectable faults, in all combinations, will in fact be detected; there is no explicit scoring of test coverage necessary.

The E-expression for Fig. 1 illustrates this point as follows:

$$E = a\bar{b}b + a\bar{b}c + a\bar{b}d + \overline{bcd}$$
$$= P_1 + P_2 + P_3 + P_4$$

In step (1) this expression is scanned product-by-product. Each subproduct, called a *growth*, is examined for *vertices* (input combinations that should result in an output of 0), which are the tests for that growth. For example, the product P_1 has the growths $\bar{b}b$, $a\bar{b}$, and ab. Because none of the vertices in the cube ab should result in a network response of 1 (since $z = a\bar{b} + \bar{b}\bar{c}\bar{d}$), any input combination with $a = b = 1$ is a test for the growth ab. However, the growth of P_1 to $a\bar{b}$ is harmless; therefore, no test for it can be made. The growth $\bar{b}b$ is similarly harmless, since this subproduct contains no vertices. In this manner, all the growths in all the products are examined and, where appropriate, tests for them are entered in a list. One way of completing step 1 is to examine in the same manner, the complement of the E-expression, also written as a sum of products. In step 2, one selects from among the tests on the list, a subset such that every harmful growth is detected. It can be also shown that the selected tests do indeed detect all possible stuck-at fault combinations and no further effort is required to find multiple-fault detection tests.

A significant feature of this method of developing

II. State table for up-down counter

Count	Up	Down
A	B, O	D, borrow
B	C, O	A, O
C	D, O	B, O
D	A, carry	C, O

tests is that while it leads to a test set that is guaranteed to be complete, the effort expended in finding it is probably no greater than in any other approach, including even those that are restricted to single faults. However, the procedure does not lead directly to the smallest possible test set. Where a premium is placed on keeping down the size of the set, additional effort has to be expended to remove tests that are unnecessary for fault location.

Testing sequential networks

While present algorithmic methods for finding tests are adequate for purely combinational networks, this is not so for *sequential* networks. In the latter, the response at time t is determined not only by the input at that time, but also by past inputs, possibly reaching all the way back to the time when the network was placed into its initial conditions by a resetting operation. Sequential networks have memory, and therein lies the root of the difficulty in testing: *it takes an entire sequence of inputs to detect many of the possible faults in a sequential network.* Moreover, the mechanization of the memory function, often achieved through complex feedback loops, makes the effect of certain faults rather difficult to determine and there are no really simple ways of investigating them.

At this time, there are three major ways of finding tests for sequential networks: by verifying functional characteristics, by translating the given network into the related iterative circuit; and by verifying the state-table for the given network. Of these, only the first two have gained practical acceptance.

It has become popular to speak of "highly sequential" networks, meaning networks in which flip-flops (or other memory elements) comprise a substantial portion of the circuitry. The logic circuits in an electronic watch are examples of a highly sequential network. For networks of this type, current practice is to specify test sequences on the basis of functional considerations to verify that registers can be cleared, certain flip-flops can be toggled, counters behave as expected, etc. The effectiveness of tests is then judged by simulation. This approach can be quite effective, providing those specifying the tests are intimately acquainted with the details of the network and sufficient time and resources are available for repeated simulations runs, as refinements are made to successively improved tests. Therefore, implicit in this method is great confidence in the simulation. However, adequate simulation of sequential circuits requires time-consuming programs that are costly to run, and so there is a great temptation to restrict the number of times one goes through the analysis-simulation loop. Or perhaps worse, there is a temptation to simplify the model used in the simulator, so that its judgment becomes impaired.

The need for algorithmic techniques applicable to sequential networks is only partially met by approaches described so far in the literature. Foremost among these is the transformation of the *time*-sequential case into the related *space*-sequential case. Consider the block diagram of a sequential circuit shown in Fig. 2A, where the memory function is implemented by means of a set of flip-flops. The flip-flops are assumed to be clocked, such as is done with J-K flip-flops.

Operation of the circuit proceeds as follows: The contents of the set of flip-flops together with the input determine the output. When the clock line changes from 1 to 0, the contents of the flip-flops are revised on the basis of the inputs at that time and the contents of the flip-flops, just prior to the update signal from the clock. No change in the flip-flop contents, collectively called the *state* of the sequential network, occurs until the next demand signal from the clock. This type of sequential network is frequently called *synchronous* because updating of its state only occurs when specified by the clock, so that its operation is in step with the clock signal. Note that the inputs described can change only once for each occurrence of the clock signal that updates the flip-flops. One can assign serial numbers to the successive clock pulses and to the inputs that exist whenever updating occurs. Thus, we have the input sequence $x(1), x(2), x(3), \ldots, x(t)$.

Suppose there are c copies of the same sequential circuit and these were so interconnected that the state of the first copy was communicated to the second, the state of the second communicated to the third, etc., and the state of the (c-1) copy communicated to the c copy, as shown in Fig. 2B. Furthermore, assume that each network makes its output depend on its input and on the state information it receives from its neighbor on the left, rather than the state update that it performs. If input $x(1)$ is applied to the first circuit, input $x(2)$ to the second circuit, $\ldots$, input $x(c\text{-}1)$ to the (c-1) circuit, and input $x(c)$ to the final circuit, the ith circuit would make the same output decision and perform the same memory update that it would make if it had the form of the circuit in Fig. 2A and had been given the sequence $x(1), x(2), \ldots, x(i)$. Note that in the revised network, the entire sequence $x(1), x(2), \ldots, x(t)$ might as well be supplied to all of the networks at once.

The possible time delays in the execution of state updates have no effect, providing the entire set of inputs $x(1), x(2), \ldots, x(n)$ persists long enough. Therefore, the network in Fig. 2B behaves very much like that in Fig. 2A. The difference is that in 2A an *input* time sequence becomes an *output* time sequence; in 2B a linear *input* space sequence becomes a linear *output* space sequence. If in the network the number of copies (called *cells*) is sufficient, then 2B can do the work of 2A.

Suppose that, in 2B, the identical logic change is made in *every* cell. Then the input/output spatial transformation performed in 2B will still be the same as the input/output temporal transformation performed by 2A, if that logic change is also made in 2A. Conversely, the effect of a fault in 2A will be the same as that in 2B when every cell has that fault. But 2B is a combinational network, since it consists of a cascade of identical cells, with each cell having a pair of inputs (to receive x signals and information from its neighbor on the left) and a pair of outputs (to supply information to its neighbor on the right and to the network outputs). Thus, the transformation indicated in Fig. 2 allows one to replace the problem of finding tests for a synchronous sequential network by the problem of finding tests for the related *iterative combinational* network.

Note that each fault in the sequential network leads to c identical faults in the iterative network, and so the problem of fault detection in a synchronous sequential circuit can be treated as the problem of multiple fault detection in an appropriate related combinational network. As one might expect, the transformation has a substantial cost. The size of the iterative network is c times that of the original sequential network, and so it is important to have an efficient algorithm for fault detection in large combinational networks with multiple faults.

When sequential circuits are not strictly of the synchronous type, the iterative circuit approach can be modified somewhat and still serve as a reasonable model for deriving tests. However, it must be recognized that this model does not accommodate two major problems in unclocked, also called *asynchronous,* sequential circuits: those of races (coincidence timing) and hazards (false outputs). Still, some engineers use the iterative circuit technique, or a variant of it, for deriving tests in asynchronous circuits. The validity of these tests is then explored by a simulation run, and the remarks previously made on completeness in connection with functional tests also apply here.

State table approach. A fundamentally different approach to finding tests for sequential networks is based on verifying that the network under test does operate according to its *state table,* which is a convenient way of specifying the two operations of state updating and output generation. In a state table, brevity is achieved by labeling each state, not by a binary n-tuple such as would be the contents of a set of n flip-flops, but by an abstract symbol, such as A, B, C, etc. The table is therefore attractively succinct. An example of such a table is shown in Table II for a scale-of-four counter.

The procedures for deriving tests that verify that a sequential network satisfies the state-table specifications are based on at least two assumptions: that the network in the presence of faults has no more states than are listed in its specification and that for the fault-free network there is an appropriate input sequence which can transfer from each state to every other state. Because the state table is expressed in abstract symbols, the test procedure is entirely independent of hardware fault models. Therefore, the test can detect any fault that does not violate the above two assumptions and still lets the network be described by a state table. The approach based on state tables can also be extended to asynchronous sequential networks.

It has become customary to refer to the process of verifying that a given sequential network does operate in accordance with its state table, as conducting a checking experiment. Although techniques for developing checking experiments were first considered 20 years ago and have been refined and amplified since, they have apparently failed to attract practitioners, primarily because the task of compiling a state table, given only the network block diagram, can be arduous. For example, a network with ten flip-flops, requires a table with $2^{10} = 1024$ rows. Therefore, checking experiments are a form of functional testing that can be complete, but at the expense of considerable tedium.

Improving testability

The fundamental cause for the difficulty in finding tests for sequential circuits is the fact that the state variables are not available for inspection when a sequential network is tested. If state variables were observable, testing would not be significantly more difficult than for combinational networks. So an immediate remedy suggests itself: bring out the state-variable leads. From the test point-of-view, this is indeed a good solution. (It is always helpful to make more of the network signals observable.) But bringing out more leads is directly contrary to modern layout and packaging objectives, and so the instant cure is not often implementable. But it does illustrate that there is much one can do in design and layout to ease the testing function—both to make tests short and to make them simple to develop.

Various other alternatives for making testing easier are open to the designer. He can sometimes implement what might be called continuous testing, where the equipment is duplicated and the outputs of the pair are compared. But in all except the most critical real-time applications, this approach is not economically feasible. A variant of this scheme, however, can be highly advantageous and arises when the second unit need not be a full-fledged replica of the first, but need do only a rather simple related task.

Prominent examples are the use of parity checks in memory applications and residue checks in the execution of arithmetic. One normally considers the replica an integral part of the main unit and the entire augmented equipment is called self-checking. Unfortunately, the number of functions amenable to self-checking is severely limited. However, for most logic networks, continuous monitoring is not feasible or even necessary, since periodic testing usually suffices.

Major architectural features can deeply affect testability. For example, microprogrammed machines lend themselves well to function tests, particularly if data paths can be divided into small sets. Small additions to hardware, if wisely chosen, can be very effective in achieving good diagnostics. Even the choice of technology has a profound influence on testing. For example, dynamic memories are harder to test than static ones and shift registers are usually more diagnostable than less structured sequential networks. However, testing considerations are usually of secondary importance in the selection of these features.

Assessing the state of the art

Although significant progress has been made in the field of diagnostics for logic networks, substantial shortcomings still exist in the current state of the art. In particular, intermittent faults are still not handled much better than they were many years ago. But even for permanent (also called solid) faults there is much room for improvement. Some hold that for the case of combinational circuits, present techniques for finding tests are entirely adequate. But are they really so for LSI? There is a need for better fault models for LSI technologies, and it is far from clear that tests derived in accordance with currently established procedures are in fact adequate. There also is need for progress in dealing with networks that contain memory elements.

Past practice has been to develop diagnostics after the design has been completed. This is clearly a poor procedure, and diagnostic considerations need to be incorporated into design rules. An easily diagnosable system may have higher initial cost, and yet it might be justifiable from a long-range point of view. The minimizing of logic elements is certainly an outmoded criterion for logic network design, but testability is not.

Alfred K. Susskind (SM) is currently Professor and Chairman of Electrical Engineering at Lehigh University, Bethlehem, Pa. Since 1970, his research there has concentrated on diagnostics for digital networks, and in 1970 and 1971, he organized and conducted workshops in this field. His teaching at Lehigh has been in the area of logic design and switching theory.

From 1948 to 1968, Prof. Susskind was at M.I.T. in various capacities: first, as a research assistant; then, as a research engineer; and after 1955, as a faculty member of the electrical engineering department. His major research interests at M.I.T. were in computer design, numerical control of machine tools, information retrieval, and switching theory.

Mr. Susskind received the B.E.E. degree *summa cum laude* from Polytechnic Institute of Brooklyn in 1948 and the S.M. degree from M.I.T. in 1950.

TRANSFER FUNCTION TESTING

by
J.P. McCarthy and P. W. LaClair
The Singer Company
Librascope Division

INTRODUCTION

Transfer function testing is a method by which the characteristics of a device or a system can be determined, even in the presence of noise, and under excitation by an input signal which has largely arbitrary characteristics. Once the transfer function is known, the response of the system or device can be determined for any input signal as long as an assumption of linearity is valid and, conversely, changes of transfer function with amplitude, frequency or time can be related to the occurrence of nonlinear phenomena such as distortion or component aging. Measurement of transfer function, which can often be performed without even removing the device under test from its normal service, can provide information that could only be obtained by many different individual tests with conventional test methods.

The transfer function of a device is defined as the ratio of output to input usually expressed as a function of frequency in terms of the Fourier (or, equivalently, Laplace) transform of the output and input. The transfer function concept has been widely used in design of control systems and synthesis and analysis of electrical networks but up to now it has not been used in testing. This has been because of the difficulty of acquiring the necessary raw data and the difficulty of calculation of the Fourier transform from the raw data. However, recent developments in computer and data acquisition hardware and computational methods have significantly reduced these difficulties, making transfer function test methods based upon Fourier analysis practical for many test situations.

FOURIER ANALYSIS

In the real physical world, things happen in the time domain. All physical processes occur in an irreversible, sequential way. In electrical control or communication systems, signals consist of voltage levels which vary with time either in a very simple, regular way as a sinusoidal wave or in a complex, irregular way as with modulated, information-carrying signals. Unfortunately, the mathematics required to describe and analyze complex signals as functions of time becomes so cumbersome that analysis of all but the simplest cases is practically impossible.

Fourier Series and the Fourier Transformation is a method by which complex functions of time may be transformed into sums of simple, sinusoidal functions at different repetition rates or frequencies. Thus, to determine the response of a certain system to a complex input, we can break the input down into a sum of sinusoidal elements, analyze the system response to each element, and - for a linear system at least - sum the individual responses to obtain the system response.

Such an approach is called analysis in the frequency domain and is probably the most widely known and used design and analysis procedure for all types of electrical engineering problems. The Fourier Transformation is the mathematical tool which permits translation between the time domain where all physical processes occur and the frequency domain where analysis is possible.

Fourier analysis is used in practically all phases of engineering design and analysis. Typical applications are the use of frequency response characteristics to design or specify an amplifier, specification of harmonic distortion or phase shift characteristics of an electronic device, or use of Bode plots (amplitude and phase vs frequency) in the design and analysis of control systems.

Once the transfer function of a system is known it is possible to determine the response of the system to any and all possible input signals by calculating the inverse Fourier transform of the product of Transfer Function and the Fourier transformed exciting signal.

There are two methods for determination of transfer function, analytical determination based upon thorough understanding of physical processes involved or direct measurement, i.e., calculation based upon simultaneous measurement of input and output.

The method of analytical determination works quite well in the analysis of passive electrical networks, even those of great complexity and is in fact widely used in the field of network analyses and synthesis. This is possible because of the fact that we are able to construct electrical components which are almost mathematically perfect devices. However, as we consider more and more complex devices involving interaction between several process variables, this method becomes less practical. While we can analytically determine a transfer function for a control valve or a servo motor, we find that the results are not nearly so reliable as they are for the electrical network, and for any practical industrial process or system this method is of little value.

So we are left with the alternative of direct measurement and numerical solution of the Fourier equation. Until recently this approach was completely impractical because of the enormity of the numerical calculations. Even as recently as the middle 1960's, a typical Fourier transform calculation would require several hours on the largest digital computers. But in 1965 Cooley and Tukey described an algorithm(1) (2) (3) which reduced the number of arithmetic operations in calculation of a Fourier transform by literally orders of magnitude, making direct numerical calculation a feasible procedure. The two curves of Figure 1 illustrate the magnitude of this reduction for a typical range of sample values.

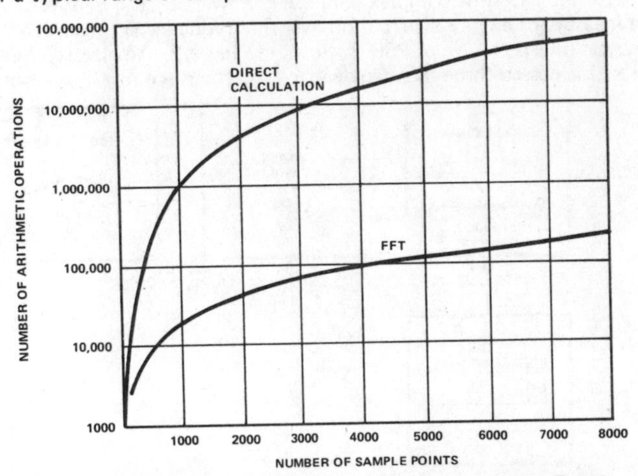

Fig. 1. Comparison of arithmetic operations for fast Fourier transform (FFT) algorithm and direct calculation

Before proceeding further, it will be helpful to define a few terms used in Fourier analysis. These are the direct and inverse Fourier transform, power spectral density, cross spectral density function, and coherence function. The transfer function itself will also be defined and the method of calculation from measured quantities indicated.

Fourier Transform

The Fourier transform of a continuous function is defined as

$$X(\omega) = \int_{-\infty}^{\infty} x(t)e^{-i\omega t}\, dt$$

and the inverse transform is

$$x(t) = \int_{-\infty}^{\infty} X(\omega)e^{i\omega t}\, d\omega$$

where x(t) represents the time function, (ω) the frequency function, and $i = \sqrt{-1}$.

Reprinted from *IEEE Auto. Support Systems Symp. Rec.*, Nov. 1972.

In terms of a discrete data set, rather than a continuous function of time, the transform pair takes the following form

$$X(j) = \frac{1}{N}\sum_{k=0}^{N-1} x(k)e^{-i2\pi jk/N}$$

and

$$x(k) = \sum_{j=0}^{N-1} X(j)e^{i2\pi jk/N}$$

$$j = 0,1,\ldots,N-1 \qquad k = 0,1,\ldots,N-1$$

The fast Fourier transform (FFT) ([1]), ([2]), ([3]) is a very efficient algorithm for evaluating the discrete Fourier transform when the number of samples is an integral power of 2.

Power Spectral Density (PSD)

The power spectral density function is defined in terms of the Fourier transform as

$$G_{xx}(\omega) = X(\omega)X^*(\omega)$$

where * indicates the complex conjugate. Power spectral density is the measure of energy distribution over the frequency spectrum. A physical interpretation of PSD is shown in Figure 2. An electrical signal x(t) is passed through a narrow band filter and the total power in

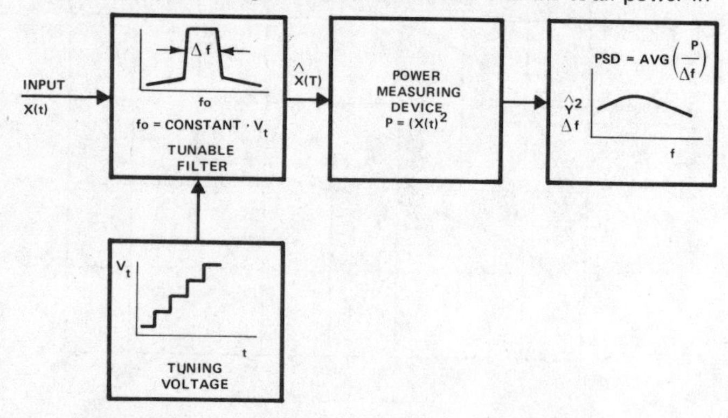

Fig. 2. Power spectral density measurements

the filter output is measured. If this measured power is normalized with respect to the filter bandwidth, the result is a point on the PSD curve. By tuning the filter over the entire frequency band of interest, the complete PSD curve is generated. The approach to PSD measurement outlined by Figure 2 is quite feasible for implementation by analog methods and is, in fact, the basic method used by a large class of instruments variously known as wave analyzers or spectrum analyzers.

Cross Spectral Density

The cross spectral density is defined from the Fourier transforms of the two inputs by

$$G_{yx} = Y(\omega) \cdot X^*(\omega)$$

The physical interpretation of the Cross Spectral Density function is shown diagramatically in Figure 3. The situation is analagous to the interpretation of power spectral density except that two signals are simultaneously processed through identical filters and multiplied together. If the x(t) and y(t) inputs both have components at a particular frequency, the cross spectral function at that frequency will have a magnitude equal to the product of the magnitudes of the two components times the cosine of the angle between them. Thus, unlike the power spectral density function which is always real valued, the cross spectral density is a complex function having both magnitude and phase.

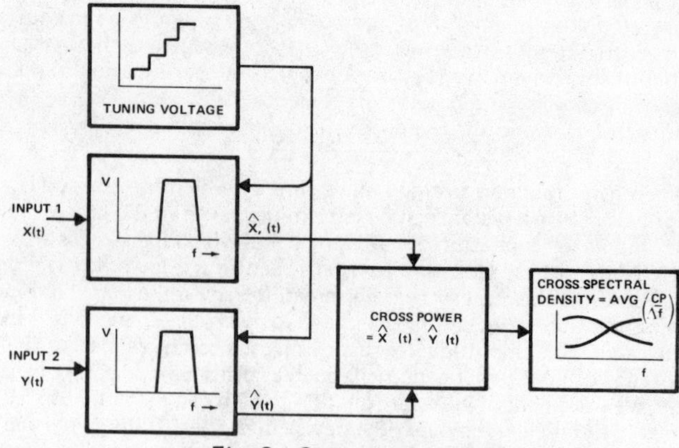

Fig. 3. Cross spectral density

The cross spectral density is a measure of the degree of correspondence between the two inputs. Where both inputs are identical, it reduces to the power spectral density. If both inputs are completely uncorrelated, the average cross spectral density is zero.

Transfer Function

The transfer function of a device is defined by

$$H(\omega) = Y(\omega)/X(\omega)$$

where $X(\omega)$ is the Fourier transformed input to the device and $Y(\omega)$ is the Fourier transformed output. This may also be expressed in terms of the spectral densities by

$$H(\omega) = \frac{\overline{Y(\omega)\,X^*(\omega)}}{X(\omega)\,X^*(\omega)} = \frac{G_{yx}(\omega)}{G_{xx}(\omega)}$$

To compute the transfer function of a device, simultaneously sample input and output, calculate Fourier transforms, and compute the transfer function. If either of the signals consists entirely or partly of noise, the effect of the noise on the measurement can be minimized by taking the average of several determinations of transfer function. This comes about because of the fact that the average cross spectral density of noise tends toward zero.

Coherence Function

The coherence function is a measure of degree to which the output from a system is related to the input. A coherence value of one implies that the output is entirely a consequence of the input while a coherence value of zero implies no relation between input and output, i.e., output is all noise. The coherence function is defined by

$$\gamma^2 = \frac{\overline{|G_{zx}{}^2|}}{\overline{G_{xx}}\ \overline{G_{zz}}} = \frac{\overline{G_{yy}}}{\overline{G_{yy}} + \overline{G_{nn}}}$$

where the horizontal bar signifies an average value and the subscripts x, y, z, and n refer to Figure 4.

The signal to noise ratio as a function of frequency may be related directly to the coherence function by

$$SNR = \frac{\gamma^2}{1-\gamma^2}$$

Application of Fourier Testing

Consider the generalized measurement problem shown in Figure 4. A "black box" with transfer function $H(\omega)$ is excited by an

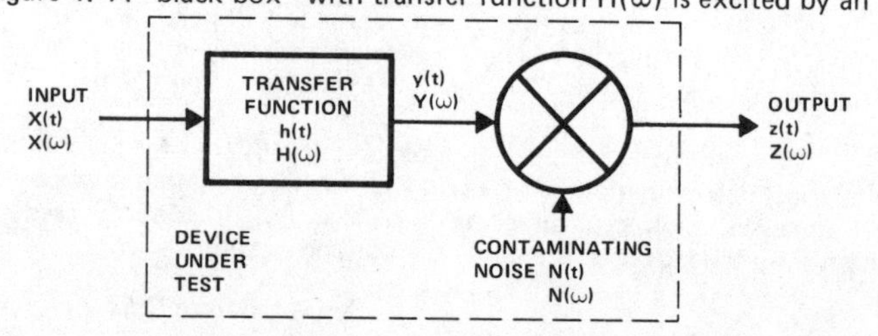

input signal x(t) whose Fourier transform is X(ω). Internal to the system, an unknown amount of contaminating noise n(t) is introduced. The measured output of the system is z(t). To determine the characteristics of this system, we must measure the transfer function in the presence of contaminating noise and separate those components of the output attributable to the input signal from those generated by the noise.

At first thought, these may seem to be simple problems. We might, for example, "short circuit" the input to determine the noise components or apply various test signals with known characteristics at the input. But this assumes that the sources of signal and noise are completely separate and distinct and that it is possible to control the input signal. What we would like to do is perform the measurement of transfer function and the noise contamination for any arbitrary signal and noise characteristics. If this can be done, it is conceivable to think of measuring and analyzing the performance of a device or even a whole system without interrupting its normal operation by making use of naturally occurring inputs and outputs. This can, in fact, be done. The transfer can be measured in the presence of contaminating noise by taking advantage of the fact that the cross spectrum between the signal component of the output and the noise component will tend toward zero in the average. If we measure the transfer function as an average of several determinations, the effect of the noise can be made vanishingly small. The measurement of coherence function gives a direct indication of whether the output being measured is a result of the input signal or comes from some other source (noise).

To illustrate the power of this approach consider a system with multiple inputs and outputs as in Figure 5.

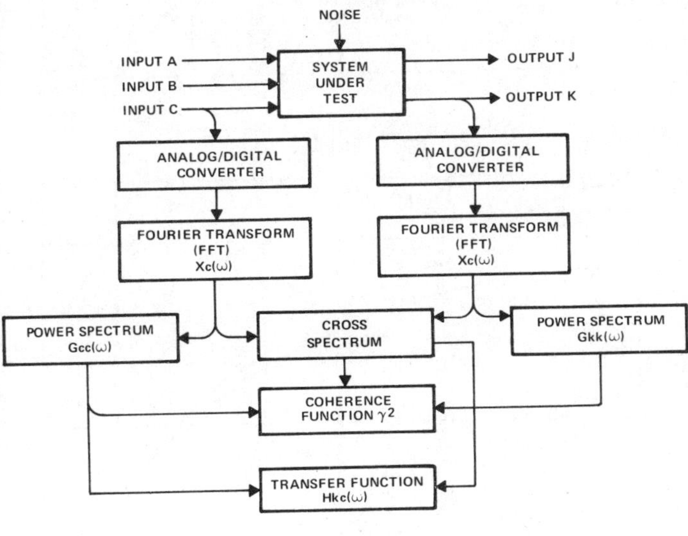

Fig. 5.

It is desired to determine the relationship between input C and output K in the presence of inputs signals at A and B as well as internally generated noise. The first step is to provide some signal to excite the system. The requirements for the excitation are remarkably loose, the only one being that the excitation signal possesses energy over the entire frequency range of the measurement. Random noise or any other signal or mixture of signals with the required energy distribution will serve this purpose. If the normal system input meets this condition, it can be used and the system can be tested while in normal service. Analog to digital converters will simultaneously sample the input and output signals and the discrete Fourier transforms calculated by the FFT algorithm. From these, the power spectra of input and output and the cross power spectrum are calculated.

Next, before trying to evaluate the transfer function, it must be determined that the observed output is in fact the result of the input C signal rather than one of the other inputs or the internal noise. This will be done by evaluating the coherence function. A low coherence value will indicate little relationship because the output and input or, looking at it another way, successive evaluations of the transfer function may differ radically since input and output are not strongly related. Once coherence has been established, a meaningful transfer function may be calculated.

A major problem that arises in the application of transfer function test methods is the question of how to relate the transfer function tests to other more widely accepted test criteria such as gain, leakage, linearity, etc. This question must be considered separately for each case. Results of three particular applications will be shown for illustration. The following principles will aid in interpretation of transfer function measurements.

1. The transfer function will completely determine the functional characteristics of any device for which a linear approximation is valid. Once the transfer function is known, the output to any specified input (impulse, sine wave, square wave, ramp, etc.) is known. Two devices with the same transfer function are entirely equivalent from an operational standpoint.

2. A change in transfer function is an indication of nonlinearity in the device under test. For example, assume that curve A in Figure 6 represents the gain transfer function of a particular amplifier measured at a low level broadband input. Now if the signal level is increased the transfer function will remain unchanged until the effects of saturation (a nonlinear phenomenon) become pronounced. At that point the gain transfer function will shift towards curve B where the magnitude of gain is reduced by output limiting and higher frequency components are apparently enhanced because of harmonic distortion of the output.

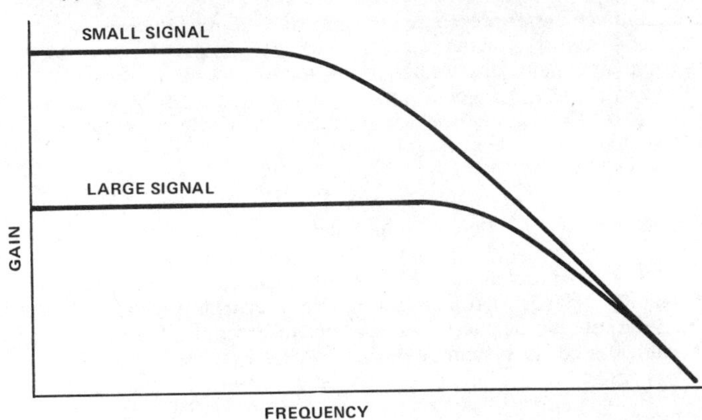

Fig. 6. Measured transfer function of an overloaded amplifier

3. The transfer function of two devices whose individual transfer functions are known is the product of the individual transfer functions.

Application Examples

To illustrate the use of the transfer function test method in practice, consider three specific applications.

Application No. 1 - - Oscillator

The R-C coupled oscillator of Figure 7a is required to be tuned to a frequency of 30 kHz. Resistor R_1 and capacitors C_1 and C_2 are

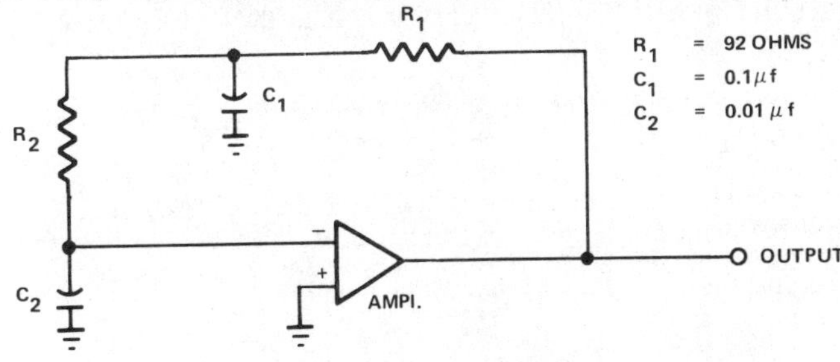

Fig. 7a. Oscillator circuit

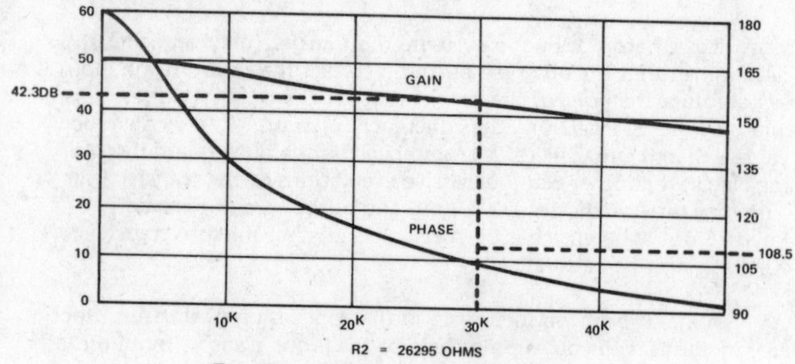

Fig. 7b. Gain/Phase amplifier No. 1

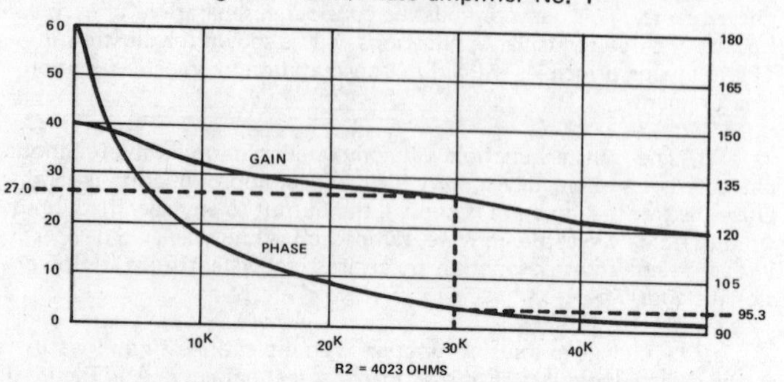

Fig. 7c. Gain/Phase amplifier No. 2

fixed and resistor R_2 is to be used for tuning. But there is one serious flaw to this circuit, namely that small changes in the gain and/or phase shift characteristics of the amplifier result in large changes in the value R_2. During fabrication, it will be necessary to tune the circuit by an expensive and time consuming "cut-and-try" procedure to select R_2.

A much better procedure would be to determine the gain and phase characteristics of the amplifier by means of a transfer function measurement. Once this has been done, the correct tuning resistor value to match the individual amplifier characteristic can be calculated. Figures 7b and 7c show the characteristics of two different amplifiers with the resulting tuning resistor values.

Application No. 2 – Power Supply Stability

A major problem in the design of regulated power supplies is instability. This is even more serious if the instability occurs after the power supply has been in operation for a period of time since it will cause costly system downtime and field repairs. Usually this comes about because the circuit was only marginally stable to start with and component aging or a particular load configuration was enough to drive the power supply over the threshold of instability. Oscillation usually occurs at some frequency well above the power supply's operating range, frequently 1 megahertz or more.

The situation is illustrated by Figure 8a which shows a typical power supply regulator circuit. Depending upon the characteristics of the feedback control loop (consisting of Q1, R1, R2, and the difference amplifier), the output voltage (Vout) will be regulated to equal the reference voltage (Vref), or the circuit will be unstable and will oscillate. The performance can be predicted from analysis of the transfer function of the control loop. To do this, a broad-band test signal would be injected at point A and the return signal around the loop measured.

If the returned signal at any frequency has zero phase shift and a gain of unity, or greater with respect to the exciting signal, the circuit will oscillate.

Three different control loop characteristics are illustrated by Figures 8b, 8c and 8d. In each case, the DC gain is the same, but relatively small differences in high frequency characteristics are critical for stable or unstable operation. Figure 8b is clearly unstable since the phase curve crosses zero with a gain of +8dB. In this configuration, the regulated power supply would go into immediate oscillation.

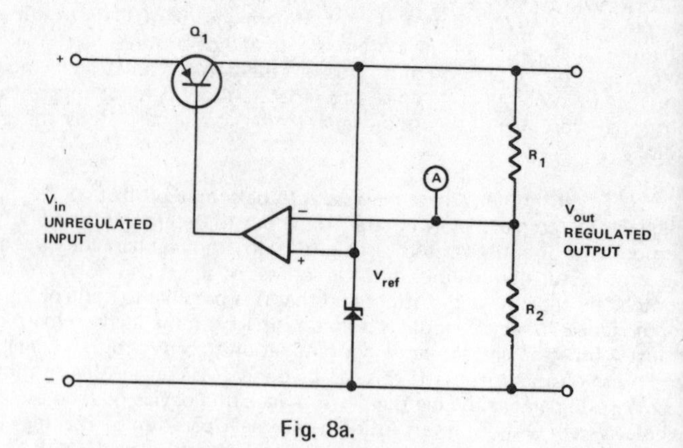

Fig. 8a.

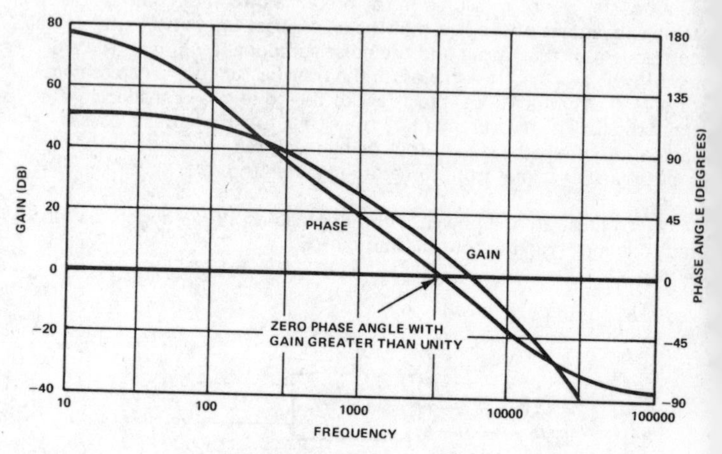

Fig. 8b. (Unstable system)

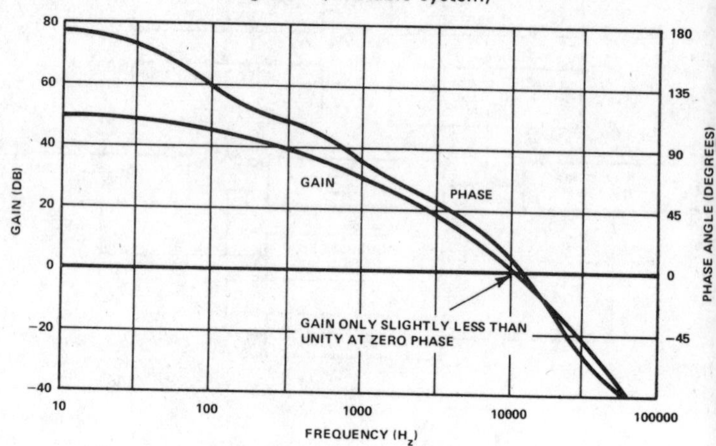

Fig. 8c. (Barely stable system)

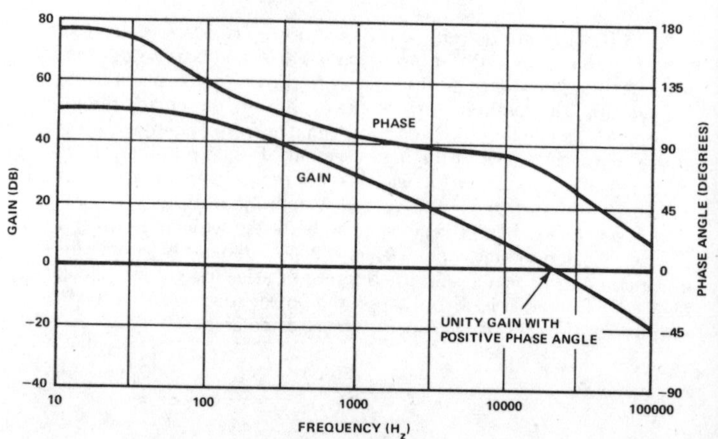

Fig. 8d. (Stable system)

The system of Figure 8c illustrates the advantage of the trans–r function test method. Although this system is stable and will ot oscillate, only a very small change in operating conditions would e required to cross the threshold of instability. This power supply quite likely to become unstable at some time in the future because f component aging and application of reactive loads. The transfer nction shows this condition very clearly, while ordinary tests ould pass this unit.

Finally, Figure 8d shows a completely stable system which is nlikely to cause any problem, even after component aging.

pplication No. 3 – Communication System

For reasons of preventive maintenance, it is required to peri–dically monitor the transmission characteristics of long distance ommunication systems. This monitoring process normally requires emoval of the system from service, resulting in loss of operating evenue or additional costs to provide an alternate system. Use of ourier analysis can eliminate downtime costs for testing and, at the ame time, improve test quality.

Figure 9 shows a generalized communication system, with mes–age traffic arriving at a sending terminal and being transmitted to a istant receiver. For testing purposes, the receiving end signal is "looped back" to the sending end for comparison with the original ignal. Noise enters the system at some unknown point. It is re–quired to determine the transmission quality of this system.

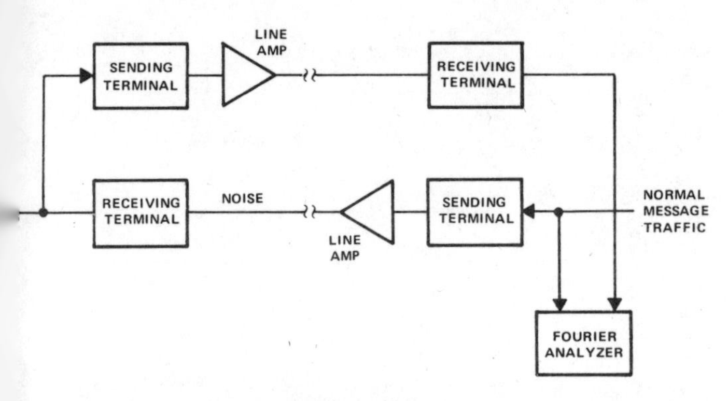

Fig. 9. Communication system

The first parameter measured is the coherence function to establish the correlation between the input and the received signal. If Figure 10 is the measured coherence function, it is seen that at frequencies above 0.7 MHz, a large part of the received signal con–sists of uncorrelated noise. However, at this point it is not clear

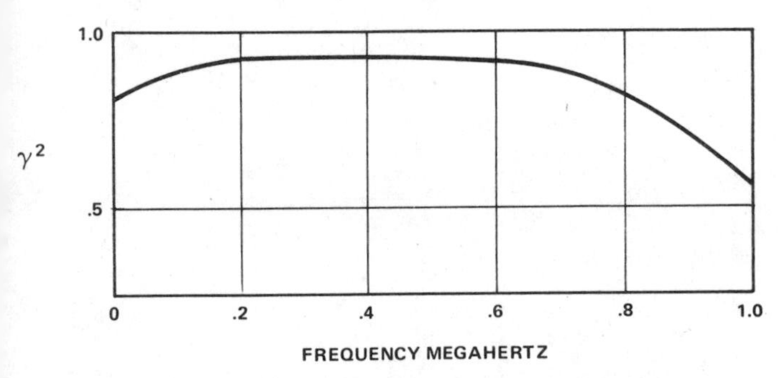

Fig. 10. Coherence function for system of Fig. 9. Above 0.7 MHz effect of noise is significant

whether the low coherence value measured is the result of particu–larly high noise level in that part of the band or a normal noise level with a low signal level. Fortunately, this distinction can be made based upon the system transfer function. By using an averag–ing time long enough to eliminate variations due to the noise, an accurate transfer function measurement can be made. Two possible results are shown in Figure 11. In one case, the system gain trans–fer function is found to be within normal limits, indicating that abnormally high noise is entering the system. In the other case, the system gain is definitely below normal at the high end of the band indicating a possible amplifier or equalizer problem.

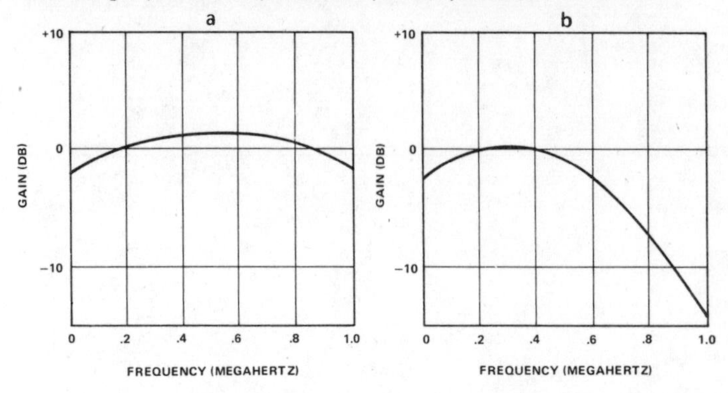

Fig. 11a. Normal transfer function indicates that low coherence results from abnormally high noise level
Fig. 11b. Loss of gain at high frequencies accounts for low coherence

CONCLUSIONS

Use of the Fourier transfer function for testing and diagnostic purposes offers very significant advantages over conventional test methods through its high information content, use of realistic test signals and speed and simplicity of testing with automatic test equip–ment. To obtain these advantages, a degree of reorientation will be required on the part of test equipment designers and users as well as in the way that equipment is specified. In spite of this, the authors believe that the benefits to be gained from the approach discussed here are so great that transfer function testing will replace other test methods in a great number of applications. Especially where high reliability requirements and/or high cost of down time dictate the necessity for very extensive testing, it may well be the only economi–cally feasible test method.

REFERENCES

1. Cooley, J. W. and Tukey, J. W., "An Algorithm for the Machine Calculation of Complex Fourier Series," Math of Comput, Vol 19, April 1965

2. Cochran, W. T., et al., "What is the Fast Fourier Transform?" IEEE Trans on Audio and Electroacoustics, Vol AU–15, No. 2, June 1967

3. Bergland, G. D., "A Guided Tour of the Fast Fourier Transform", IEEE Spectrum, Vol 6, July 1969

4. Roth, P. R., "Effective Measurements Using Digital Signal Analysis", IEEE Spectrum, April 1971

5. Bendat, J. S., and Piersol, A. G., "Measurement and Analysis of Random Data", John Wiley & Sons, 1966

6. Roth, P. R., "Digital Fourier Analysis", Hewlett-Packard Technical Journal, June 1970

PART IV: Test Languages and Program Preparation

Just as a computer is useless without an operating program, so too is an ATE without a test program. As essential as test programs are to ATE, the ingredients and preparation process somehow eludes the understanding of everyone except those who have spent considerable time solving the many problems and frustrations involved in test program preparation. Conceptually the process is quite simple, but until one actually experiences the problems involved in the interrelationships between UUT, tester, and its control and production software, there is not a full appreciation of the engineering analysis involved. This part provides some insight into the program preparation process and the relationship of the test language to that process. Of course, this is not a suitable substitute for actual test programming experience, but the papers do about as well as can be expected to convey a complex experience obviously familiar to the authors.

The first paper, "Program Preparation for Automatic Test Equipment," analyzes and describes the various steps and functions involved in the process from initial conceptual design to sell-off of a validated program. To describe each facet of the process it is necessary to treat each task as a separate function. Of course, in the actual performance of these functions, the separation is not so evident. In fact, one might go through the process often unaware that transitions occur in the nature of the tasks being performed. How structured and formal the various tasks should be performed depend on how large a programming effort must be undertaken at the time. As with any complex process, there are natural advantages to compartmentalizing tasks according to workers skills and specialties. In structuring a large scale effort, however, it must be remembered that an effective end product depends on continuity among all of the related tasks. The person who must assure that continuity is the test design engineer. He may have various skilled personnel to help, but unless he is given overall responsibility throughout the process, excessive costs will accrue and the product will be shoddy.

Just as in other computer applications, ATE technologists have concluded that a problem-oriented language is required to enable users to prepare test programs without having to spend several years as a programming apprentice. Since the test language is used throughout the process of test program preparation, many people have concluded that a good test language is the key to good economical program preparation. The paper entitled "The Test Language Dilemma" tends to destroy this conception by putting the test language role into its proper prospective against the total problem of test design, programming, and validation. The paper briefly reviews five test language examples showing that there are many effective languages to choose from with the hope that industry will abandon the futile search for the "perfect" language and get on with the real problem of learning how to design good tests cheaply. Even though finding the best test language is not really the solution to effective program preparation, it is an important consideration. The language which seems most promising today is Atlas. The next paper, "Atlas—Abbreviated Test Language for Avionic Systems," introduces this language. The reason this language seems most promising is not because of any inherent quality or beauty, but rather because it is based on a collective effort of many experienced users. It must be remembered that languages are not adopted by people for intrinsic beauty but simply because there is a need to communicate with many people. Thus the most effective language is one that has proven itself by substantial usage and evolution. The prime weakness of Atlas is that it allows users to develop "adapted" Atlas languages and processors which, if one takes all of the liberties allowed, can result in a new language bearing little resemblance to a complete Atlas or even an Atlas subset. Hence the buyer of a system that provides an Atlas capability had better look carefully at the Atlas capability offered if he is really concerned about language standardization.

The paper entitled "Software Automation in Automatic Support Systems" takes a production-oriented look at the program preparation process with an eye toward automating as much of it as possible. Recognizing that there will always be a need for original thinking in the process, the idea is to reduce the more mundane tasks to computerized processes for the purpose of labor cost reduction and standardization. This analysis is intended primarily for large-scale program preparation processes. Those having only an occasional need for a new program would not find such automation practical. But for really large-scale efforts, such techniques provide the only practical way to optimize the relatively scarce commodity of test design engineering labor.

The fifth paper, "Automatic Generation of Test Programs," offers an interesting approach to on-line preparation of test programs. By using a semiautomatic test station to manually set up each test, the test engineer has the means for trying out his test design as he generates it. Not only does this approach concurrently validate the program produced, but it avoids the issue of test language by reducing man-machine communication to physical actions such as setting knobs and reading meters. As promising as this approach is, it seems curious that it has not been widely used.

PROGRAM PREPARATION FOR AUTOMATIC TEST EQUIPMENT

By A. M. Greenspan*

Summary
Quality is a product characteristic which can be designed into a test programming effort by means of deliberate control in all phases of conception, design, production, validation, and demonstration. A satisfactory level of quality can be achieved only by a planned effort with specifically defined tasks and objectives.

It is management's responsibility to assure efficient and effective results in the ATE programming effort. The management team, when operating properly, should detect promptly and correct design practices which could result in defective program sets. Technical data, design policy or other elements of contract performance which could create system support problems must also be identified and modified. However, in order to accomplish these ends, it is necessary for management to be able to see the programming job in the context of the overall effort as well as to understand each of the sub tasks comprising this effort. Once this is achieved, development of a successful ATE programming function is possible through timely application of proven management procedures such as quality assurance plans, design reviews, and other controls.

1. Introduction
The programming task for an automatic test system is the key to efficient utilization of manpower and time in an operational environment.

The Automatic Test Equipment (ATE) program bears the same relation to the ATE system as a detailed step-by-step instruction manual bears to a human tester. Preparing either a step-by-step instruction manual or an ATE program can require considerable engineering knowledge. However, since the ATE system can neither think for itself, and most systems cannot yet learn from experience, the ATE program must be more detailed and specific. It must anticipate beforehand all of the eventualities that the tester will encounter. Thus, the ATE program requires considerably more engineering and anticipation of possible occurrences than a manual step-by-step instruction procedure.

The ATE programming process as performed at RCA in Burlington, Massachusetts, consists of four basic phases:

1) Analysis and Program Design
2) Coding Assembly and/or Compilation
3) Validation
4) Demonstration

The portion of effort typically required for each of these phases relative to the overall process is shown in Figure 1.

***RCA Aerospace Systems Division, Burlington, Mass.**

Reprinted with permission from *IERE Joint Conf. on Auto. Test Systems Rec.*, April 1970.

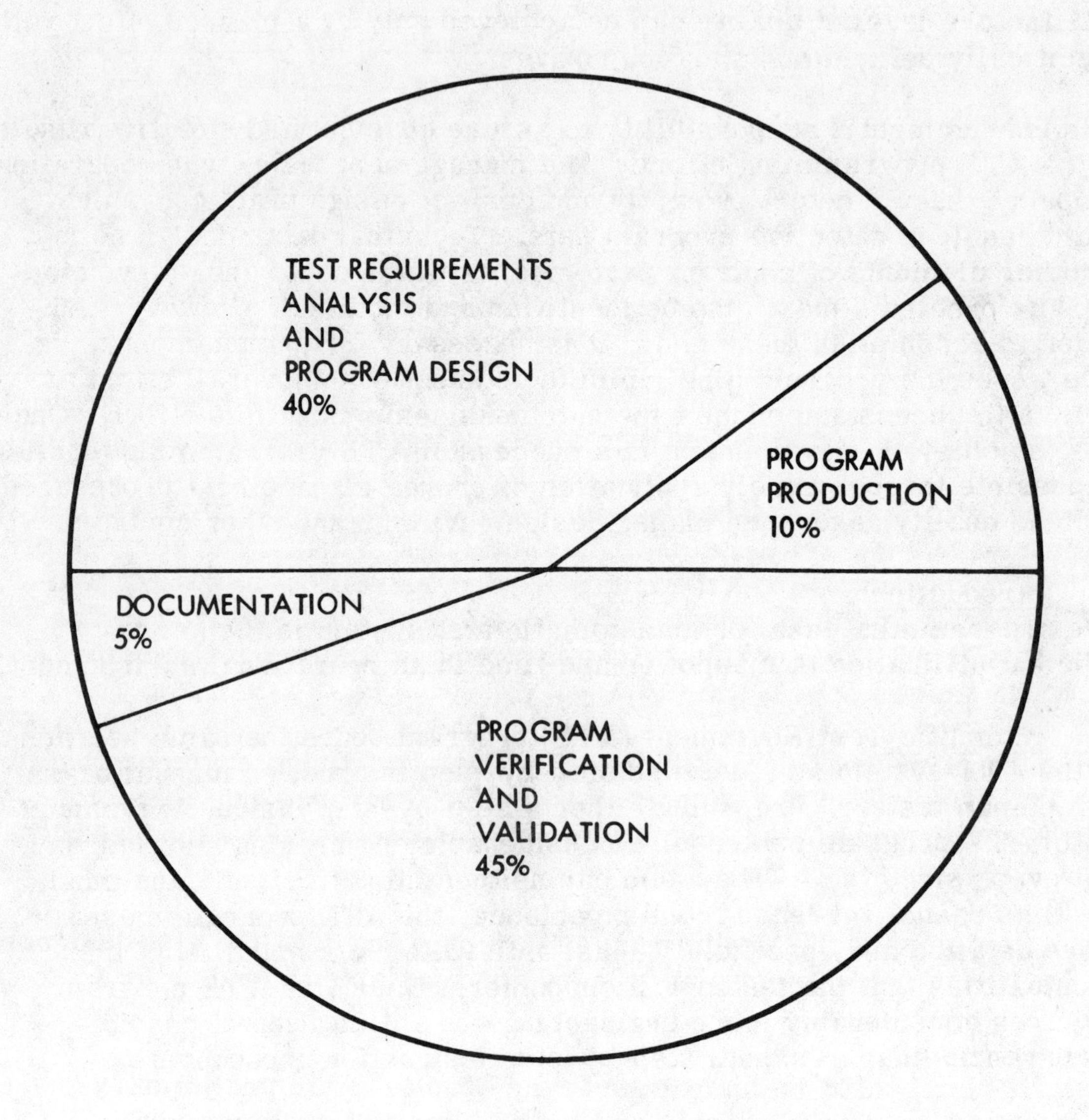

Fig. 1 Distribution of Programming Effort

2. Analysis and Program Design

The ATE programming procedure begins with a basic UUT (Unit Under Test) test specification. This specification is drawn up by the UUT design engineer, possibly with support from a technical writer. This test specification, describing the UUT operation, characteristics and tolerances, is then turned over to a Senior Test Designer along with other UUT documentation such as schematics, wiring, diagram, manufacturing drawings, factory test results, failure mode analysis and other pertinent information. Usually, however, documentation is not this complete when received by the Senior Test Designer. As a minimum he must have the test specification and schematic; any other materials available will serve to make the Test Design programming job that much easier.

Upon receipt of the UUT documentation the Senior Test Designer will begin his preliminary test design. First he must study all of the source documentation and guidelines available to him in order to develop an efficient test plan. The test plan is a general concept as to the best method for testing the UUT. While engaged in test planning the test design engineer will anticipate possible testing problems and attempt to devise means of minimizing these through utilization of special test techniques, segmenting the manner in which testing is performed, or design of special adapter or signal conditioner to be used as part of the UUT ATE interface. Another aspect of the preliminary test design is performance of the Test Requirements Analysis (TRA). The purpose of the TRA is to identify (based on the UUT documentation) any incompatibilities (performance + physical) between the ATE and the UUT. The test designer will also at this time establish preliminary criteria for specifying Go and No-Go limits.

During the process of preliminary test design the design engineer should have available to him the support of a quality assurance engineer, a hardware design engineer, the UUT design engineer and a representative of the eventual user of the ATE program.

Upon completion of the preliminary test design and TRA the test designer will prepare a formal test plan in a format adaptable to review by a concept review team.

This test plan should indicate clearly the results of the preliminary design, point up problem areas and suggest possible solutions or approaches to minimize or eliminate these problems. The test plan may be a simple report, flow chart, tabular listing, or comprehensive design document depending upon the UUT complexity and ATE compatibility. In preparing the test plan the design engineer requires the support of a technical typist, in order to prepare the results of his analyses in a formal manner, to be reviewed by a concept review team.

When the test plan is complete the design engineer will request a concept review. The primary function of the concept review is to evaluate the proposed test plan and provide direction as required. A secondary but also important function is to consider the impact of the prepared test approach on documentation, spares, training and other maintainability requirements.

The participants in a concept review usually are made up of:

1) The Test Program Design Engineer.

2) An ATE Systems Engineer (a specialist in ATE applications and design).

3) A Senior Test Program Design Engineer (a specialist in test techniques other than the Program Design Engineer).

4) A UUT Systems Engineer (expert in operational and support requirements of the UUT).

5) A representative from the contracting agency for whom the program is written.

6) A representative from Quality Assurance and Program Management.

Copies of the test plan should be made available to all participants of the concept review a number of days before the review takes place in order that all may familiarize themselves with the UUT and proposed program approach. Thus, when the concept review takes place the reviewers will have gathered material to help resolve problem areas which have been indicated, or to suggest subtle problems which may have been overlooked by the program designer. Alternative approaches may also be suggested which could possibly enhance the results obtained by the program. Each participant will, of course, be primarily concerned with his own area of interest and will be expected to be particularly helpful in lending support in his own specialty. However, the cross fertilization of ideas that takes place between all participants at the concept review often results in a whole effort which is greater than the sum of its parts. An example of a typical test plan is shown in Figures 2 through 5.

2.1 Detailed Test Plan

Subsequent to the concept review, the test design engineer will begin to prepare his detailed test design. First he will resolve any action items which have resulted as a consequence of the concept review. This may involve resegmenting of the UUT for test, change of test approach, addition or deletion of tests, variation of preliminary limits, or addition or deletion of signal conditioning. Each action item correction will require approval of the concept review team member into whose area the specialty falls. Clearance of all action items will result in authorization for the actual detailed test design to begin.

The amount of work involved in achieving the final test design in the form of an English Language Program will be influenced by the amount of detail already in the basic test plan. As a rule, a definitive test plan simplifies the final test design task.

1. DESIGN DESCRIPTION

1.1 Test Plan Coverage

This Test Plan covers Radio Receiver-Transmitter XX-XX/X, which is a modularized HF transceiver operating over the frequency range of 2 to 30 MHz. In service, it is normally used in conjunction with Remote Control Unit YY-YY/U and Whip Antenna ZZ-Z.

The remote control unit and the whip antenna will not be covered in this test design. Rather, the transmitter will be terminated in a dummy load, and remote control signals will be simulated by ATE stimuli.

This particular transceiver has been in service over a number of years and is well documented. However, there are discrepancies between the factory test specifications and current field maintenance practice. These discrepancies must be resolved.

1.2 Source Information Summary

1. Next Higher Assembly		
2. Next Lower Assembly		
3. Manuals Technical	Pages	Date of Issue
Organizational	Pages	Date of Issue
Maintenance	Pages	Date of Issue
4. Drawings Schematic	Revision No.	
Parts Lists	Revision No.	
Cable Diagrams	Revision No.	
Test Specification	Revision No.	
Outline Drawing	Revision No.	

Fig. 2 Typical Design Description Page

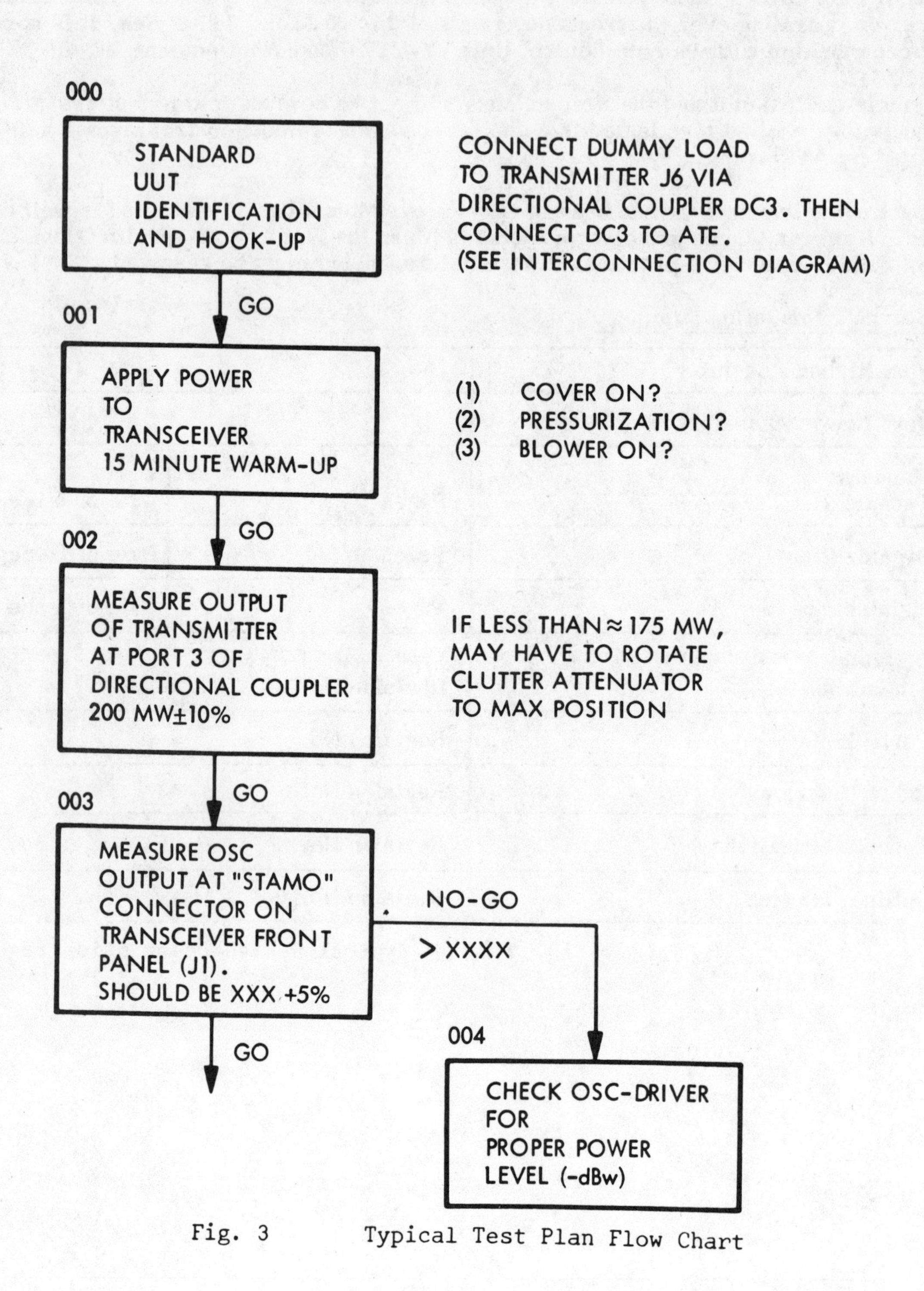

Fig. 3 Typical Test Plan Flow Chart

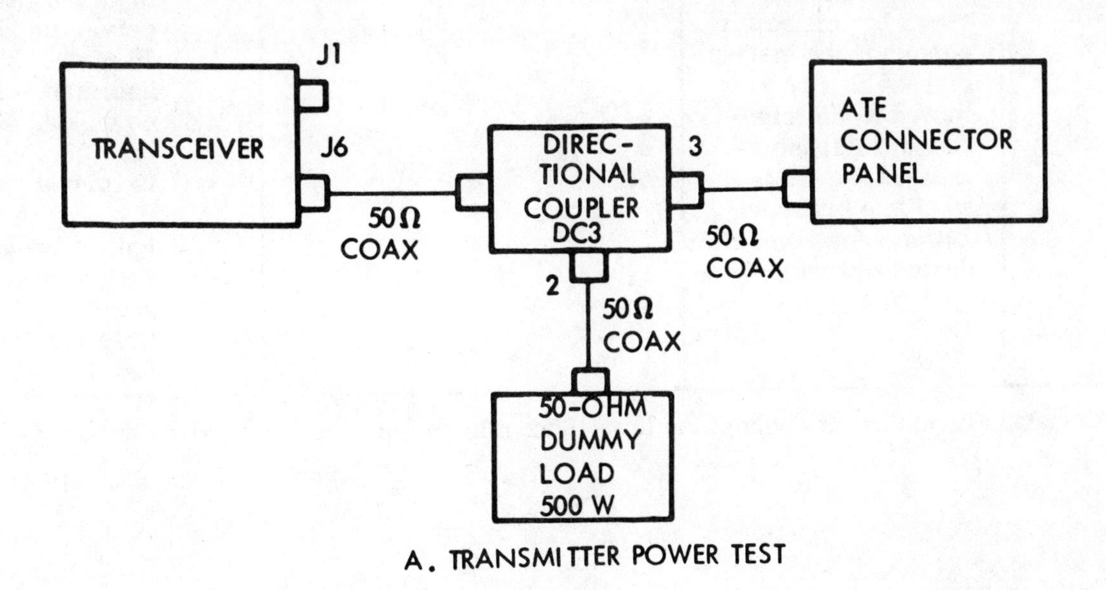

A. TRANSMITTER POWER TEST

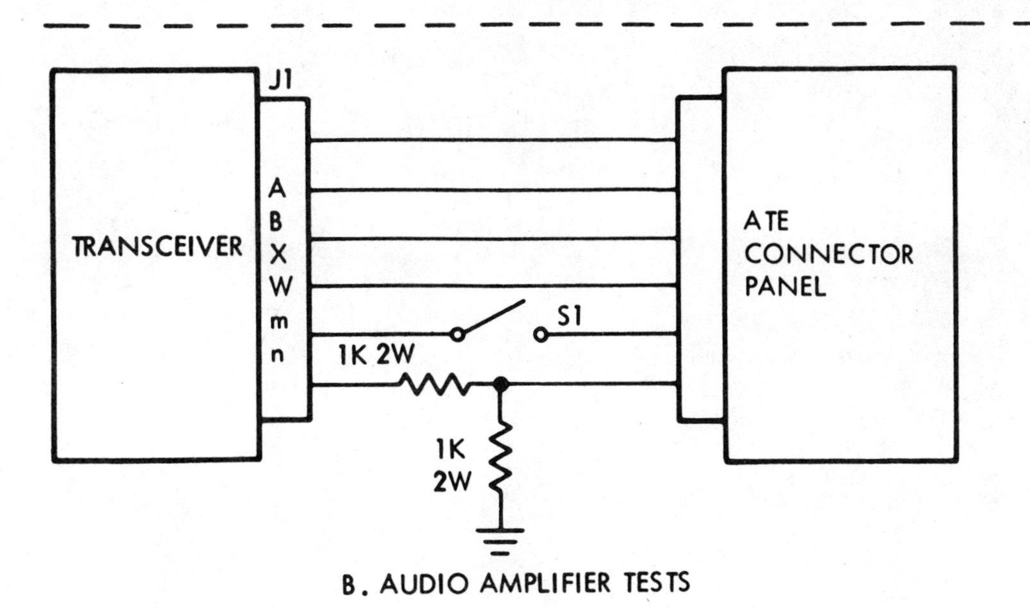

B. AUDIO AMPLIFIER TESTS

Fig. 4 Typical Preliminary Interconnection Diagrams

Test No.	Problem	Recommended Solution	Action
000	No dummy load furnished with UUT.	Requisition GFE Model XX or equivalent. In meantime, use Load Box, Power XX-XX/X from stock.	*Contracting officer to furnish GFE by 1 April 1968. Stock item XX-XX/X approved for preliminary testing.
001	Should transceiver cover be left on for performance testing, or should it be removed in anticipation of fault isolation testing? Also, what are pressurization requirements during testing?		*(1) Mount transceiver in Test Fixture XX-Y and then remove cover when fixture cooling air is turned on. (2) Ignore pressurization requirements at this time (30 January 1968).

*Information furnished at concept review. Shown here only for illustration purposes.

Fig. 5 Typical Problem Report

The approach recommended by the writer in designing a flow diagram representing the detailed test design is to first utilize static tests to verify that the UUT can be powered safely and also to quickly eliminate the possibility of any gross catastrophic failure easily found in a power off condition. Next the dynamic performance tests are added. These are the detailed designs of the performance tests outlined in the test plan, and the function of these tests is to check the UUT performance requirements under dynamic operating conditions. The static and dynamic tests described above represent the UUT "Go" chain. When the program is run on an ATE, should all of the above tests pass or be within the limits specified by the test designer, the UUT would be considered good and a statement to that effect would be printed or displayed by the ATE at the conclusion of the final test.

The third step in the creation of an ELP is to add the "No Go or diagnostic portion of the program to the chain. This involves analysis of the UUT failure modes in order to establish the manner in which UUT failures manifest themselves as well as determine the nature of the ATE fault isolation steps which will be necessary to establish the point of failure.

The final step in creating an English Language Test Document requires a thorough check of the test program design, interface, diagrams and documentation to insure that they are all complete and compatible.

The detailed test design is usually documented as a flow diagram. The flow diagram form is much preferred to a tabular listing, especially for diagnostic testing, due to the ease with which the testing sequence can be tracked in this form. This flow diagram will describe each test in the sequence in which they occur, the stimulus and measurement requirements for each test, special switching to connect or disconnect interface equipment, test limits and tolerances and, finally, the consequences of "high", "low" or "go" branches. The flow diagram is usually written in English language oriented terms to allow persons not familiar with the ATE to follow the test design logic, and when complete, the flow diagram serves as a detailed description of the testing function as it will be performed by the ATE. The persons involved in the writing of the English Language Program (ELP) are the Senior Test Designer with technical support, when necessary, from the UUT Design Engineer, and clerical support from an engineering aide. A typical ELP flow diagram is illustrated in Figure 6. For ease of preparation the test design should be provided with a template containing standard flow chart forms and graph paper upon which to draw the flow diagram. The template form used at RCA, Burlington, is seen in Figure 7.

2.2 Interface Design

Once the ELP is complete and the testing process is clear, the design engineer must turn his attention to design of the interface to be used between the UUT and ATE. The interface is used to resolve compatibility problems. Currently, all multipurpose ATE systems utilize some sort of general purpose interface to adapt the UUT to ATE and augment the ATE capabilities. The use of interface adapters is so widespread that they are often considered a part of the ATE system, but actually the interface design is as much a part of the software preparation process as the test program itself.

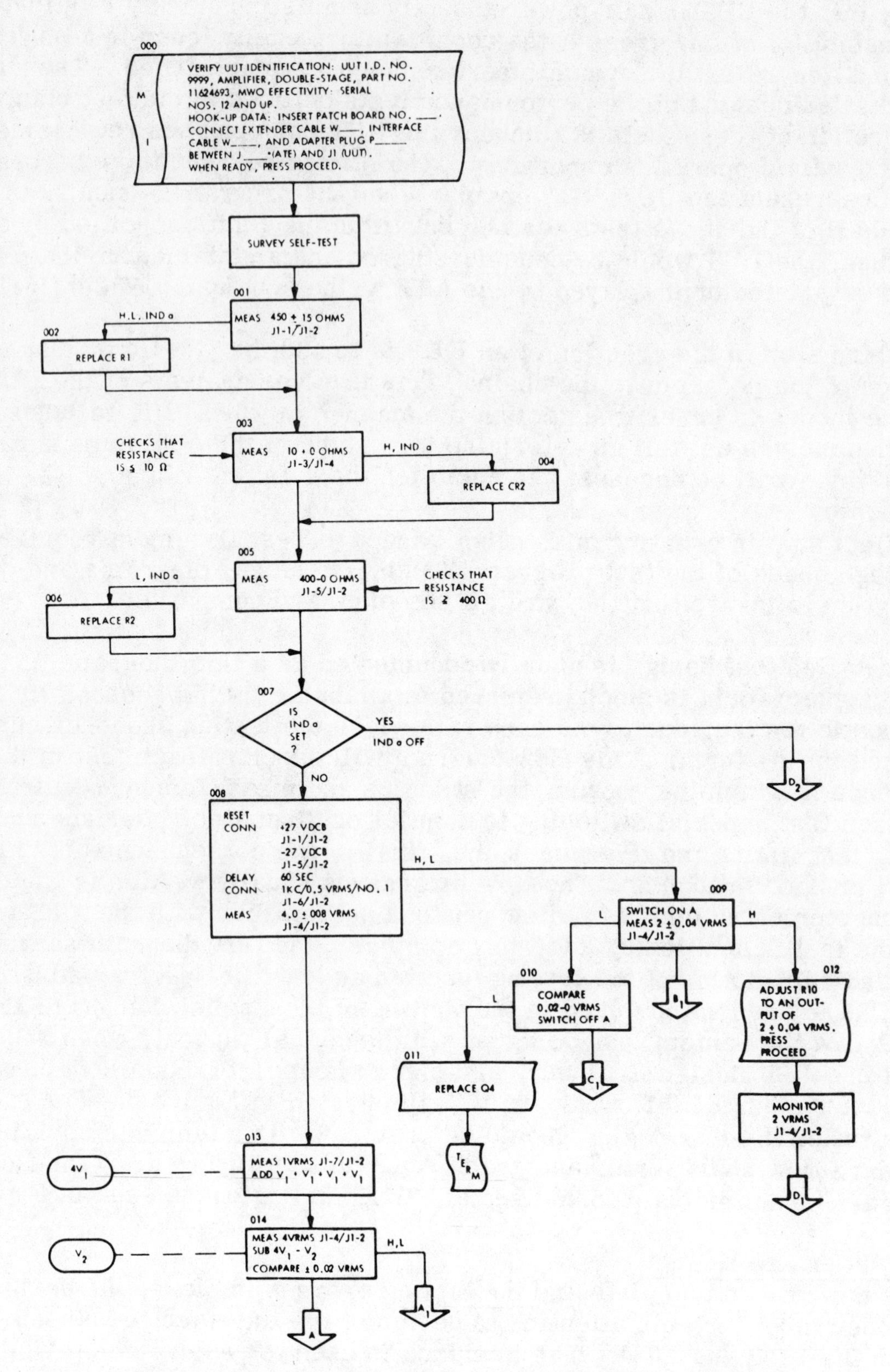

Fig. 6 Example of Test Program Flow Chart

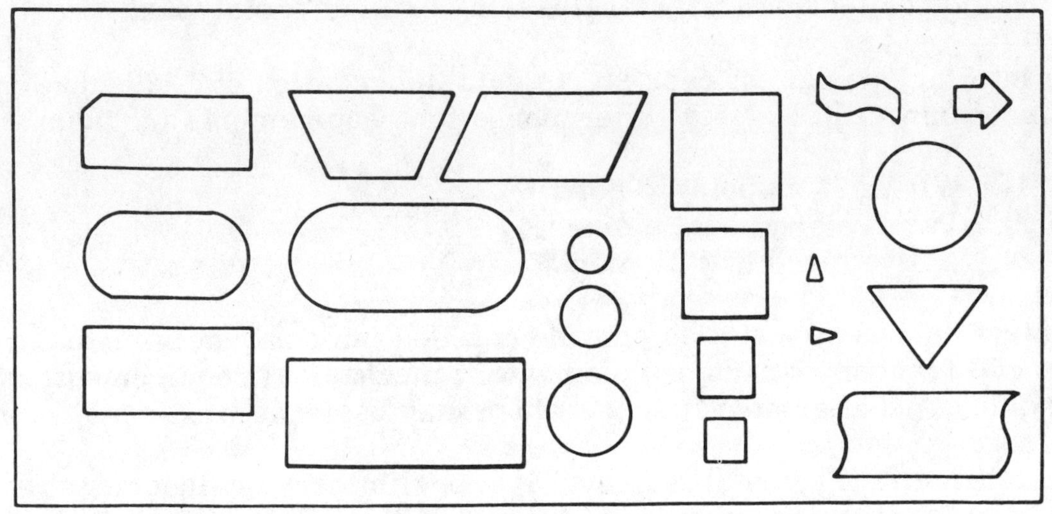

Fig. 7 RCA-ATE Mechanical Guide Outline

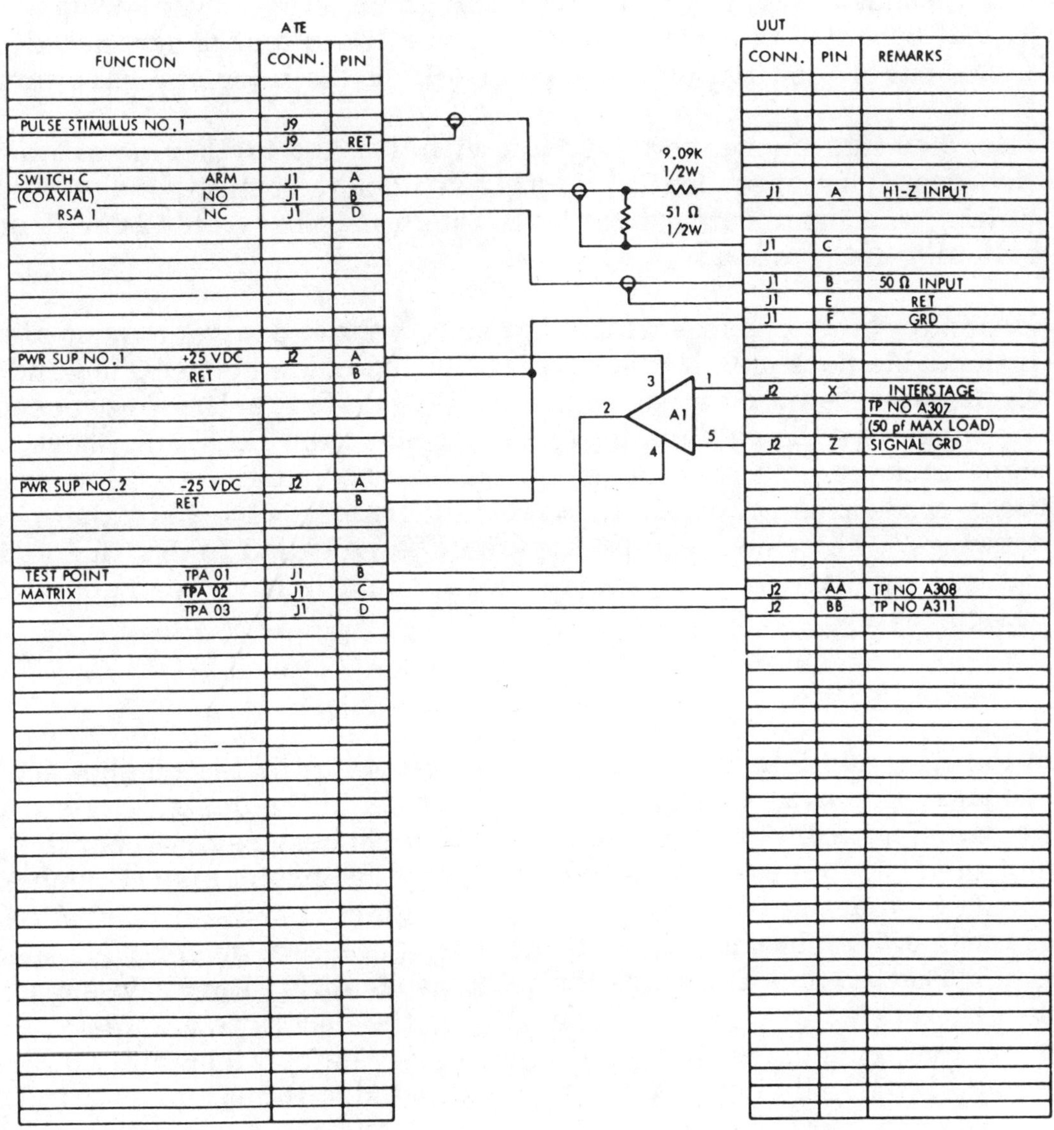

Fig. 8 Typical ATE/UUT Interface Connection Diagram

Design of the interface usually requires support from mechanical engineering, the UUT design engineer, the drafting department, and an engineering aide or technician for engineering build of prototypes.

The interface design process starts with the creation of a schematic diagram defining all electrical interconnection requirements including:

1) Wiring from point to point
2) Passive electrical elements
3) Active electrical elements.

The test designer must also provide detailed information in textual or pictorial form concerning supplementary mechanical requirements such as holding and alignment fixtures which may be necessary.

The schematic (Figure 8) is basically a preliminary engineering drawing developed in the process of test design. The format should be simple to make for ease in maintenance and updating of the interface adapter and for debug of test logic during validation. Marginal notes, signal identifiers, and logical grouping of circuits and wiring will aid test program validation and check-out of the first production models of the interface hardware.

The interface interconnection diagram will also add further detail to the ELP by identifying specifically, routing switches, stimuli, measurement test points, and input/output terminals which may have only been described symbolically in the ELP.

In addition to the interconnection diagram, the test designer must also detail the cable hook-up diagram indicating the cable connections and/or switch placements which must be made in order for testing to proceed, as well as formalize the cable point to point connection lists. In short, the interface package must include all necessary information to allow drafting to create production schematics, wiring diagrams, wire lists and mechanical drawings. This interface package will be released to drafting and/or mechanical engineering after all test design details have been approved at the design review.

2.3 Design Review

When the ELP and interface package are complete the test design engineer will request a design review. The participants in the design review should be the same persons as were present for the concept review, but the purpose of the concept review was to detail potential problems and smooth the test design, whereas the purpose of the design review is to insure that the program is of high quality and will meet the goals set by contract specifications. That is, it will ensure that, the program will provide support to the level specified by contract stipulations, the design is complete, the program test techniques conform to previously determined optimum practices and, finally, the program can be validated with maximum efficiency.

Experience has shown that an analytical sampling process provides good engineering confidence regarding the quality of the program design. The method used at RCA for checking program design is as follows.

Hypothetical or "paper" faults are selected by one or two members of the design review team. The members of the team selecting these faults should be familiar with the UUT or with similar units, in order that the faults picked will be in the area of common failure modes of the UUT, as well as failure modes suspected of being the most difficult to detect. Each fault selected must be of a type that will manifest itself as a detectable UUT failure. The symptom or symptoms which should occur as a result of the hypothetical malfunction must, thus, also be analyzed by the person making the fault selection.

The number of faults selected are a function of the size of the UUT to be tested. The number of faults must be sufficiently large to establish a reasonable confidence in the program's ability to find faults and yet not be so large as to be inefficient.

When the design review takes place the faults are presented one by one to the program designer. After each fault is presented the program will be examined step-by-step, to verify that the fault will be detected and/or correctly isolated.

The criteria used to judge the program is merely an exercise in inferential statistics in which the number of faults successfully found and the number of faults not found serve as the basis for accepting the program as being of sufficiently high quality as to be ready for validation or for rejecting the program and sending it back for further rework.

The design review plays a very crucial part in the program preparation process. If it is done properly, validation time can be kept to a minimum. This is critical as machine time is very precious and there are typically many conflicts between personnel attempting to obtain as much time as possible on a limited number of machines. Also, validation is typically the greatest time consumer in the program preparation process. Steps taken to optimize machine utilization by insuring efficient validation will pay great dividends in saved dollars and reduction of pressures caused by tight schedules.

2.4 Program Production

Successful completion of the design review leads to the program production phase of the program preparation process. Program production can be defined as the process of coverting a completed UUT English Language Test Procedure into a machine coded program ready for validation on the ATE. This includes transformation of the program into the medium used to operate the ATE, i. e., magnetic tape, paper tape, mylar, etc.

Because of the many demands on the test designer, test program production should be reduced to a simple, straightforward routine handled largely by programming assistants. Of course, supervisory control must be retained by the test designer but his direct involvement should be minimal.

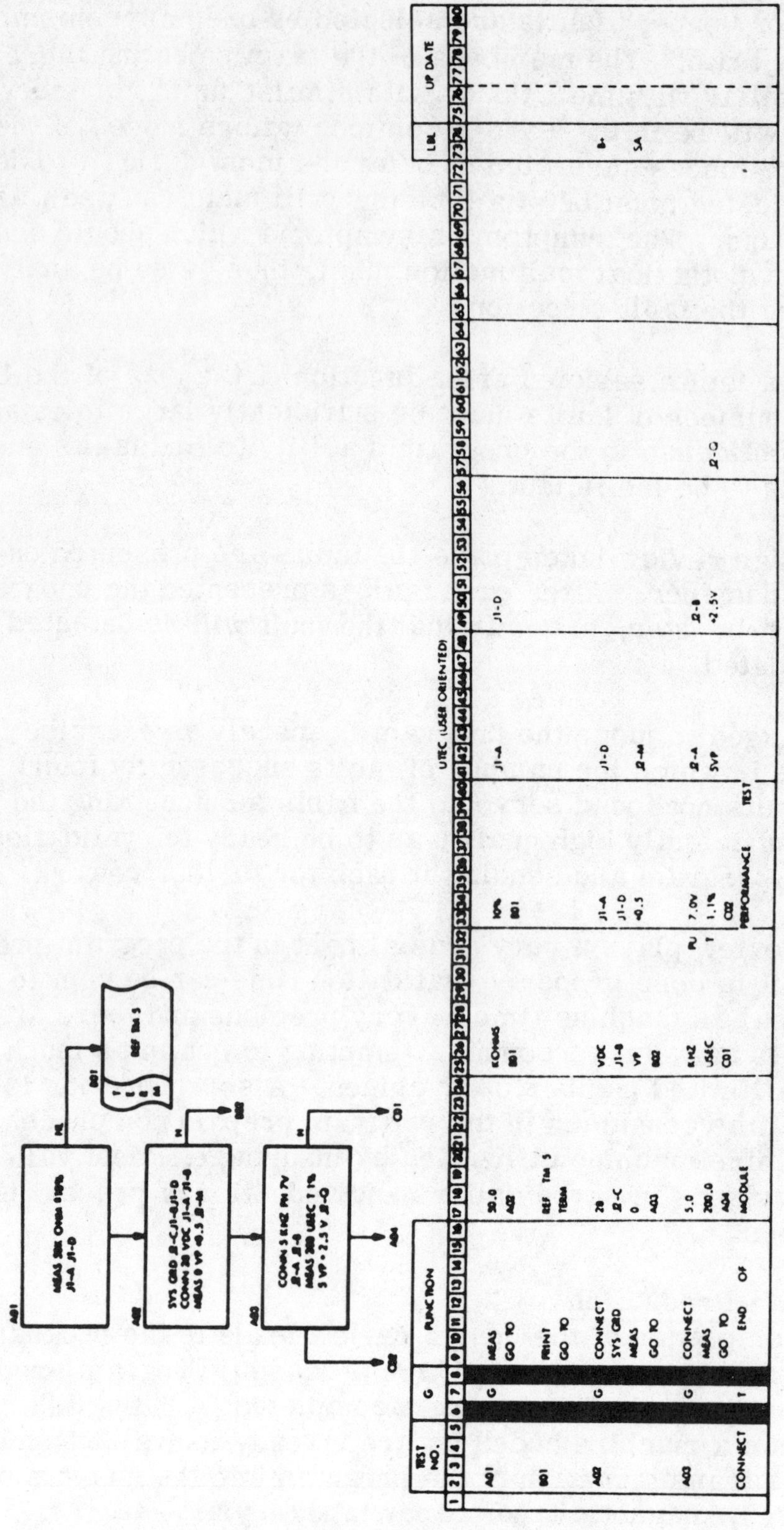

Fig. 9 Typical Program Segment

The production process discussed in the following paragraphs assumes an ATE procedure in which a high level source language will be converted to a program tape through use of an off line general purpose support computer for computation, assembly and tape generation. The basic principles discussed should be adapted to meet specific circumstances.

Program production begins with transformation of the tests described in the ELP into special ATE coding forms. The individual who transcribes or translates the ELP test program statements into alpha-numeric line entries on the ATE coding form is called a "coder". This job should be handled by a programming assistant. Coder skill level requirements vary with the degree of ELP interpretation necessary. The simpler the translation, the more routine the coding, and consequently the lower the technical requirements. Figure 9 shows the coding necessary for a program to be run on the RCA UTEC (Universal Test Equipment Compiler). Figure 10A gives an indication of the increase in complexity in the coding if the program were to be assembled in ASEIP (An RCA Assembly Language) which is a lower or less user oriented programming language. Figure 10B demonstrates the even greater complexity of attempting to translate and code directly into machine language. Regardless of the ATE language utilized, close liaison is often required between test designer and assistant to achieve proper coding for special test techniques, or requirements which are not easily put into the ELP.

Once the program has been coded, the coding sheets will be sent for keypunching. Keypunching of coded UUT test programs is substantially the same as any other keypunch operation. Standard machines are used, and the keypunchers used for ATE keypunching need no special skill other than that required for normal keypunch work.

Once the program has been keypunched onto computer cards the deck is verified by having another keypuncher retype each card on a verifying machine which will compare the two cards and identify those cards which are not identical, after which necessary corrections are made.

The keypunched test program should contain all of the tests specified in the ELP for a particular UUT. Standard control cards peculiar to the support computer being used to compile the program are then added to the deck and the card package is submitted for compilation and/or assembly.

The job of submitting the coding sheets to keypunch, collecting the keypunched deck, adding control cards and submitting this card deck to the computer should be handled by engineering aides, preferably by the aide who has been giving support to the test design engineer since the start of the programming process. Of course, the design engineer must retain supervisory control and be certain he is kept informed of progress, problems, and/or conflicts.

The output of the support-software system used to compile and/or assemble the UUT test program is generally called the object program. It is the coded machine instructions used as input to the ATE system

LANGUAGE LEVEL	LANGUAGE SAMPLE
ASEIP SYMBOLIC LANGUAGE (COMPROMISE BETWEEN ATE DEVICE AND USER ORIENTED)	PSI 146 DAI 155 SGN 100, 10000 TPA 1, TPB 5 MDC 3 DLY 50 MEAS

FIGURE 10A

Fig. 10 Programming Language Levels

LANGUAGE LEVEL	LANGUAGE SAMPLE	
MACHINE LANGUAGE (ATE DEVICE-ORIENTED)	15006317 55100125 55501469 02107440	INSTRUCTIONS, INTERNAL DATA, I/O, T.E. INSTR.

FIGURE 10B

during the testing process. Actually there are two main outputs that will be of interest to the design engineer. These are: the medium carrying the program which he will use on the ATE, be it paper tape, mylar tape, magnetic tape, etc.; and the computer listing. The computer listing is a copy of the input source language statements read into the computer and the corresponding machine language output which was generated as a result of these inputs. Usually other information will also be available on the listing, such as memory assignments, interface wiring lists, and error flags. The extent of additional information is dependent on the sophistication of the computer program under which the ATE cards are processed.

When the computer outputs are received by the test design engineer he will start verifying the results of the program run. To do this he must carefully examine the listing in order to detect errors which may have been introduced inadvertently into the program by the test designer, coder, keypuncher, computer, or combination of all of these. The effort necessary to check the listing will depend upon the ability of the support computer to create attention flags statements which detect and point out the various errors which may exist. In any case, even under the best of conditions a careful check should be performed by the test design engineer to insure against test conditions which could damage the UUT, ATE or both.

When the test designer is satisfied that the program can be utilized safely, validation can begin. There is some debate as to whether a program with known errors should be utilized to begin validation or whether it is best to correct a program through an update (recompilation) and start with a program in which no known errors exist. In this writer's opinion, the best approach is to make as much use as possible of each program run. Even if the results of the first run contain errors, if they are not dangerous to the UUT or ATE the test designer should use this tape (if a tape input is used) to obtain whatever further information is possible on the ATE itself. In this way he will be able to derive additional benefits from the second computer run.

3. Validation

The validation process involves exercising the UUT on the ATE under control of the test program. This is the only means for conclusively proving that the program set is ready for field use.

The person validating the test program is usually the original test designer, although this is not mandatory. The validator may be another person, as long as he is appropriately familiar with the UUT, the ATE and the test program.

The validator will start the validation procedure by connecting the UUT through the interface to the ATE, following explicitly the hook-up instructions which are included as part of the program set. Any errors or vague instructions will be noted and corrected. The examination of hook-up, cabling and printed instructions as well as mechanical orientation of parts and ease of operator actions is as much part of the validation process as the actual running of the program on the ATE.

When the hook-up is complete a UUT known to be good is connected to the interface hardware and the ATE takes control of the testing procedure.

If the ATE is operating properly, all coding is correct, and all the programmed tests have been set up with proper limits, delays and switching (test design is correct); the program should run through to the end and show that all tests are "Go".

Practically speaking, this is never the case with a program run for the first time. Experience has shown, especially on newly developed ATE systems, that one hour (on the average) is necessary for validation of each test in the program. Further, 90% of the validation time will be spent on 10% of the tests. These tests are, of course, the particularly difficult and critical tests within the program.

Each performance test that fails should be analyzed by the validator to determine the cause of failure before proceeding further. Once the cause of failure is known, testing can continue, but the validator must be alert to the possibility that the error causing the initial failure may cause failures in subsequent tests. If the error is one which overstresses ATE stimulus generators or routing switches, or which could create conditions in succeeding tests which might induce additional failures in the ATE or UUT, the error must be corrected before testing can continue.

The entire series of "Go" chain tests are validated first in the sequence in which they normally occur in the programs. The results of each test should be recorded. This recording process should preferably be automatic through control of the ATE program. If this capability is not available in the ATE upon which validation is taking place, the recording process should be done manually. Availability of the recorded readings is important to the validator for it is through these that he can analyze the testing process as it is taking place on the ATE relative to what the expected test results were when the program was written. For example, test results may be barely within limits because of unanticipated offsets or response times. This type of situation will usually necessitate a change in stimulus value, a shift of limits, or additional settling time delay.

In cases where tests actually fail in the "Go" chain, the validator must be able to analyze the test situation and determine if the fault lies within the ATE, UUT, Interface, program or if it is an intermittent failure due to marginal allowance for settling time. Load impedance, cross-talk noise, and a multitude of other possibilities must also be considered. This is a difficult task at best and requires that the validator be a person with broad technical background as well as possess the ability to logically analyze and solve complex testing situations containing many variables. Support of ATE maintenance personnel should be available to the validator to repair the ATE when the problem is isolated to the ATE itself. If the validator and test designer are not the same person, the test designer should also be available for help with particularly complex tests. In addition, the validator should be able to avail himself of spares for failed components and even wiring specialists when wiring errors, whose repair is complex, are found. In addition, technical writing specialists should be available to help in correcting and/or modifying technical manual references, hook-up instructions, or program operator instructions.

Of all the support mentioned, only the maintenance and spares support should be immediately available while validation is taking place on the ATE. The remaining support can be obtained during program analysis when the validator is not actually utilizing the ATE.

When all of the performance test problems have been corrected and the "Go" path can be run repeatedly without failure, the validator will begin to check the test program's fault isolation logic to ensure that the programming is continuous and compatible with the ELP. Forcing the program from "Go" chain tests into their associated "No-Go" sequences can be accomplished in many ways. For example, branch instructions can, on most ATE's, be entered manually through keyboard or other peripheral devices by the operator causing the program to jump to the "No-Go" test of interest; an alternate but similar method is to manually insert an artificial measured value for a given test in a results register. When control is given back to the program, the ATE will be fooled into thinking the test has failed and should branch accordingly out of the "Go" chain. However, the preferred method for this phase of validation is to have the ATE equipped with a validation aide type of control which can be set to automatically force a high or low branch on any given test.

Once the program coding has been proven by forcing the program "No Go" paths, the task of validating the No-Go still remains. In order to do this it is necessary to induce faults into the UUT and see if the program isolates these faults correctly.

The number of faults which must be inserted into a UUT will usually be dependent upon the program confidence which must be obtained before the program can be released for demonstration to the customer.

The selection of faults to be inserted into the UUT should be made by someone other than the original test designer, preferably a Senior Test Design Engineer. The fault selector should choose faults which allow all, or as many No-Go branches as possible to be exercised relative to the amount of validation time and program confidence expected of the test program. The type of fault selected should be consistent with the maintenance philosophy under which the program is to be used.

Since it is best to simulate field use of the test program as closely as possible, the ATE operator should have skills and training similar to that expected of the operator who uses the ATE in the field. This operator will run the ATE program after the fault is inserted following procedures and directions called out by the test program. The validator will act as an observer, noting any problems, operator or program, which may occur.

Validation of the Go-paths and fault insertion generally leads to changes in the program set. If the procedures and support personnel necessary to handle these changes are not carefully set up and controlled, an inefficient validation effort can be expected.

Minor program changes are initially noted on the listing in some conspicuous fashion. More complex changes may require rework of the original test design logic (ELP) and the coding of new tests. Changes should be accumulated until validation efficiency is impaired by errors in the test program. The changes are then entered on standard coding sheets and reprocessed through the production process. Control must be maintained to ensure program changes are processed efficiently by support personnel.

When the updated program is returned, the entire "Go" path should be rerun to verify that the update did not introduce errors in the performance tests. Problem areas that were the basis for the update are then checked to see that they are now corrected. Special attention must be given to the possibility of interaction of changes with areas of the program previously validated. Interaction is most likely when "Go" path changes are made to locate previously undetected faults or when the test sequence in the "Go" path is changed to simplify fault isolation.

Interface changes follow a different path. Work orders are issued to either the mechanical or production support groups, depending on the nature of the change. From the validator's viewpoint, however, the process remains much the same. He should obtain a commitment as to when the work will be ready and keep in contact with the necessary people as required to ensure that the required change action is promptly taken. The waiting time between submitting a change and having it incorporated can be utilized by the validator to update paper work or work on another program.

A validator must establish good lines of communication to expedite changes necessary in tapes and interface hardware. He must schedule events to avoid situations such as having an ATE idle because updated tapes or interface hardware are not available, or inefficient operation due to improper ATE alignment. Experience has confirmed that validation is best accomplished with one person expediting the overall effort, including ATE maintenance and test validation.

4. Demonstration

The final step in the program preparation process is demonstration. It is at this time that the test program's ability to meet system support requirements, and function as an integrated set (i.e., tape, operating documentation, and interface hardware) is proven.

The demonstration procedure generally begins with the running of the ATE system self-test programs in order to establish a base line that can be used for judgment of the test program. That is, the demonstration is of the program set's capability and thus an attempt is made to remove from consideration any other variables which could affect the program results.

Once the system's operational capability has been verified, the demonstration procedure typically follows a field operating pattern allowing no deviation from documented or displayed operator instructions.

The program tape interface hardware and UUT are connected to the ATE in accordance with the instructions contained within the program set.

When the initial hook-up is complete the program is run end to end. Success in this phase of the demonstration is predicated upon the program's ability to verify that the known good UUT being utilized is actually good, or that all performance tests have been satisfactorily passed. The ability of the program to accomplish this tells the program acceptance team that program tolerances are correct insofar as not indicating false alarm failures for good UUTs.

The next step in the demonstration requires a specified number of faults to be inserted into the UUT, one at a time, to demonstrate the program's fault isolation capabilities.

The faults that will be inserted by the program acceptance team are kept as confidential information until the actual insertion takes place. The purpose of this is to avoid program design which is biased in the direction of specific faults. The physical insertion of the fault is done under careful scrutiny to insure that the fault is that which has been specified as well as to guard against accidental damage to other components. Once the fault has been inserted the program is run again, from the beginning, and acceptance is qualified on an automatically terminated run identifying the faulted component.

The number of faults to be inserted in any given assembly is determined by a statistical process used to ensure contract target quality. Faults are selected on the basis of failure data pertaining to the UUT or class of UUTs. The demonstration of a long complex program will generally require a greater number of inserted faults than a short program, because sample size is based on the number of decision points in the test program. Following fault insertion the program set is inspected and verified prior to acceptance and delivery.

5. Conclusion

This paper has attempted to describe the ATE programming procedure in order to help provide both an overview of the entire task and an insight into some of its more critical aspects. The procedures described have been developed at RCA in Burlington, Massachusetts, over years of trial and error. It is our hope that others may now benefit from this experience in order to help the field of Automatic Test Equipment realize the widespread system support potential promised by this discipline.

The Test Language Dilemma

F. Liguori, Naval Air Engineering Center, Philadelphia, Penn.

Accompanying the spreading popularity of Automatic Test Equipment (ATE), there has been a proliferation of test languages. These languages attempt to bridge the communications gap between the hardware (test equipment) oriented engineer, who must devise the procedure for testing a Unit Under Test (UUT), and the ATE system which generally is controlled by a computer-like control system. As various test languages evolved from machine-orientation through assembly languages, translators and compilers, each ATE manufacturer developed his own standards for his "ideal" language. Most have pursued the development of an ATE language and associated processor under the following delusions:

> The perfect test programming language will provide an easy bridge between engineer and computer, and
>
> Having built this bridge, all of the problems plaguing test program production will vanish.

Unfortunately neither of these expectations have been realized. Instead, the concentration on test language has diverted attention from the need to define the test problem effectively. Yet, there is much that an effective test programming language can do to improve the test production process. This paper discussed some of the test languages developed so far and then summarizes the best features of each. It also attempts to define the limitations of even the best test language in bridging the gap between man and machine.

TEST LANGUAGE OBJECTIVES

Test programming languages, like their data processing counterparts of FORTRAN, COBOL, and ALGOL, attempt to provide a vehicle for the user to program his problem in a more natural language. To be effective the source language must not only be natural to the user, but it must be oriented to the problem to be programmed. In other words, it must have all of the statements required to define the problem explicitly enough to be processed by a computer. Thus the term problem-oriented language has emerged. How well suited a language can be to both the user and a computer depends on both the ability of the user to orient his thinking to the step-by-step logic of computers and the facility provided in the translating software to re-orient the source language in favor of a computer.

Reprinted with permission from *Proc. ACM National Conv.*, Aug. 1971.

EVOLUTION AND VARIATIONS OF LANGUAGES

A close parallel to test language development is found in the data processing industry. When programming was done in machine language, as many languages were required as there were different computers. With the development of assembly languages, there was some tendency to use common terminology for identical instructions. Mnemonics were developed to enable the programmer to use short hand alpha-numeric statements for the various computer instructions. For example, CLA was commonly used for "clear and add". Since IBM dominated the computer industry much of the standardization that occurred was of IBM origin. The fact that most programmers were forced to learn IBM terminology accounts for its common usage more than any planned attept to standardize. Early programmers were not prone to standardization, nor are they really so disposed today. It is only with the advent of large scale programming efforts of recent years that they have found it mandatory to standardize some in order to interface with each other.

When compliers emerged, the first attempts were made to orient the programming language to the problem rather than the computer. Since two classes of applications dominated the data processing industry, two problem languages emerged. FORTRAN became the scientific or mathematical programming language and COBOL the business or accounting language. Yet even with standardization forced on the programming community, several versions necessarily existed because of the varying capabilities of the computers involved.

Today these languages are more standard than ever but the problem is far from settled. With the application of computers to new disciplines, the traditional separation between business and scientific applications has broken down. A new language PL1 has emerged which combines the features of FORTRAN and COBOL. The problem now is to overcome the prejudices favoring FORTRAN and COBOL in applications that are more effectively handled by other less popular languages which are often more problem oriented. From this, it should be evident that computer languages, like communication languages, will evolve as the nature of problems change. ATE programming languages are not excepted from this phenomenon. The quest for the perfect language is no more a realistic goal in ATE than it is in the data processing industry. Each user must ultimately settle for an adequate language for his application that meets most of his requirements using conversion software that he can afford.

INHERENT LIMITATIONS OF PROGRAMMING LANGUAGES

Having settled on the best language for a given appliaction and having developed the most sophisticated compiler to translate that language, the user is still faced with a fundamental problem. That is, an effective source language must be very specific, self-consistant, and unambiguous. Natural languages, like people are none of these things. It is only to the extent that people can learn to develop their thinking along these lines that they can effectively program computerized systems with even the best of problem languages.

Before effective problem languages were developed, many of the problems associated with programming tasks were attributed to the unnatural statements, abbreviations, and punctuations required for programming. Many people therefore presume that having a more natural programming language will solve all programming problems, whereas it simplifies only the expression of the problem, not necessarily its solution. The fact is that by eliminating language barriers, the real problems of logical inconsistency become more visible. That of course, is very desirable and makes the use of problem languages worthwhile. But to expect that all other problems will vanish by the use of a natural, problem-oriented language is a false hope in ATE or any application.

EXTENT OF STANDARDIZATION IN ATE

Recognizing that all ATE applications are not identical, no single ATE programming language will meet the needs of all users. Yet, within these appliactions there are varying degrees of common functions, such as the need to specify test points, and to branch from one test to another. It is therefore possible to develop a core language useable for a class of ATE systems, if not all of them. Such a language must be expandable to handle the peculiar requirements of the test problem. The more definitive the core language is, the less likely it can be fully compatible to a given test problem. Hence care should be taken in the formulation of a test language to avoid excessive intricacies. The user should be allowed to expand it with his jargon, just as tradesmen supplement english to accommodate their needs.

INCENTIVES FOR TEST LANGUAGE STANDARDIZATION

Unless an organization is constrained by contract to adopt the industry standard or there is a need to interface with other organizations, there is little incentive for adopting an industry standard. That is why there has been little enthusiasm by ATE manufacturers toward industry standardizations. The ATE buyers, however, are becoming increasingly concerned with standardization for several reasons:

They do not want to be locked - in to sole source procurements for test programs.

They may use ATE systems by several manufacturers and would like standardized documentation for their programs.

They may wish to develop their own test programming capability based on a single language.

For data processing programming languages, the Government has been instrumental in influencing standardization. As the largest user of computers, it has assured the acceptance of FORTRAN and COBOL as standards by insisting that programs procured by the Government be encoded in those languages if at all possible. While the Government has not settled on a standard test programming language yet, the time is ripe for such a decree. It is up to the leaders of the ATE industry to have a worthy candidate language to offer and exercise technical influence for its adoption or a less than desirable language will be selected. A bad choice of standard language for ATE could seriously impare full realization of the ATE potential for many years to come.

EXISTING ATE LANGUAGES

As with early computer assembly languages, almost every ATE manufacturer has his own test language. Of the many in existance, a few are particularly noteworthy because of substantial usage or desirable characteristics. A discussion of a few should suffice to point up features and shortcomings of the available ATE languages. Brief examples of some test languages are give in Figures 1 through 5. Each example expresses the same test requirement, namely, stimulating the UUT with 28 volts, dc, and measuring the UUT output for between 26.5 and 29.5 volts, dc.

ATLAS (See Figure 1)

Abbreviated Test Language for Avionics Systems. This language is still in the process of being defined and refined under the auspices of the ARINC Airline Electronic Engineering Committee. It is the only test language to date approved for use by a group of user organizations. On 1 June 1969, ARINC Specification 416-1 was issued defining ATLAS in substantial detail. Meanwhile, the committee continues to develop and modify ATLAS.

A full description of ATLAS is beyond the scope of this discussion, however, a few key characteristics are listed here for further discussion and evaluation:

The language is abbreviated, using mnemonic contractions of words to express many of its functions rather than spelling out the words. (For example ERRLMT is used for "error limit".)

It is functionally oriented to the test problem and ignores the characteristics of the test system in its statements.

It presumes to be equally applicable to manual test definition and to ATE.

It relies heavily on punctuation symbols to separate elements of each test statement.

It provides for full ATLAS compliance or use of subsets or "adapted" ATLAS.

It attempts to select "a level of encoding which will result in the least expensive compiler while still retaining a readable quality".

Although ATLAS has been heavily debated and defined in substantial detail, there is little evidence that it is being used to any extent by organizations preparing substantial quantities of test programs.

PLACE (See Figures 2 and 3)

Programming Language for Automatic Checkout Equipment. This language was developed along with a compiler, by Battelle Memorial Institute under the sponsorship of the U. S. Air Force. Some of the key characteristics of PLACE are:

It was developed to provide engineers with a language that can be used to program a variety of test systems.

Rather than defining a specific language, PLACE defines a structure in which the ATE user defines his own test statements.

Along with the PLACE Processing Language, a processing program was developed.

The basic language is very cryptic and depends heavily on punctuation symbols.

The processor has been proven adaptable to user-defined, problem oriented, source languages.

The language and associated processor have had substantial usage on at least three military test systems, AN/GJQ-9, AN/APQ, and AN/GSM-204(v). Each of these have found significant use as Air Force, Depot-Level, Maintenance Tools for testing aircraft avionics systems. In the AN/GSM-204(v) application, extensive use has been made of the macro capability to define standard test functions, called System Preferred Methods (SPM's). The PLACE processor provided the necessary flexibility to apply the SPM approach to test design as well as the generation of an output formatter to generate machine language code. An example of SPM adapted PLACE is shown in Figure 3.

The use of the macro concept through SPM's enables the test programmer to express his testing instructions at the testing function level rather than in fundamental actions related to a particular ATE. This reduces the level of knowledge required of a test programmer and also results in more standard use of the ATE because the compiler calls out the sequence of fundamental ATE actions rather than the programmer. The latter benefit has a substantially favorable impact on program debugging and validation by reducing it to the time needed to prove out the test design rather than the design and coding technique.

ELATE AND DIMATE (See Figures 4 and 5)

These are examples of problem oriented languages developed for a specific ATE application. ELATE was developed by Hughes Aircraft Company for their VATE (Versatile Automatic Test Equipment) and the DIMATE (Depot Installed Automatic Test Equipment) language was a derivative of an earlier RCA development for their MTE (Multipurpose Test Equipment). Both languages are readily readable but do require some programming experience for effective utilization. Since these languages were developed with a specific ATE application defined, they have the operational advantage of being relatively simple. On the other hand, they are less flexible for adaptation to other ATE applications.

LANGUAGE TRADEOFFS

In selecting or developing a test programming language there are a number of tradeoffs involved which make the choice largely a function of the intended application. There is no perfect language nor even an acceptable one for most users.

The first tradeoff is between an extensive and flexible language, expandable to serve the needs of many ATE systems and a more restricted language that handles only the test problems anticipated for a particular ATE or family. The former approach obviously requires much more effort, but also suffers from unnecessary complexity for any given application. The latter approach is more economical and meets the needs of a given system better but may not be usable for any other ATE.

The second tradeoff concerns how closely the language should resemble english. A free-flowing language embracing a full english vocabulary and conventional punctuation is impractical for at least two reasons:

All natural languages, including english, are too ambiguous for exacting communication needs.

To develop a translator to enable a computer to process english language statements would be prohibitively expensive and such a translator would be slow to process.

Hence a restricted english sub-set is the best that can be achieved. The tradeoff must determine just how restricted the english is to be. Restrictions are of two basic types: in vocabulary and in syntax. Vocabulary deals with the number of acceptable words from which statements can be generated. Syntax deals with the acceptable use of the vocabulary with regards to punctuation and allowable combinations and sequence of words. Vocabularies are generally easy to expand, but syntax, once defined, is difficult to change and more difficult for the user to adjust to.

A limited vocabulary restricts a user's ability to do complex things, but is easier to learn. A restrictive syntax is unnatural but less likely to be misused and is easier to implement. Extensive use of punctuation clarifies syntax, but tends to be unnatural and introduces a source of error for most users.

Related to the vocabulary selection is the problem of abbreviations or mnemonics. Most programming languages make extensive use of mnemonics to reduce the amount of writing involved in a statement and to simplify translation by a computer. However, mnemonics are not very natural and can become very confusing to the non-programmer.

CRITIQUE OF EXISTING LANGUAGES

Because so much of the language selection process is based on the peculiar needs of a given application, it is difficult to make general recommendations. Also criticism of existing languages become rather subjective, depending on the evaluators experience. Nevertheless, for those who have little experience and weak biases, an evaluation of the languages mentioned earlier should be helpful.

The author favors free-flowing statements with a minimum of punctuations, special symbols, and mnemonics. While such features tend to make the conversion software more expensive, they reduce test programming errors and attendant costs. Since software tool design is a non-recurring cost as contrasted to the recurring costs of UUT programming, the allowable investment in software design is a function of the test programming volume.

Of the languages shown in the illustrations, basic PLACE is least natural and DIMATE most readable. It also requires substantially less writing to program the equivalent function in DIMATE language. The use of fields or columns to separate words in a statement rather than commas and parenthesis tends to reduce errors and simplifies spot checking of statements for completeness.

Universality of a language is extremely important both from its design and application. A universal design allows the language to be applied to many ATE systems so that as one is replaced or a new one added, the test designers do not need to learn a new language. One simply extends the existing vocabulary to handle any new features of a given system. Application universality refers to how used a language is. The more organizations that use a language on more test problems, the better refined it becomes. The problems with the language tend to be worked out and it becomes a more practical language. That, after all, is what any language is for - to be used. That more than any other is the reason some languages become accepted, because they are used by many people.

Because ELATE and DIMATE were not designed to be universal, they have limited applications and tend to be used less. ATLAS is being designed to be universal, but so far has not been applied to any extent. As it is used, many of its shortcomings will be uncovered and may result in its demise. In the opinion of the author, it attempts to be too universal by trying to meet the needs of all test equipment rather than only ATE. There are many characteristics of ATE that differ from conventional testing which cannot be taken advantage of if the procedure must apply equally to manual testing. Furthermore, the extensive use of mnemonics and need for punctuation symbols make it far from natural to anyone except the accomplished programmer. In attempting to develop the language so that it is easy to compile, too many of the computers needs have been favored to the neglect of the test designer.

PLACE has the advantage of being both universal in design and having had substantial use. It has been used for ATE systems of different manufacturers and has been implemented on several general purpose computers, including the IBM 7090, CDC 6400, and IBM 360 series.

It is believed that more test programs have been produced using PLACE than any other language. The complex input structure and awkward vocabulary can be overcome by adapting PLACE to a users vocabulary using MACRO phrase definitions. The results of one successful adaptation of PLACE using the technique is shown in the examples of SPM adapted PLACE in Figure 3. With a more elaborate MACRO phrase library, the source language could be made substantially more free-flowing and natural.

Because of its universal design, adaptability to many ATE's and its proven usage, PLACE is probably the most practical ATE language available. Industry would do well to modernize PLACE or at least pursue its univeral design objective rather than continue to proliferate ATE languages.

CONCLUSIONS

The brief examples of the five test languages touched upon are intended only to provide a feel for the test problem and the many alternative methods of expressing the problem. Many other test languages have been developed and perhaps more are yet to come. Unfortunately, none has fulfilled its promise to solve the programming problem, namely to reduce the cost of test programs to palatable proportions. The reason is simple; the test language merely provides a convenient vehicle for test problem expression, it does little, if anything, to define the test requirements or instill in the user of an ATE, the discipline required to express his problem with the precision inherent in computer-controlled testing. The technician has too long relied on human judgment to solve the testing problem. It will take him a while to re-orient himself to the effective use of computer decisions as the medium for practical testing of complex electronic equipment.

The greatest contribution of the higher order or easier-to-use test languages has been to expose the myth that the test designer was hampered only by an unnatural ATE language. The greatest fault of higher order test languages is that they have compounded the complexity of the ATE world with many languages and dependence on complex language processors. Hence the language dilemma lies in its exposure of the problem rather than its solution. But that, of course, is not all bad.

REFERENCES

V. MAYPER
"A Survey of Programming Aspects of Computer-Controlled Automatic Test Equipment", Volumes I and II,
Project SETE
New York University, 1964

PROJECT SETE
New York University
"Problems and Pitfalls in Automatic Test Computer Programming", May 1965

R. C. MILLER
"Simplifying the Use of Automatic Test Equipment with a Compiler"
IEEE Proceedings of the 1965
Automatic Support Systems Symposium

F. LIGUORI
"ATE Designer-Programmer or Systems Engineer"
IEEE Proceedings of the 1967
Automatic Support Systems Symposium

R. J. MEYER
"Computer Controlled GPATS"
IEEE Proceedings of the 1967
Automatic Support Systems Symposium

T. A. ELLISON, L. S. O'NEILL
"ATLAS - A Standard Compiler Input Language for Commercial Airlines"
IEEE Proceedings of the 1968
Automatic Support Systems Symposium

R. L. MATTISON, R. T. MITCHELL
"UTEC - A Universal Test Equipment Compiler"
IEEE Proceedings of the 1968
Automatic Support Systems Symposium

AERONAUTICAL RADIO, INC.
"A Guide to ATLAS for Test Specification Writers"
ARINC Report 418, May 1969

Program D. C. VOLTAGE TEST
Coded By F. LIGUORI
Checked By M. MARTIN

Date 1 JUNE 1971
Page 1 of 1

C FOR COMMENT

STATEMENT NUMBER (1–5)	Cont. (6)	FORTRAN STATEMENT (7–72)
10340	3	APPLY, DC SIGNAL, VØLTAGE 28V ERRLMT +-2V,
		CNX HI J1 H LØ J1 EE $
10340	4	MEASURE, (VØLTAGE), DC SIGNAL,
		CNX HI J1-JJ LØ J1-EE $
10340	5	CØMPARE, MEASUREMENT,
		NOM 28V UL 29.5 LL 26.5 $

FIGURE 1

EXAMPLE OF A SIMPLE TEST IN ATLAS LANGUAGE

Program D. C. VOLTAGE TEST
Coded By F. LIGUORI
Checked By M. MARTIN

Date 1 JUNE 1971
Page 1 of 1

C FOR COMMENT

STATEMENT NUMBER (1–5)	Cont. (6)	FORTRAN STATEMENT (7–72)
1034		(DCM.,RESET),(TP.,ØUT=2530,IN=1936,L=1)$
		(TP.,ØUT=2031,IN=1831,L=2) $
		(TP.,ØUT=2334,IN=1330,L=3) $
		(DCPWR.,LV1=RESET,LV2=28) $
		(DCN.,R=100,MØDE=NØRM,SCS=IM) $
		(CØMP.,UL=2950,LL=2650,HOLD,CC=IM) $

FIGURE 2

EXAMPLE OF A SIMPLE TEST IN BASIC PLACE LANGUAGE

Program D. C. VOLTAGE TEST
Coded By F. LIGUORI
Checked By M. MARTIN

Date 1 JUNE 1971
Page 1 of 1

C FOR COMMENT

STATEMENT NUMBER (1–5)	Cont. (6)	FORTRAN STATEMENT (7–72)
1034		CØNNECT MULTIMETER $
		TP J1-JJ, TPRET J1-EC $
		USE LINE 3 CONNECT DC SUP LV2 TØ J1-11 $
		(DCPWR, LV1=RESET, LV2=28) $
		MEAS PLUS 28VDC PLUS 1.5 MINUS 1.5 $

FIGURE 3

EXAMPLE OF A SIMPLE TEST IN SPM-ADAPTED PLACE LANGUAGE

Program D. C. VOLTAGE TEST
Coded By F. LIGUORI
Checked By M. MARTIN

Date 1 JUNE 1971
Page 1 of 1

C FOR COMMENT

STATEMENT NUMBER	Cont.	FORTRAN STATEMENT
103403		APPLY DC PWR SUP, VØLTAGE=28V UPLIM=+2V, LØLIM=-2V
103404		MEASURE DVM
103405		LIMCK DVM NØM=28V, UL=11.5, LL=-1.5V

FIGURE 4

EXAMPLE OF A SIMPLE TEST IN ELATE LANGUAGE

Program D. C. VOLTAGE TEST
Coded By F. LIGUORI
Checked By M. MARTIN

Date 1 JUNE 1971
Page 1 of 1

C FOR COMMENT

STATEMENT NUMBER	Cont.	FORTRAN STATEMENT
T1034		CØNNECT +28 VDC J1-H J1-EL
		MEASURE +28 VDC +-1.5 J1-JJ J1-EE

FIGURE 5

EXAMPLE OF A SIMPLE TEST IN DIMATE LANGUAGE

ATLAS - ABBREVIATED TEST LANGUAGE FOR AVIONIC SYSTEMS

T. A. Ellison, Manager of Navigation Engineering
United Air Lines, International Airport, San Francisco, California

INTRODUCTION

ATLAS is a language easily readable both by men and machines. It further is a language which expresses the stimulus and measurement requirements of a unit under test (UUT) in terms oriented to the parameters, pins, and physical constants as seen from the UUT. It is thus independent of the ATE or manual test facility being employed. The above characteristics make ATLAS a language which can be, and is being employed either directly as a documentation language for manual or automatic test requirements, or with minor adaptations, as a compiler input language.

Used for test documentation, it helps avoid problems of ambiguity, incompleteness, or inaccuracy that are notorious pitfalls in normal English procedures. As a compiler input language, ATLAS provides easy readability for debugging, verification, and management or control reviews, and allows a close correspondence between documentation and the automatic test program.

Developed under the sponsorship of the U.S. and European airlines, with assistance of many participants from other industries, ATLAS is available without proprietary restrictions to all who find its use advantageous.

DEVELOPMENT OF ATLAS

The ATLAS programming language was developed under the sponsorship of the Airlines Electronic Engineering Committee (AEEC) of Aeronautical Radio, Inc. (ARINC), to provide for exchange of test requirements between airline avionic equipment suppliers and airline engineering and maintenance staffs. Need for a single designated language was recognized by the airlines to avoid coping with a variety of languages from the individual ATE complexes and philosophies of various suppliers.

An Automatic Test Equipment Subcommittee of AEEC began activities to select or develop an optimum language for airline avionic test needs in January of 1967. Participation in the activities was open to any interested parties and representation of a broad segment of the airlines, avionics supplier, and ATE suppliers formed the core of the Subcommittee participation. Approximately 100 participants attended the first Subcommittee meeting, but more common working group size in later meetings was from 30 to 40 participants.

Most of the participants in the Subcommittee had extensive background in development or application of other test languages. ATLAS acknowledges many benefits from the experience of these participants and from the innovations that resulted from synthesis of their background with airline pressure for a UUT oriented language. After evaluating objectives to be reached, it was concluded that a new language would be required, and the group began work to devise a language structure and vocabulary that would achieve these objectives. As a result, in June of 1969, ARINC Specification 416 was issued which defined the ATLAS language in sufficient scope to cover analog, air data, and CW radio equipment test applications. Work has continued since that time on extension of the language to cover requirements of digital testing, and expansion and refinement of the analog portion of the language.

DESCRIPTION OF THE ATLAS LANGUAGE

Documentation

The ATLAS language is defined by ARINC Specification 416. Revisions and additions to this document since its initial release are issued as supplements. At present, two such supplements have been completed and formally approved, and are incorporated in the March 1, 1971 reprinting, designated as Specification 416-1. A third supplement is awaiting final review and approval at the next ARINC Automatic Test Equipment Subcommittee meeting. In addition, proposed changes or additions to the language in various stages of finalization are documented in the working papers of this Subcommittee, which are circulated by ARINC to the working members of the Subcommittee. These working papers, referred to as ATE Newsletters, and numbered sequentially, provide both a history of the past growth stages of the language, and, in the active papers, a forecast of the probable lines of future growth of the language.

ARINC Report 418 provides a simplified introductory guide to the language and is usefull for a general overview of ATLAS. This document is not intended to be a complete or rigorous treatment of the language, and should not be used for other than introductory purposes.

Reprinted with permission from *WESCON Conv. Rec.*, Aug. 1971.

A variety of privately prepared ATLAS descriptive material exists in both the United States and Great Britain. These publications have been prepared generally as Training or Programming Manuals for an individual user and are arranged and edited to clarify or simplify the presentation according to the users needs. All, however, depend on the ARINC material noted above as basic information source and controlling documents. Some representative examples are listed in the References.

Language Structure

Program Organization. ATLAS programs are divided into two sections. The first, "preamble", section provides definition of signals, sensors, sources, loads, and procedures for subsequent reference and use in the actual testing. Statements in this portion of the program do not actually cause execution of any testing but do simplify and make more readily interpretable the text of the second or "procedural" section of the ATLAS test procedure. Functions or procedures defined in the preamble are assigned labels by which they are referenced in the procedural section. The procedural section provides the actual statements required to execute the test procedures desired. In these statements, sources, sensors or loads may be expressed by labels created in the preamble or by direct use of the electrical or physical parameters involved.

Statements in an ATLAS test procedure are sequentially numbered beginning with the preamble and continuing through the procedural section. Gaps in the numbering may be provided, if desired, in anticipation of later program expansion needs. An example of a very simple preamble and test procedure is shown following:

```
000300 DEFINE, 'BIAS', SOURCE, DC SIGNAL,
VOLTAGE -6V ERRLMT + - 0.1V,
CNX HI ( ) LO ( ) $

010102 APPLY, 'BIAS', J1-1, J1-2 $
```

Statement Format. Individual statements of an ATLAS program have a combination of fixed variable length fields. The format of ATLAS statements is shown diagramatically in Figure 1. The statement begins with the first character, a flag field, which uses a letter designator in certain statements to assist in defining the character of the statement. The second field is a statement number, which is required for each ATLAS statement in order to allow positive identification and reference to any portion of the test program. The remaining fields are variable in length and are separated by commas between fields. The first variable field contains a verb, which describes the type of action prepared by the statement. The next field defines the general characteristic to be measured such as voltage, pressure current, and so on. In the next field a noun, in combination with modifiers in the following fields, describes the exact characteristics of the stimulus or measurement. A number of modifier fields may be used in the statement characteristics section. The final field is a connection field which establishes the reference points where the stimulus is to be applied or the response measured. A statement terminator, $, marks the end of the ATLAS statement. The statement format and vocabulary of ATLAS allow the language to be a self reading, English-like language. This makes for easy readability with a minimum of personnel training for use of the language as manual test documentation, and for easy program generation, debugging, or verification when used as a compiler input language.

Vocabulary. The latest issue of ARINC Specification 416 defines 238 language elements, of which 42 are verbs, 35 are nouns, 80 are modifiers, 14 are connection field entries, and 67 are miscellaneous statement characteristics such as flag signals, punctuation elements, calculation symbols, and words having some specially restricted usage. (Nouns, verbs, etc., are used in the ATLAS sense rather than English sense in this paper.) This vocabulary quite well covers the full set of physical parameters needed to describe electrical, mechanical, and pressure testing sources, sensors and loads. Although conceived to cover avionics testing in these areas, the vocabulary needed for this test domain is large enough that extension of language to other applications, such as testing of internal combustion engines, for example, appear to require only minor modifications or additions to the language. The chief area not covered in the present vocabulary are the complex relationships and data sets associated with digital testing. It is here that the bulk of the present language development is centered.

Syntax Diagrams. To define the language elements and their relationships with enough detail and clarity to allow generation of compiler, it was necessary to find something more precise than verbal descriptions. To meet this need a diagramatic approach to definition of the language was developed. These syntax diagrams or as they are sometimes called, Atlagrams, are contained in Specification 416 and are the final authority as far as rigorous definition of the ATLAS language. A syntax diagram is provided for each Verb in the ATLAS language in Specification 416. These diagrams, in turn, use a subdiagram to simplify major diagrams. The subdiagrams are referenced by their title and enclosed in dotted boxes. The full set of subdiagrams is also provided in 416. Some of the subdiagrams are complex enough to, in turn, require subdiagrams for clarity. Examples are provided of syntax diagrams for the ATLAS verb GO TO, which covers branching, for the subdiagram condition which covers the branching conditions in the verb GO TO, for the verb MEASURE, and for the subdiagram of the connection field, in the accompanying figures.

The description provided here covers highlights of ATLAS. More complete documentation is readily available in the referenced ARINC documents, including an explanation of the syntax diagram symbols employed, vocabulary, and general language background. There documents may be obtained at nominal cost upon request to ARINC.

PRESENT DEVELOPMENT AND IMPLEMENTATION STATUS

Development

The initial thrust in the development of ATLAS was to provide a language with "analog" capability. Analog in this context includes electronic flight control equipment, air data equipment, communications and navigation radio equipment, audio equipment, and all other equipment whose processes are basically analog in nature. The digital equipment in service in the airlines, until recently, was simple enough to be covered by the basic logic statements included in the presently defined ATLAS. More sophisticated airline digital equipment is rapidly being introduced. Need obviously exists to expand ATLAS or redefine the language to allow processes such as memory core loading, matching of data sets, and filling of data lists. It has been discovered that existing language provides considerable digital capability, and much of the recent work has been devoted to minor extensions and redefinitions of existing language in order to allow consistent use of the presently defined language in digital applications. The other major work area has been to create the vocabulary and syntax required for the handling of list definition and manipulation. The work in this area has not been formally approved and exists primarily in the ATE Newsletters and their attached working material for the year 1971. It is anticipated that formal approval of much of this material is very near, and that the later ATE Newsletters and working material can be considered closely representative of the final outcome of the extensions of the language for digital testing.

The complexity and variety of digital testing is such that creation of a very specific language to cover all the possible operations and variations associated with the present variety of digital hardware would be a massive undertaking and an unwieldy language. At present the concept instead is to create a simple basic digital language structure, and then to devise representative subroutines using this language as an example and as a reference library. Those subroutines which are determined later to have a wide application, may later be covered by a standardized higher level representation to provide a more economical means of writing programs. A analogous procedure was provided in development of the analog language where initially, to obtain working agreement and a stable language base, simple "single action" verbs sets were developed to cover stimulus and measurement processes. These were combined subsequently into higher level words where obvious frequent usage justified it.

Implementation

Language Utilization. The most widespread present use of the language is as a documentation language for test procedures on equipment, where use of automatic test is current or considered probable within the line of the equipment. United Air Lines, at present, has a policy of requiring ATLAS documentation on all communications, navigation, electronic flight control, and electrical equipment that we procure unless our Engineering Division specifically determines that the equipment is not suitable for automatic test. The Air Transport Association maintains a specification for technical documentation, ATA 100, which is referenced in documentation requirements in virtually all procurement contracts by airframe manufacturers and airlines. In August of 1970 a requirement was added to ATA Specification 100 for electronic test specifications in ATLAS for all complex electromechanic units which accept electrical or pneumatic inputs and have electrical outputs. Specific requirements for ATLAS documentation have been appearing in individual contracts from various airline and airframe manufacturers since as early as 1968. As a result, most of the major electrical electronic airline equipment suppliers in the United States and Europe either have prepared or are preparing documentation in ATLAS.

Within the United States, there is also application of ATLAS on the part of the military. The VITAL language associated with the Navy ATE applications has drawn on the ATLAS activity to some extent. The Frankford arsenal of the United States Army is planning to implement ATLAS, with extensions, (ATLAS-X), as a test language for land vehicles, including engines and systems, and has been actively working with the language.

At the last ARINC ATE Subcommittee meeting it was reported that an informal United States Army ATE language standardization working group had been formed. A recommendation from this group to use ATLAS as a basic for a standard test procedures language was planned to be forwarded to the high command.

United States Air Force has had representatives participate in the ATLAS language activity and expressed interest in its potential use, but continues to rely on PLACE for all present applications. Many of the military aerospace suppliers have been participants in the language development, and have in-house activities and awareness of ATLAS.

There is definite awareness of ATLAS, and some utilization by non-airline industries in the United States, as well as the airline and military activities.

In the United Kingdom, the adoption of ATLAS for civil and military usage is even more prevalent. As reported at the ARINC ATE Subcommittee activity meeting, in London in March of this year, ATLAS is in use within Britain in a number of industries and has been supported by both the British Ministry of Aviation and Ministry of Technology as the preferred documentation language for electronic equipment. ATLAS documentation is being required for all appropriate systems for the European multi-role combat aircraft (MCRA) project. Several British technical colleges have run courses in ATLAS, attracting pupils not only within Britain but from other countries in Europe. Two two-day lectures on ATLAS, sponsored by the British Aviation Ministry, were given in November of 1970, and one more was given in Munich at the end of March.

Compilers. The high level of the ATLAS language would require a very sophisticated compiler to cover the entire scope of the language and incorporate all potentially compilable features. Consequently, all implementation of compilers, at this time, operate in subsets of ATLAS appropriate to the needs of the user.

There are at least nine ATLAS compliers either operative or in preparation; of these, six were prepared with testing of airline avionic equipment as the objective, one was prepared under military sponsorship, and two are in preparation for internal testing needs by the companies involved. Four of these compilers were generated by the staff of the ATE manufacturer, three were prepared by software consultant, and two were prepared by the staff of the equipment manufacturer. A table below, of the vocabulary actually implemented, approximately indicates the subset size involved in four of the airline oriented compilers.

	ATLAS Vocabulary	Compilers A	B	C	D
Verbs	42	37	30	44	32
Nouns	35	10	20	60	12
Modifiers	80	25	**	*	40 ***
Connections	14	*	14	*	*
Misc.	67	56	40	*	*

* Information not Available. ** Not compiled.
***12 required, 28 optional.

Compilers began being implemented approximately one year to two years ago, so none of the compilers have operated for a long period of time. In some cases, the compilers operate on-line in the ATE computer and, in some cases, they are off-line on a separate computer. The computers size used for these compilers range from 8,000 word machines to very large general purpose computers. At least one of the compilers operates in a real time mode and allows direct control of the ATE complex by ATLAS instructions inputted by the operator, when desired.

In addition to the question of sub-set size of the compiler, there is the question of degree of adaptation or variation from ATLAS defined procedures. This is difficult to assess and compare, but the following comments can be made. Most compilers appear to follow the vocabulary and syntax of ATLAS as defined very closely in permitted operations, but many introduce restrictions over and above those specified in ATLAS. There is no compiler that (to my knowledge) provides for automatic allocation of the ATE source and sensor resources, for example. Use of the ATLAS preamble and define conventions appears to be extensive in all of the compilers. Several of the compilers were developed as interface translators to an existing language, to utilize existing compiler software and ATE instruction codes.

In summary, utilization of ATLAS, both for documentation and compiler input, appears to be generally satisfactory to the users and is showing rapid growth.

FUTURE DEVELOPMENTS TRENDS

Utilization of the language, both as a documentation language and as a compiler input language, is obviously still in early phases. As this utilization increases, problems and restrictions of the language will become more apparent and it will be necessary to continue to polish and, to some extent, extend the existing analog ATLAS language. This work, plus the additions to the language to enhance its digital process control capability, should continue to occupy developmental activities of the ARINC ATE Subcommittee for perhaps a period of two years.

At some following time, the growth rate of the language should drop substantially and the emphasis will shift to minor housekeeping and extensions of the language. As the definition phase of the language draws to a close, more attention may be both necessary and possible to review of the utilizations that exist, to compare the adaptations and subset limitations involved, to exchange information on compiling techniques, and possibly to find some way to share the development effort required to support software costs for a more extensive compiler of the language. Joint efforts on compilers might be undertaken to improve syntax checking capability, or automatic review or allocation of ATE hardware resources for a given test program. Other compilable features of the language which, at present, cost and time limitations prevent us from fully utilizing, might be made available in similar efforts.

For the airlines, it seems certain that the present days of program generation by the airline, or under contract by the ATE supplier, will tend to be replaced by provision of the ATLAS program by the original equipment manufacturer. The basic desire to preserve relationship was a primary airline motivation to develop the language. There will undoubtedly continue to be internal programming efforts for extension of ATE to existing equipment, and to modify programs already prepared. The airline programming effort will tend, however, to concentrate on modification and extending diagnostics rather than original development of programs.

Continued extension of military and non-airline industrial use of ATLAS also seems probable. The chief obstacles to this expansion are twofold. First, there is some apparent reluctance to utilize a language which is titled "Avionic Systems". At first glance, the more general applicability of the language is not apparent. There have been suggestions to change the title of the language from reference to "Avionics Systems" to reference to "All Systems". The airlines to date have been reluctant to make the change because this would imply an ability and willingness to provide the staff work necessary to keep up with a very broad application of the language. This brings us to the second obstacle. The airlines are, or at least would like to be, profit-making organizations and must place limitations, particularly at present, on overhead activities that do not have quite an immediate return in airline activities. The support of ATLAS meetings from other industries and organizations is also not directly funded, and subject to the impact of budget restrictions. This means that the development rate of the language may not be as rapid as individual users need at times. This would particularly apply to early applications to new technology such as the military might make.

There appears to be several alternative responses to problems of rapid expansion of the size and utilization of the language. One would be for the airlines and the voluntary ATLAS activity participants to limit their sponsorship to those areas where immediate benefit can be seen and allow individual areas of special need to move forward without any coordination, before or after the fact. To some extent this is unavoidable and has happened already. However, it has the obvious disadvantage that standardization is compromized and the benefit of cross-utilization of others efforts is rapidly lost in this way. The airlines feel some reluctance to see this happen. First, because of our general desire to progress, and second, because our ability to predict is subject to error, and the growth of the language in areas outside the airlines may be an indication of applicability within the airlines.

A second alternative is to continue airline sponsorship of the effort but in some way increase its support, particularly of the secretariat and coordination functions of the ATLAS Subcommittee. This added support might come from some form of military support, or from within the airline industry through some means of adding stature and priority to the long range aspects of the activity.

The third alternative is a combination of the first two. The existence of divergent "trail breakers" is recognized and accepted, but, after the fact, coordination is encouraged.

A fourth alternative would be to transfer sponsorship of the language to some other organization that appeared to have greater resources for carrying on the development. Some caution is recommended in considering alternative, because the language is still evolving rapidly and could not be administered in the same way as more conventional standards.

The current practice falls somewhere near the third alternative. The extent to which ATLAS can accommodate and encourage rapid extensions to accommodate individual needs depends largely upon finding a successful way to allow those with special needs to give special support to the activity. Some possible approaches to this problem have been explored, but no effective way to accomplish this has yet been established.

In summary, ATLAS appears to be firmly established both in the United States and Europe as the preferred language for test specifications for commercial aviation electronics. It shows every sign of extension to a much broader base of usage in military and general industrial applications. Some fragmenting into related families of language has occurred and will continue to occur in expanded utilization of ATLAS. Retaining language standardization benefits without restricting this rapid growth, and the coordinated development of an effective language for digital testing are the two principal challenges in the ATLAS activity at present.

References

(1) ___, "Abreviated Test Language for Avionic Systems," ARINC Specification 416-1, Aeronautical Radio, Inc., 2551 Riva Road Annapolis, Md. 21401, March 1, 1971.

(2) ___, "A Guide to ATLAS for Specification Test Writers," ARINC Report 418, Aeronautical Radio, Inc., May 15, 1969.

(3) T. A. Ellison and L. S. O'Neil, ATLAS - "A Standard Compiler Input Language for Commercial Airlines," Proceeds of the Automatic Support Systems Symposium - 1968. St. Louis Section, IEEE.

(4) Mayper, Victor, "What is ATLAS," Jacobis Systems Corp., Paper.

(5) ___, "Abreviated Test Language for Avionic Systems," IPG/TS/126, British Aircraft Corp., Stevenage, Hersfordshire, England, November 13, 1970.

(6) ___, "ATE Newsletters," Nos 1-44 Aeronautical Radio, Inc., Annapolis, Md. (Includes letters issued from December 14, 1966 through May 20, 1971. Letters published during most recent six months would generally include all active working papers.)

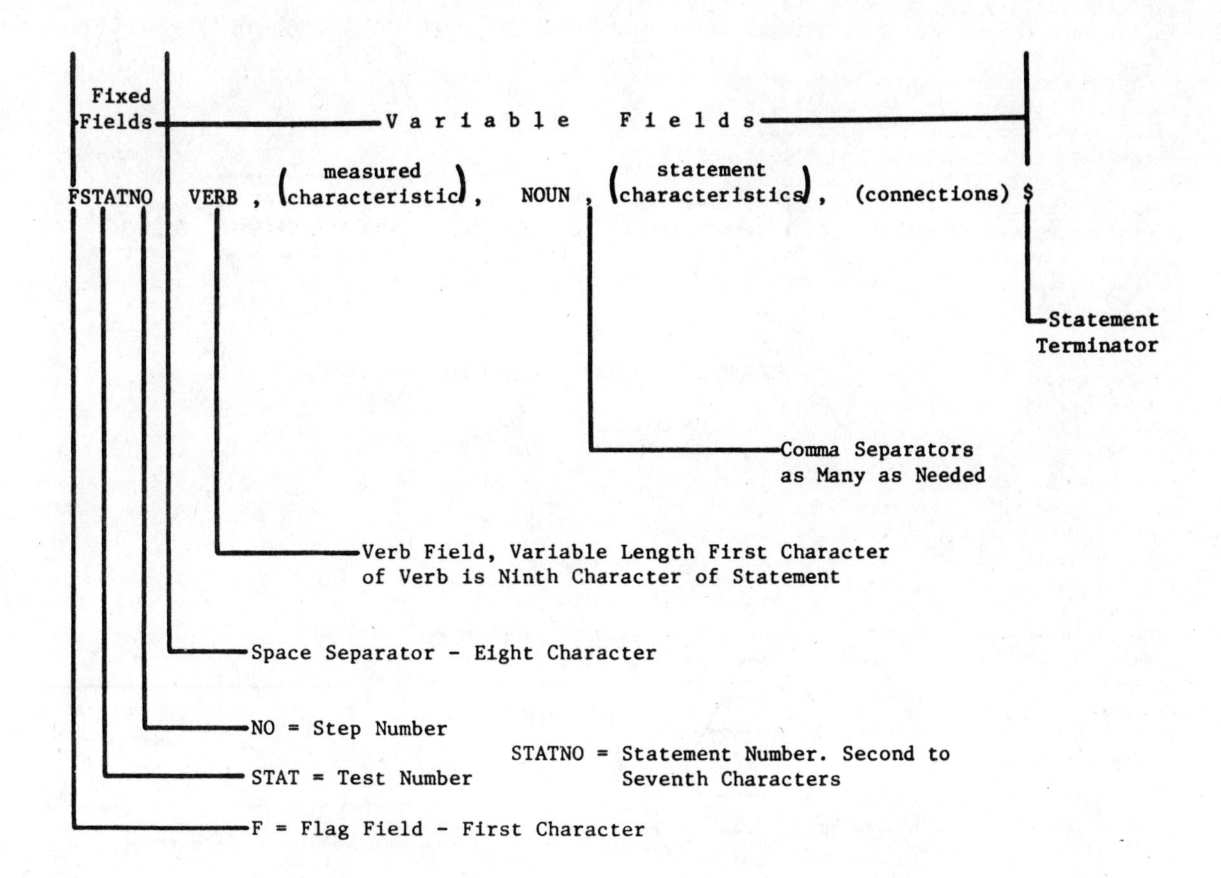

FIGURE 1 - ATLAS STATEMENT FORMAT

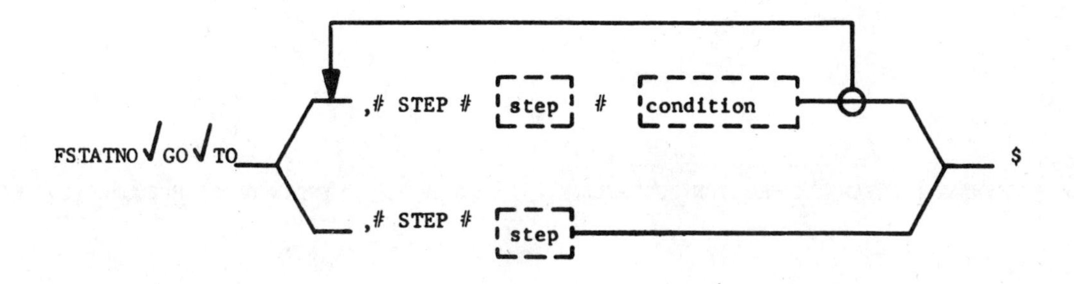

FIGURE 2 - SYNTAX DIAGRAM FOR GO TO

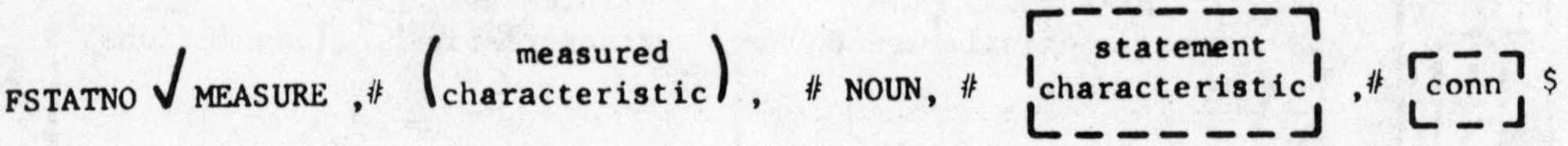

FSTATNO √ MEASURE ,# (measured characteristic) , # NOUN, # statement characteristic ,# conn $

FIGURE 3 - SYNTAX DIAGRAM FOR MEASURE

condition
IF #
HI
LO
EQ
NE
NOGO
GO
boolean expression

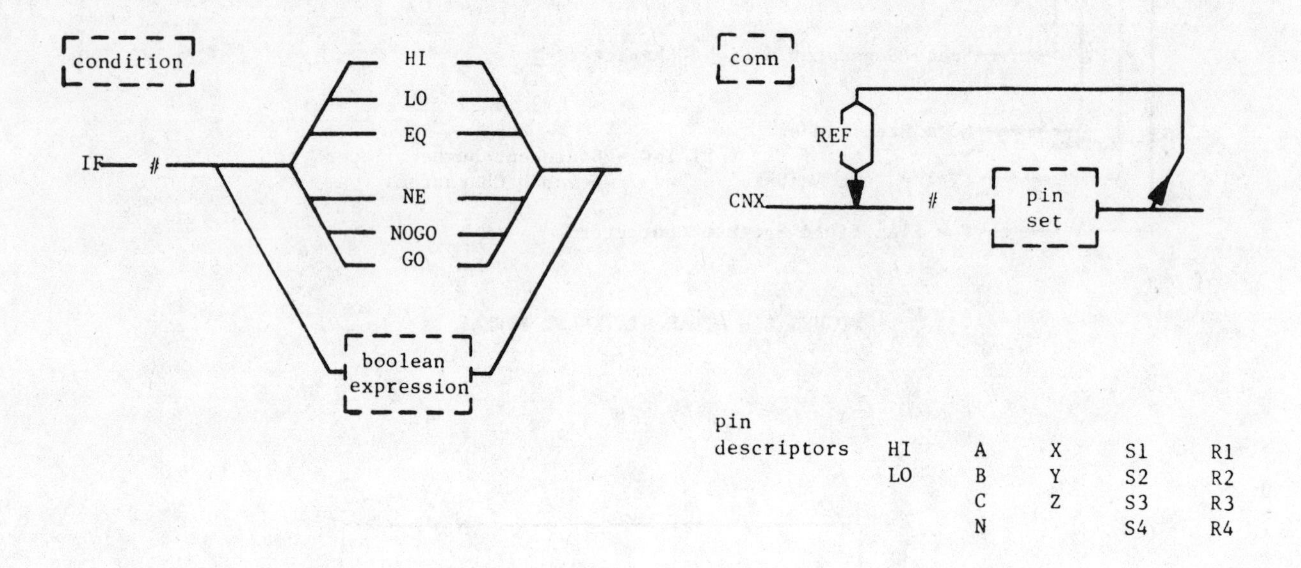

FIGURE 4 - SYNTAX DIAGRAM - CONDITION SUBFIELD

conn
REF
CNX
#
pin set

pin descriptors	HI	A	X	S1	R1
	LO	B	Y	S2	R2
		C	Z	S3	R3
		N		S4	R4

FIGURE 5 - SYNTAX DIAGRAM CONNECTION FIELD

SOFTWARE AUTOMATION IN AUTOMATIC SUPPORT SYSTEMS

by

F. Liguori, Manager
Software Systems Engineering Laboratory
Emerson Electric Company
St. Louis, Missouri

INTRODUCTION & SUMMARY

The emergence of GPATS from prototype development to operational status, with multiple installations for maintenance support of a number of avionic systems, represents a significant departure from the traditional maintenance concept of utilizing special purpose hardware to service each fielded system. While multipurpose automatic test systems, such as GPATS offer numerous operational advantages over traditional maintenance techniques, they add a dimension of complexity by making software (in the form of test programs) a vital operational element of an already complex maintenance system. Not only has software become vital to the maintenance of numerous aircraft avionic systems, but the magnitude of software generation and control requirements make it the key factor in the operational success of the automatic test equipment (ATE)concept.

Emerson has recognized the vital role of software from the earliest days of ATE and has been progressively developing its software generation capability to improve quality and production volume of avionic test programs. Substantial progress has already been made toward accomplishing these objectives by two basic means:

1. Development of mechanized methods of program production using computerized data processing techniques.

2. Development of software management and preparation procedures covering various aspects of operating an effective test programming center.

To date, a number of meaningful achievements have been realized in the mechanization effort, including:

1. Adaptation of the PLACE "Compiler" as an aid in the GPATS program preparation task.

2. Development of System Preferred Methods (SPM's) to simplify the test setup programming requirements for commonly occurring functions of GPATS.

3. Development of a mechanized method of producing user documentation using standard preferred comments (SPC's) for operator message generation and computerized photo-composition techniques for generating repro-negatives.

4. Origination and issuance of a series of control specifications covering all aspects of test program preparation and management.

5. Establishment of a teleprocessing configuration, with associated software, which provides telephone line linkage between Emerson, St. Louis, and remotely located programming facilities.

The most recent Emerson developments in test program preparation and handling are:

1. The full definition of GPATS system losses and accuracies as they pertain to the test programming task.

2. Investigation of advanced compiler techniques and programming language which will further advance the state-of-the-art of test program preparation using automated techniques.

3. Investigation of techniques for more efficient validation (proving out) of programs on the GPATS system, including simulation techniques.

All of these accomplishments and on-going efforts of improving test program preparation and handling are significant and in many areas represent innovations in engineering applications of computers. The rate of program preparation and validation required to meet current military commitments to ATE call for a full scale effort toward maximizing test program automation.

Reprinted from *IEEE Auto. Support Systems Symp. Rec.*, Nov. 1969.

Emerson efforts are currently addressed toward a total process solution of the programming process consistent with the predicted but unprecedented rate of test program production. The approach is essentially an expansion of the current techniques of software mechanization. Whereas software mechanization to date has concentrated on program preparation techniques. The current "software automation" plan not only expands software mechanization of production, but also deals with two equally important aspects of software production, namely:

1. Computerized aids to validation.

2. Electronic linkage of remote users and the central program preparation facility.

This paper discusses the concept of software automation and develops both accomplishments and the anticipated results of this ambitious effort. Each element of the total process is being developed under control of individual work statements for each separately definable product. The need for full implementation of all tasks in order to successfully meet the software challenge of ATE cannot be over-emphasized.

PROBLEM DISCUSSION

Test Design Programming Pecularities

Test design programming is the process of developing programs which automatically control the complex events involved in performance and fault isolation testing of electronic units. This process is functionally illustrated in Figure 1.

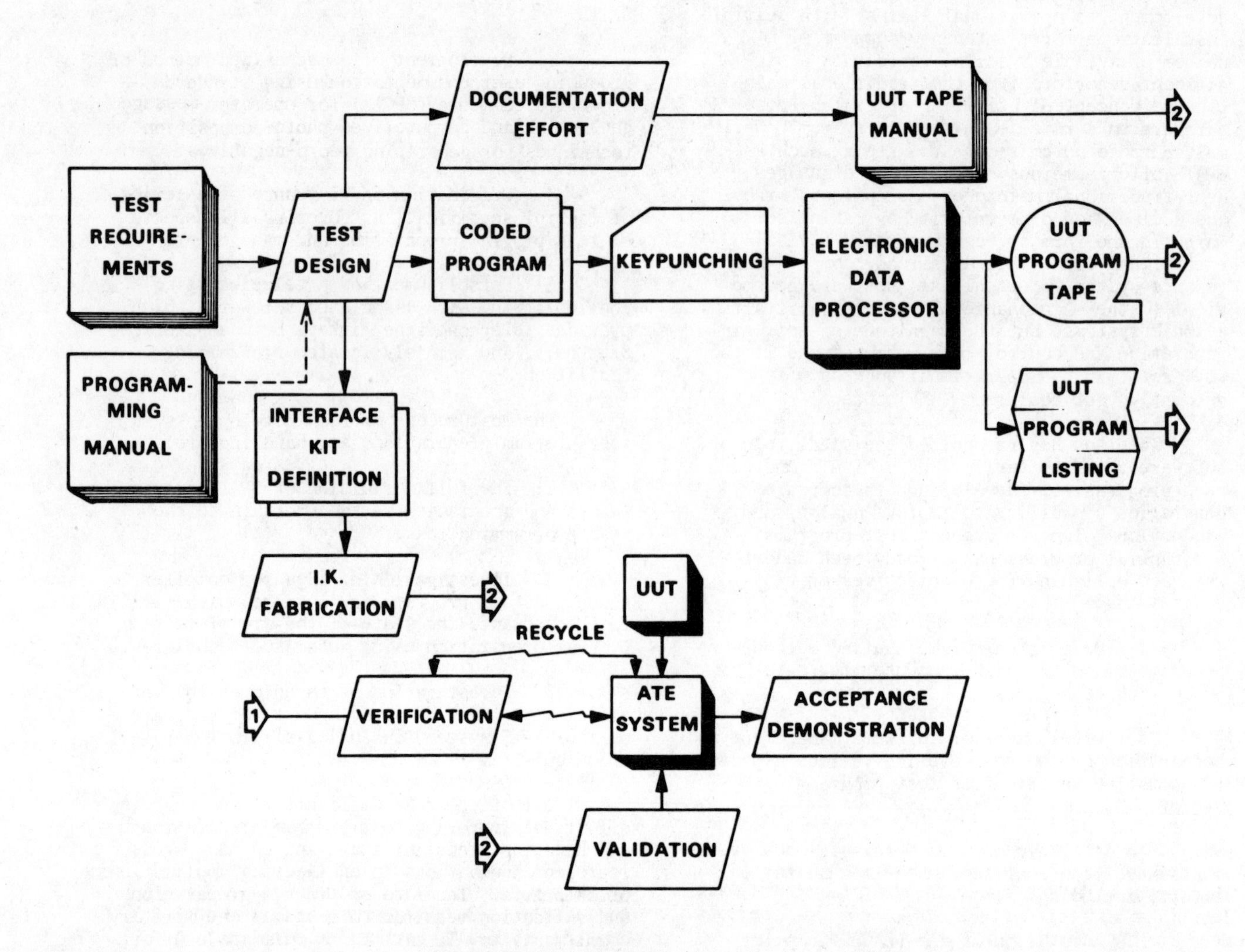

CONVENTIONAL TEST PROGRAM PRODUCTION PROCESS

FIGURE 1

While operating as computer programs in principle, test programs differ significantly from conventional programs in two ways:

1. Test programs operate a complete network of electronic signal generating and measuring devices contained in an automatic test system such as GPATS. Programming such a system is substantially more involved than simply manipulating data in a computer.

2. New test programs are continually being produced to test an ever increasing complement of new units under test (UUT's). Thus, unlike other systems where program development is required only for initial system design, ATE requires test programs to be produced throughout the operational life of the system.

The Changing Complexion of the Problem

The electronic control aspect of test programs require that test design programmers be competent electronic systems engineers as well as adept programmers. Training engineers in programming has been the approach used in developing needed personnel, but the short supply of capable, interested engineers has been a problem.

The continuing need for new program development poses a second serious problem. The heavy demand for ATE operating time to validate (prove out) new programs conflicts with production testing of UUT's with operational programs.

In the early days of ATE, it was not too difficult fulfilling manpower requirements because:

1. Several engineers were sufficient to prepare the few programs needed to prove the concept of ATE, so they could be discriminently selected.

2. Because of the novelty, test design programming was challenging and intriguing to engineers.

Availability of ATE machine time during the development phases was not a serious problem because there was no production schedule for testing units to conflict with validation time requirements.

As test program requirements increased, it became increasingly clear that the generation process became less inventive and more productive. Yet the "stored intelligence" characteristics of a program continues to require sophisticated engineering effort in its production. To further complicate the situation, a new problem was created by the volume programming requirements. That problem is configuration control of software. Inventive as test designs are, there is a serious need for standardization and adherence to original test requirements definition and close attention to detail throughout the program production process. The solution to these apparently conflicting requirements lies in the concept of "software automation".

THE SOFTWARE AUTOMATION SOLUTION

Concept and Advantages

Software automation is the basic concept of developing a functional system of men, materials and mechanisms for efficient, large scale production and validation of test programs. The concept is gradually being implemented at Emerson based on a continuing analysis and improvement of the test generation process. The technique involves reducing the complex process to well defined functional elements, and incrementally replacing manually performed tasks by programmed machines wherever practical. Each incrementally definable automation "tool" is developed under control of a definitive work statement with operational requirements and implementation schedule fully defined. The reason this concept is so effective is that the very tasks engineers dislike or do poorly are the ones data processing equipment do best. Thus optimum use is made of the engineer's analytical and inventive talents as well as the ability of data processors to do tedious, routine operations with great speed and accuracy. Thus, a Software Automation System (SAS) can be developed from this man-machine team which efficiently produces large quantities of new programs of consistently high quality.

The incremental implementation approach realizes several advantages:

1. Elements of the manual system are replaced with "automated" building blocks as developed, without disrupting the on going production effort.

2. The most important or cost effective elements are developed first.

3. Maximum advantage is taken of state-of-the-art advancement, such as new test language developments as they become available.

4. Developmental and training costs are spread-out as a continuing effort involving a minimal staff, thereby achieving best development efficiency.

5. The full impact of each added improvement can be evaluated individually, leading to further improvements.

Lessons from Data Processing Industry

Although test program design requires system engineering skill seldom called for in more conventional programming efforts, there are similarities in production problems. In developing the SAS, maximum advantage is taken of the lessons learned at substantial cost by the data processing industry. First, it is recognized that development of a compiler and other mechanization tools is a costly, but non-recurring process. This large investment has to be weighed against small, recurring individual savings in test program preparation. In so doing, it is apparant that a substantial volume of programs has to be produced to justify company investment prior to breakeven. For a company seriously committed to ATE Technology this investment is mandatory.

Rather than requiring test engineers to learn an unnatural machine language dependent on complex rules, software aids are being developed which allow the test designer to express his requirements in a more natural test oriented language. The use of system preferred methods (SPM's) and PLACE language has shown the soundness of this approach, but the current tools must be expanded to meet intended production rates.

Finally, it should be recognized that program preparation, using the automation concept, leads to a centralized facility serving all remote users because of the huge initial investment and virtually unlimited productive capability of the system. This conflicts, however, with the requirements of remote facilities. Thus, the need exists for a network of data processing equipment which provides direct access to the centralized facility.

Elements of Software Automation

Although the Emerson SAS embodies many separately defined incremental elements, they reduce to four functional areas:

1. Software Mechanization
2. GPATS Simulation
3. Teleprocessing
4. Configuration Management

Software mechanization embraces those software automation elements which, together, develop a means of improving production of test programs executable on GPATS. Software mechanization involves the process from Test Requirements Document (TRD) review to GPATS machine tape output with documentation. Since mechanization involves personnel as well as software and data processing equipment, training and procedure development is included.

The GPATS simulator is a program executable on a defined general purpose data processing system. It is defined both in terms of operational requirements, and a detailed design definition embodied in a definitive specification. The object of the simulator program is to model all operational characteristics of GPATS, as well as the operator action and nominal UUT response to complete any test run. The completed simulator program will provide a means for checkout of UUT programs prior to validation on the GPATS system. This checkout process called verification, will greatly reduce on-line GPATS time required for program validation and sell-off.

The established Teleprocessing Network provides a means of direct communications, over common telephone lines, between remotely located GPATS facilities and the St. Louis Data Processing Facility used in test program processing and reprocessing. Plans call for adding an on-line network of computers to the teleprocessing network to provide even greater facility in program production and configuration management.

Concurrently with established SAS element implementation, state-of-the-art developments are in process which promise to further improve the SAS in the near future.

Current Status of Software Automation

A number of the elements of the software mechanization system have already been effectively implemented at Emerson, and others are in the process of development. Because of the incremental approach, what is invisioned today as the "ultimate" system may never be realized because improvements are always in the offing. However, a working system is in use today at Emerson. The elements of this system are described in the paragraphs to follow to convey a realistic indication of the practical application of the concept. The elements described herein have been developed to working status. To date, enough of the automated elements have been utilized to prove the validity of the concept for large scale production of programs.

Functional Block Diagram of Software Preparation Process

To set the software automation elements into perspective a brief description of each element is discussed with reference to the program generation process given in Figure 2. This is the system currently used at Emerson, St. Louis, in conjunction with its remote groups. A summary of SAS accomplishments and current developments is given in Table 1.

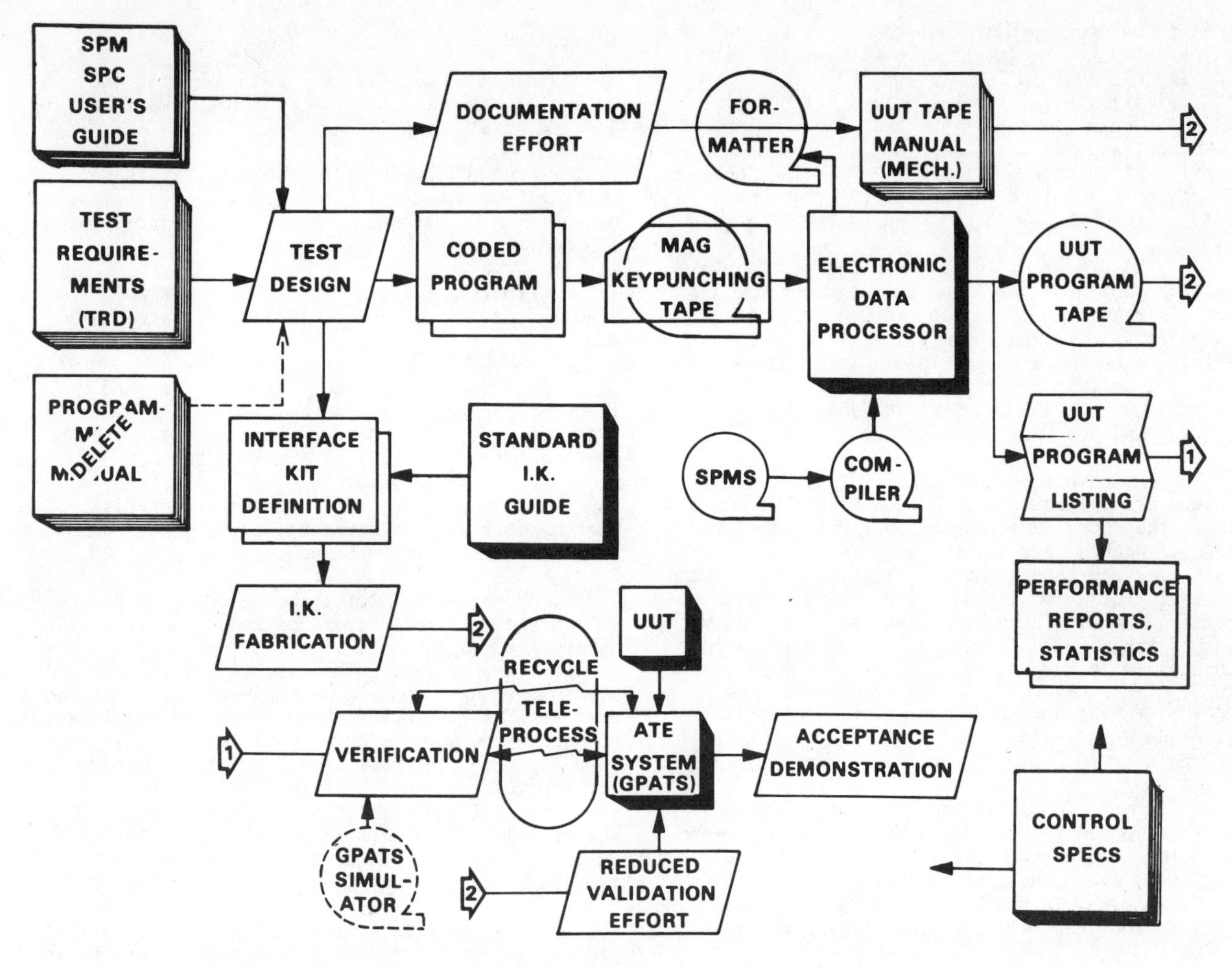

TEST PROGRAM PRODUCTION USING THE SOFTWARE AUTOMATION SYSTEM (SAS)
FIGURE 2

Input to the process is the Test Requirements Document (TRD) containing avionic Unit Under Test (UUT) testing data. The TRD is prepared by the UUT manufacturer, but he must be given guidance in its preparation. That guidance is provided by a TRD preparation specification which is supplied by Emerson but soon to be a military specification. This is one of the series of program control specifications which collectively cover all aspects of test program preparation and software management.

The supplied TRD contains all testing requirements at the interface, independent of any specific test equipment. That data must then be translated to specific functions provided by GPATS. This is the Test Design function performed by the test engineer. The translated document is called the Test Implementation Document (TID). This translation formerly involved a team effort including an experienced programmer. Now, simple programming instructions are supplied by the Compiler User's Guide which is easily understood by the test engineer.

Concurrently with TID preparation, the test design engineer develops an interface definition. This schematically defines what wiring and occasional components are required to make the UUT and GPATS electrically compatible. This definition is then translated into a practical design by an engineering design activity. The Interface Kit Design Guide (IK) helps standardize these designs to minimize costs and required interface hardware. Fabrication is the final step in interface preparation prior to checkout on an operational system. Fabrication is done in standard fashion by any manufacturing or model making activity. The TID information is developed directly on coding forms for keypunching. Currently, a hybrid source language is used comprised of SPM (Standard Preferred Method) statements and PLACE (Programming Language for Automatic Checkout Equipment) statements. SPM statements are preprinted on forms which require only minimal writing on the part of the user. They are in engineering-oriented format, so no programming experience is required for their use. The other advantage of SPM's is that functional requirements may be called out rather than having to decide what specific paths the signals must follow through GPATS to set up electrical signals or make desired measurements. This not only simplifies the test programming task, but assures that the standard, optimum configuration circuitry is used each time the same function is required. Peculiar functions not defined through SPM's are coded in PLACE language.

The SPM/SPC files contain the test implementing routines that allow the engineer to call out his requirements in the problem oriented language. Upgrading these files is an on-going effort which requires programming of higher skill levels than test programming. However, once defined new SPM's and SPC's are added to the library and need not be re-defined.

The completed SPM forms and place coding sheets are collected into a package for keypunching. The coded test program is first translated into a punched card set known as the source deck. This deck represents the UUT program in a language suitable for automatic data processing and conversion into GPATS machine language. Currently an updating procedure is operational which allows changes to be keypunched directly onto magnetic tape. These changes are then merged with the source program during a recompilation cycle. It is planned to replace card punching completely in the near future.

A more sophisticated data processing network is currently being designed employing remote computers linked to a large scale processor dedicated exclusively to test program production at St. Louis. In the interim, it is more economical to rent central data processing service as required.

The output of the Data Processor and its peripherals consists of the GPATS machine language program on perforated tape. In the SAS, the program listings and GPATS operator instructions are formatted by the central data processing run for direct inclusion in the deliverable tape manual. As the system is expanded, additional output data can be derived from the computer run. Magnetic-tape-formatted, tape manual text suitable for automatically producing military specification quality master-plates for printing is already being generated. Some configuration management data is already being generated as well as information for performance reports.

Residual Problems

Having reduced the problem of volume production of test programs to manageable proportions through mechanization, two residual problems become the limiting factors:

1. Availability of GPATS machine time for program validation (engineering evaluation).

2. Accessing the mechanized production capability from remote locations during program validation.

Validation Problems

Based on analyses of GPATS efforts, validation problems relating to test programs fall into two basic areas:

Type (1) Logic or programming errors resulting from misuse of GPATS capabilities, errors of omission, programming manual misinterpretations and coding errors.

Type (2) Complex testing problems which result from interaction of GPATS, interface Unit and UUT.

Although unable to quantify occurrence frequency or relative time devoted to the above problem areas, it is agreed that type (1) accounts for the larger quantity of problems but not necessarily the larger time element. It can be concluded, therefore, that both problem types must be delt with if substantial reductions of validation time are to be achieved.

A GPATS simulator is being designed to detect virtually all errors of the type (1) category through an off-line (not on GPATS) verification process. The type (2) category problems require individual engineering evaluation analysis. It is very doubtful that any simulator, no matter how sophisticated, will directly solve such problems. However, the simulator program can generate engineering tools to assist in type (2) problem solutions. These aids, in the form of UUT program flow charts and GPATS signal tracings will substantially reduce the problem analysis time, much of which is done "on-line" using GPATS.

Thus the simulator will be a very valuable aid in validation, reducing both engineering labor and GPATS machine time by two means:

1. Direct identification of logic and programming errors incurred in UUT program preparation.

2. Providing valuable documentation to assist the validating engineer in his solution of complex testing problems.

Accessing Mechanized Production Capability

The final requirement of a SAS is that the advanced "tools" of program production and validation be accessable by all users. Since program validation is generally done in the field at remote sites, direct communication linkage for data processing service is essential. The current teleprocessing system meets the immediate need via a network of Mohawk systems. In time, this network will be replaced by a highly sophisticated network of directly linked computers currently being configured by a state-of-the-art definition effort. When operational, that system will provide an automated solution to the next most pressing problem in ATE technology; namely, software configuration management.

TABLE 1

BRIEF FUNCTIONAL DESCRIPTION OF SOFTWARE ELEMENTS OF SAS

A basic translator program developed by Battelle Memorial Institute to allow user organizations to adapt their mnemonic language to a machine code generator.

An elaboration of the basic PLACE compiler adapted specifically to the GPATS family usage. It checks input language statements and compiles PLACE language statements into GPATS machine language. It also generates diagnostics and prints out source and object program listings.

Originally developed as a users document calling out System Preferred Methods for commonly required GPATS functions. Later it became a family of specially-printed code sheets allowing the engineer to call out complex GPATS functions by English statements. Each code sheet then provides input (via keypunch) to the PLACE compiler which will then generate a complete series of machine codes to perform the desired testing functions on GPATS.

A handbook to assist test design engineers in the design of interface kits to adapt the unit under test to GPATS. It both reduces design effort and fosters use of standard, shareable hardware rather than peculiar interface kits.

A program linked to the PLACE processor which selects appropriate source language call statements and generates output code suitable for automatically generating UUT tape manuals. These are the operators'manuals that enable a technician to hook up a unit for testing and perform the necessary manual operations involved during checkout of a unit. The object file of a PLACE run is developed on a magnetic tape which is later translated into the "Standard Preferred Comments" (SPC's) that comprise the text of the manual. The translator operates a computer controlled photo-composition system with a resident GPATS format and SPC file.

TABLE 1 (Continued)

BRIEF FUNCTIONAL DESCRIPTION OF SOFTWARE ELEMENTS OF SAS

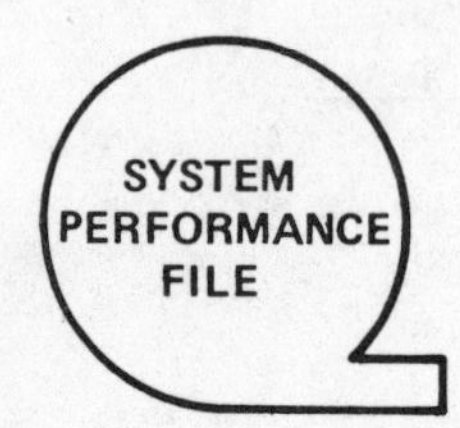

Originally developed as a comprehensive definition of GPATS performance as a system in its many variable modes or usage. Formulas developed have led to the ability to predict losses in the system and compensate for measurement inaccuracies. This information is being incorporated in the machine code generator of the compiler so that the user will not have to calculate and compensate for system degeneration of signals when he calls for GPATS functions in the SPM language.

A formal set of control specifications covering all aspects of test program generation. It is divided into two subsets: (1) Specific direction to test design engineers for each function involved in program generation and checkout, and (2) management control procedures for handling software management functions. Most controls will be embodied in a computerized software configuration management system.

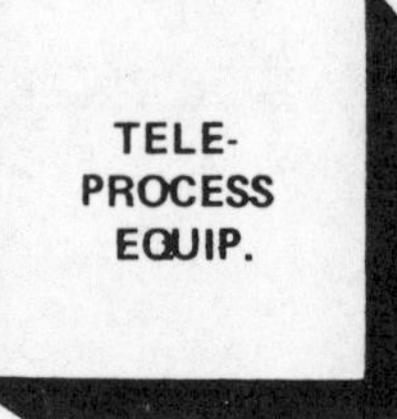

Currently consists of a network of commercial teleprocessing equipment linking remote programming centers with central facilities at St. Louis, Mo. It provides off-line keypunching directly onto magnetic tape and magnetic tape driven telephone modems and hard copy printers enabling fast access to compiler facilities for all remote user groups.

A program, currently under development, which will model the GPATS system on a general purpose computer. By inputting a GPATS machine language program, an automatic evaluation is made of function compliance with stored ground rules for testing, check on GPATS capability for a specified configuration and other verification functions. It is also capable of producing machine generated flow charts of the UUT program and signal tracing diagrams for each test in the program being evaluated.

BIBLIOGRAPHY

T. A. Ellison, L. S. O'Neill, "ATLAS - a Standard Compiler Input Language for Commercial Airlines", IEEE Proceedings of 1968 Automatic Support Systems Symposium.

R. L. Mattison, R. T. Mitchell, "UTEC - A Universal Test Equipment Compiler", IEEE Proceedings of 1968 Automatic Support Systems Symposium.

F. Liguori, "Software Management Through Specification Control", Software Age, May 1968.

F. Liguori, "ATE Designer-Programmer or Systems Engineer", IEEE Proceedings of 1967 Automatic Support Systems Symposium.

F. Liguori, "Conceptual Design for an Automatic Test System Simulator", Technical Program Proceedings, 1969 ACM National Conference.

V. Mayper, "A Survey of Programming Aspects of Computer-Controlled Automatic Test Equipment", Volume I and II, Project SETE, New York University, 1964.

AUTOMATIC GENERATION OF TEST PROGRAMS

Philip A. Hogan
Aerospace Division, Honeywell Inc.
Minneapolis, Minnesota

SUMMARY AND INTRODUCTION

The high cost of writing and debugging test programs is well recognized today as one of the major deterrents to the successful application of automatic test equipment. It is an unfortunate fact that software costs often exceed the hardware costs. This cost ratio increases as test systems become more general purpose in capability since they are able to test a greater variety of devices. In recognition of this fact, Honeywell has recently completed development of an ATE system which places major emphasis on features which reduce test programming time and cost. The system is general purpose in design since it was developed for internal use at Honeywell for production testing on a broad range of electronic devices.

This paper is presented in three parts. The first part describes features provided in the ATE system which lead to reduced costs for conventional test programming. The second part discusses the advantage of a comprehensive manual control capability in an automatic test system as it relates to test programming. A computer assisted method of achieving manual control is described. This leads to the third part which describes an experimental technique for automatically generating test programs which completely by-passes the conventional approach. Other techniques have been developed for eleminating the normal test programming process, but they are applicable only to continuity or digital logic testing. The technique described in this paper is applicable to testing complex analog devices.

TEST SYSTEM DESCRIPTION

The ATE system is a general purpose, multi-station, computer-controlled test system. In order to discuss its test programming features, it is first necessary to briefly describe the hardware system.

A typical test station is shown in Figure 1. Each station contains an integral control-computer dedicated to that one station. The computer, a DDP-516, is a small 16-bit process control computer with 8K core memory. In Figure 1, the computer can be seen immediately to the left of the three-bay console and behind the test station teletype. In all future stations the DDP-516 will be replaced by the H316 which is a newer generation computer having an instruction repertoire identical to the DDP-516's. It is also compatible with the same peripheral equipment options, and uses the identical I/O circuitry. The H316 is less than half the price and size of the DDP-516, requiring only 14 vertical inches of rack mount space.

Figure 1. ATE Test Station

A patchboard can be seen on the right hand bay. It provides the means to configure the test station signal routing for a specific UUT. It also provides the umbilical connection between the test station and the UUT, and serves as a convenient mounting fixture for adapter circuitry. Small devices such as modules and cards plug directly into connectors mounted on the patchboard adapter. Larger devices are connected through an umbilical cable to the patchboard. A minimum length path for stimuli and signal routing is provided by mounting general purpose switching relays and signal routing crossbars directly behind the patchboard receiver.

Test stations operate autonomously by reading blocks of test program instructions into the computer core memory through the high speed paper tape reader. In addition, each station can also be connected to a Central Data Station (CDS) on a high speed, party-line, data transmission line which operates serially at a 250 kilohertz bit rate. Each test station then receives blocks of test program instructions from the CDS, and transmits blocks of logged data back to the CDS. Operation with the CDS is the normal mode of operation and is illustrated in Figure 2. A test station can be located up to 1000 feet from the CDS.

Reprinted from *IEEE Auto. Support Systems Symp. Rec.*, Nov. 1969.

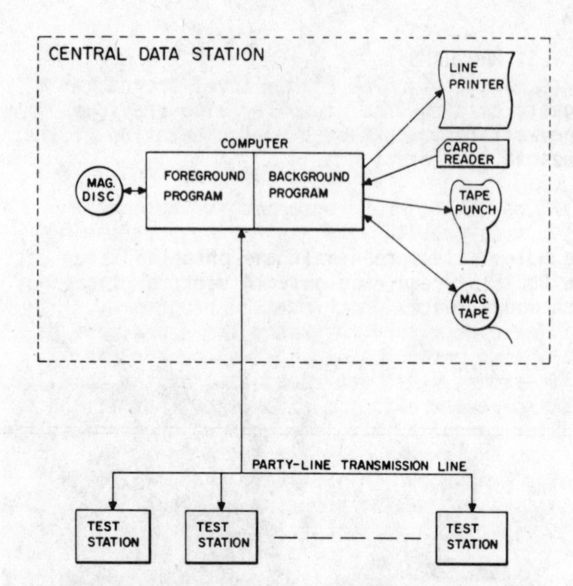

Figure 2. Operation With Central Data Station

The primary purpose of the CDS is to share a 3.5 million word moving-head disc file between test stations. The disc file contains test programs and logged data. In a secondary or background mode, the CDS also shares the other peripheral equipment. Due to modern digital computer technology the dedicated computer in each station costs little more than some of the standard commercial instruments. But the peripheral equipment is more expensive and, therefore, time-shared. The CDS computer is also a DDP-516. It can be seen along with the disc and teletype in the background of Figure 1.

Central Data Station Software

The Central Data Station software system consists of a number of separate functional packages. The CDS operates with foreground and background level programming to insure maximum system utility. The primary (foreground) operating mode of the Central Data Station is to share the magnetic disc file among the separate test stations for test program retrieval and data storage. Test stations are serviced on a request basis using a common interrupt line. When a test station requires a new program segment from the shared disc, it sends a short message to the CDS identifying the particular segment needed. Likewise, when the data buffer in a test station core memory is filled with the logged results of many measurements, it transmits a buffer record which is added to the disc file for that station. A test station request is immediately filled by the CDS before it honors a request from another station. The foreground program is contained in 4,000 words of core memory.

In addition to the foreground tasks, the Central Data Station is also used to perform numerous background jobs. Only one background job is performed at a time, but it is done simultaneously with the foreground task. The dispatcher portion of the foreground program is used to queue background job requests from the test station. Existing background jobs are performed in a second 4,000 words of core memory. The background jobs are as follows:

a) List all the logged data for a given UUT on the high speed line printer.

b) Read an Adapted ATLAS language test program from punched cards or magnetic tape onto the disc.

c) Compile and assemble a test program segment which is on the disc into an Operating language test program and store it back on the disc, available for test station call.

d) List a compiled test program segment on the high-speed line printer.

e) Accept and insert corrections to an Adapted ATLAS language test program stored on the disc.

f) Transfer an Adapted ATLAS language test program from disc to magnetic tape for permanent storage.

g) Transfer logged data from disc to magnetic tape for permanent storage.

All background jobs can be requested from the CDS teletype. In addition, tasks a), c), d), and e) can be requested from any of the test station teletypes.

The foreground tasks are handled on a top priority basis. That is, background jobs are momentarily stalled when foreground functions are to be performed. Background jobs are handled on a two level priority basis. Background jobs involving the line printer are momentarily interrupted to perform high speed background jobs which involve only the disc.

In addition to the foreground and background programs, there are other functions which are performed off-line at hours when the test stations do not require service. The off-line jobs are as follows:

a) Copy the complete disc onto magnetic tape for backup storage. This provides insurance against alteration of the disc contents.

b) Punch paper tape test programs for operating test stations in the autonomous mode. The tapes are punched by selecting a series of binary test program segments from the disc.

TEST PROGRAM DEVELOPMENT

The ATE system features described above allow the use of an efficient test program development process for reduced programming costs. That process, shown in Figure 3, includes on-line compiling in the CDS. The purpose of the compiler is to convert test instructions in a high level programmer's language to the lower level Operating language read by the test station computer. The source language for a test program segment is first read from punched cards onto the disc. After compiling, both the source language and the object binary program are retained on the disc. Only the binary test program is transmitted to a test station for execution, but the source language is still available for editing and rapid disc-to-disc recompiling. After a program segment is completely debugged, the compiler source can be permanently stored on magnetic tape and erased from the disc.

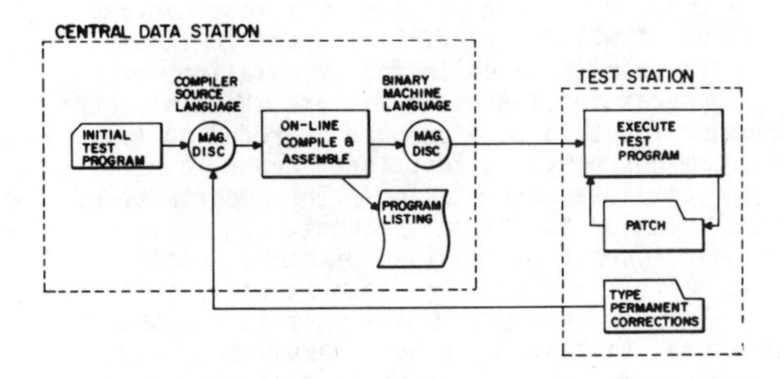

Figure 3. Test Program Development Process

Program editing can be performed from any test station teletype as corrections are found necessary during the debug process. Corrections are referenced to either source language line number or to program tags. Edit instructions are accumulated in the test station computer. After all instructions have been accumulated they are shipped to the CDS where the edit takes place as a background job.

The request for a recompile and for a program listing on the high-speed line printer can also be made from the test station. While editing, compiler source language can by typed on a station teletype for assistance in making corrections. Only the desired section of a complete source language program is requested for typeout due to the slow speed of the teletype. The test program listing from the high-speed line printer is a side-by-side listing of compiler source language, resulting assembly language, and corresponding binary object program.

Compiler

Test program compiling either requires a large mass memory in the form of a magnetic disc or tapes, or else necessitates a weak and restrictive programmer's language. For the broad range of electronic equipment encountered in most production and maintenance shops, it is important that the compiler be powerful enough to handle the multiplicity of test functions at minimum programming cost. The higher level programmer's language of a powerful compiler also provides the necessary understandable documentation of the automatic test procedures.

As new UUT's are programmed for automatic test, it is necessary to expand the programming language of the compiler input. New UUT's may require new types of measurements or new types of stimuli devices which must be programmed. It is also necessary to expand the repertoire of descriptive mnemonics, such as connector pin designations, which are recognized by the compiler. It has been the author's experience that if the compiler cannot be easily expanded, programmers begin to bypass the compiler and program in machine language. This defeats the compiler's advantages of reduced programming costs and improved program documentation.

An on-line interpretive compiler is one which resides in the core of the test station control computer and compiles the test program each time it is executed on the test station. Such a compiler is generally quite restrictive since it must share core memory with the test program and the operating executive program if mass memory is not available. Even if mass memory is available, an interpretive compiler can be restrictive because of the time required to recompile the test program each time it is executed.

The ATE compiler consists of two separate elements. The first is the Basic Compiler which is a character processor that cannot recognize any compiler source language input by itself. It must refer to the second element, which is a Language Definition File, in order to recognize and interpret the input language. The Language Definition File, which is stored on the disc, is in a form easily understood and easily expanded by test programming personnel to meet expanding compiler requirements. The Basic Compiler, which never changes even as the programming language expands, is actually a compiler/assembler combination. The compiler includes a table look-up capability which allows the program to be written in the functional nomenclature of the UUT. The look-up table for each UUT is included in the Language Definition File.

Operating Language

The Operating Language is the language of the test program as it is executed in the test station. The Operating Language is a machine language level. There are no interpretive operations required for the control computer to convert to machine language level prior to executing the program. Thus the test programmer has a printed record of exactly what instructions are being executed by the computer. Normally, the programmer follows through the program be reading only the compiler input language which is carried on the program listing as comment lines. But when required for debugging a particular problem, he is able to read and modify the exact computer instructions being executed until he has resolved the problem. The Operating Language consists primarily of a series of calls to standard sub-routines in the test station executive program followed by constant or variable data required by the sub-routine. The calls to the sub-routines appear as a JST operation code (jump and store). Associated with each JST is a mnemonic identifying the purpose of the particular sub-routine.

The Operating language is not restricted to those functions contained in the standard sub-routines. Where specific test system functional capabilities dictate, any of the 70 standard computer instructions are available to the test program.

Rapid Program Modification

After program corrections have been determined they are made through the editing and recompiling process. However, the debugging process to determine what changes should be made requires the ability to make rapid temporary changes to the program at the test station. Recompiling a 500 word program segment takes from 1 to 1.5 minutes. To use this process for the many experimental program changes often required to track down a single problem would be prohibitively time consuming.

On-line test program modification is accomplished using a special program called PATCH. This program resides in the test station computer core memory along with the executive program and test program. PATCH allows the operator to communicate through the test station teletype to make immediate changes to the test program. The test program instructions presently in the computer can be typed out, and changes or additions can be made by simply entering them on the teletype. This capability allows the test programmer to make immediate program changes.

The most significant feature of PATCH is that the programmer communicates in symbolic language rather than in meaningless numerical codes. Operation codes of program changes are entered in their mnemonic representation. Constants can be entered either as octal, decimal integer, floating point, or ASCII characters. Likewise, the test program presently in the computer can be typed out at the programmer's option in mnemonic, octal, decimal integer, floating point, or ASCII characters. The instructions in the computer are actually in binary machine codes, but PATCH interprets them into the representations meaningful to the programmer before they are typed.

After a change is made to a test program segment using PATCH, it is possible to retain that change permanently in the core memory until it is intentionally erased. This change can be retained even after many other different test program segments are executed. When the modified segment is again entered into core memory, the stored change is automatically re-entered into the test program. This allows the programmer to execute other program segments and then return to the segment being debugged without destroying previous changes.

It can be seen that the above on-line program modification capability greatly simplifies the program debugging task. Changes can be made quickly, in a form easily used by the programmer. Only after the changes have been verified must the permanent changes be made through the compiler cycle. This same capability is used for future test program changes in response to a UUT modification. A proposed change is first verified using PATCH, and then incorporated into the permanent documentation.

MANUAL CONTROL

The importance of the above ATE system features which have been provided for test program development will be readily apparent to those who have been involved in the tribulations of test programming. Another feature which has been provided for this same purpose of reducing test programming costs, is functional manual control of the complete test station. The importance of manual control is not so apparent. In fact, it has been found that the idea of manual control in an automatic test station often strongly offends an engineer's aesthetic sense. It is necessary, therefore, to examine rather closely the need for manual control as it relates to test program development. The key word is functional manual control which will be expanded upon later.

Problems of Automatic Control

A major reason that test programming costs have been so high is that errors from many different sources exist simultaneously at one point in the development process. That point is the so-called program "verification" phase which occurs when the test program is first run on the test system. This phase is more accurately called the program debugging phase, since the only thing usually verified is that the test doesn't work for some unknown reason. When a test program is first run it may contain human coding errors. But there may also be problems in the test system hardware, UUT hardware, or in the test programmer's interpretation of how the hardware is supposed to function and how it should be tested. In fact, with the assistance of modern compilers to reduce human coding errors, the greatest source of problems are those associated with the hardware and the intended test procedures. The isolation and elimination of these problems during the test program debugging phase is one of the most time consuming elements in the test program development process.

One reason that test program debugging is so time consuming is the cumbersome methods which the test programmer must use to trouble shoot the problems. This can be seen if we compare the troubleshooting methods with those used in developing test procedures for manual test equipment. When first testing a device on manual equipment the test engineer uses a preliminary test specification--the manual equivalent of an automatic test program. When a test does not perform as expected, the test engineer uses the classical troubleshooting technique of hypothesizing the problem and attempting to confirm or reject its existance through experimentation. Perhaps unwanted noise or phase shift is introduced by the test equipment, or the need for a settling time delay was forgotten, or perhaps the test specification mistakenly specified the wrong position of a mode switch. Probably the first thing the test engineer does when the unexpected happens is to quickly scan his various meters and switches for a clue to the problem. If the problem persists, the UUT designer can be called in to sit down at the test equipment to make his own observations.

Contrast this method with that commonly employed in debugging automatic test programs. There are similar problems in the hardware and in the preliminary test procedures which are now in the form of the test program. In addition, the programmer also has coding errors to worry about. But the big difference from manual testing is in the facilities provided to the test programmer for troubleshooting. He must constantly work within the cumbersome realm of digital codes and the digital test program. He still hypothesizes a problem, but the experimentation is often difficult. If he suspects that the wrong excitation conditions exist he must search back through the test program listing to see what relays and signals were programmed. The importance of chronology in a test program can be very confusing. The programmer may find a code in the program which set a certain relay. But he may have overlooked a subsequent code which reset that same relay, thus leading to an erroneous assumption of the state of the test station. The programmer must form a mental image of the complete test station status which is constantly changed as he reads the codes and chronology of the test program. Pity the poor test programmer with only average powers of visual perception.

Once the programmer thinks he has formed a correct image of the station status at the point in question, he may want to change that status in accordance with a proposed problem solution. Perhaps he thinks a stimulus level should be increased, or another relay set, or an instrument should be connected to a different test point. To do this the programmer must look up or calculate a series of coded commands which are keyed into the station to accomplish the desired function. It is difficult for people to think in terms of codes. The use of a resident program in a test station computer can assist in this process by allowing the programmer to enter commands through a teletype. But due to the computer memory restrictions, even these commands must be in a rigorous and coded format, such as a relay number, rather than the UUT function it controls.

During the time the programmer is trying to correctly determine the programmed state of the test station and command it to a new desired state, he is constantly worrying about coding errors. By this time he has probably completely forgotten the original problem he was trying to solve. Most automatic test systems provide a convenient control switch for this situation which, when depressed, allows the programmer to start all over again!

Advantages of Manual Control

At this point the reader may begin to suspect that the author is presenting an argument in favor of manual test equipment over automatic test equipment. This is not the case. What the author does favor is the incorporation of some of the advantages of manual test equipment into automatic test equipment in order to reduce the present high cost of programming. The expanding use of automatic test equipment in spite of the high cost of programming is sufficient testimony to its many advantages.

If an automatic test station has the same facility for functional manual control as a manual test station, automatic test program development is greatly simplified. Figure 4 illustrates the conventional procedure for developing test programs. When the test program is first executed it is necessary to simultaneously troubleshoot errors from all different sources. In addition, the vehicle for manipulating the test equipment during this troubleshooting phase is to make changes or to simulate changes in the automatic test program. This immediately forces the test programmer into the cumbersome realm of sequential digital codes.

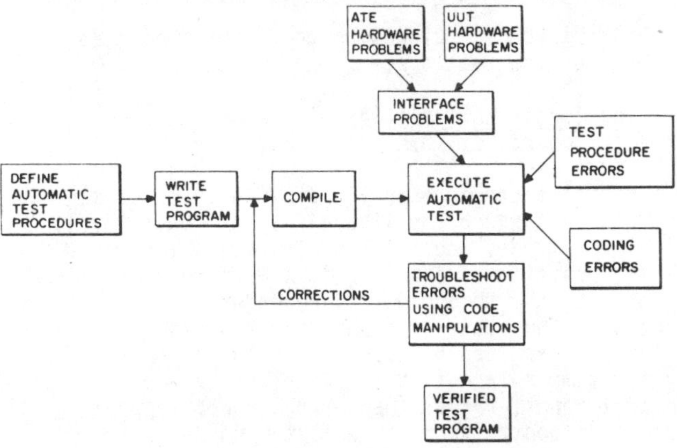

Figure 4. Conventional Test Programming

Figure 5 illustrates the test program development procedure using functional manual control. It is still necessary to troubleshoot the same sources of error. However, it is now possible to eliminate most errors in the hardware and test procedures before the test program is even written. The test programmer is able to eliminate these problems before he ever sees a digital code. Then when the test program is first executed the programmer can concentrate on coding errors and those errors which relate to the difference in speed of test execution between manual and automatic test.

There appear to be two basic characteristics of the manual control capability provided in a typical manual test station which simplifies the debugging of hardware problems and test procedure problems:

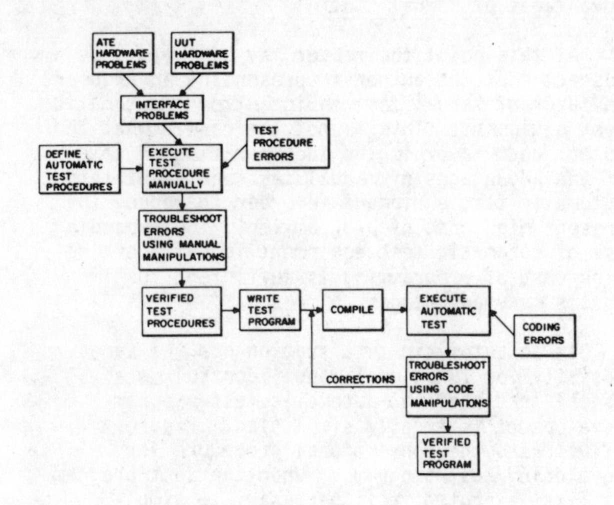

Figure 5. Test Programming With Manual Control

a) Continuous functional indication of the current state of the complete test system.

b) Direct functional control over the state of the test system.

The functional relationship is one of UUT orientation rather than ATE orientation.

Manual testers generally provide their state indication through the liberal use of functionally labeled switches and meters. The state indicated by the position of a switch is always current as opposed to the chronological dependency of a test program. The state of a specific function is quickly determined because it is always indicated by the same switch in the same placement on a control panel, rather than by the random appearance of a control word in a test program. In addition, all possible states are indicated by the switch positions on a manual test station, whereas a test program listing indicates only those states which have been programmed to be different from some predetermined quiescent state which the automatic station always assumes when "cleared". For example, if power is off on a manual station there is a switch pointing to OFF to so indicate.

Direct functional control of a manual test station is also provided by the labeled switches. There is no need to worry about what particular command codes or series of relay excitations are needed to achieve a particular function. The test engineer merely reaches to a familiar position on the control panel and turns a switch to the desired functional state.

The need to exercise a considerable amount of manual control over ATE occurs primarily when debugging the automatic test procedures during test program development. However, there are also other times when functional manual control of ATE is beneficial. This occurs, for example, when it is desirable to manually isolate faults which are detected during automatic testing, or when it is desirable to achieve correlation between manual developmental testing and automatic testing on the production line. However, these considerations are beyond the scope of this paper.

Manual Control Mechanization

A manual control capability has been incorporated into the Honeywell ATE which provides the above characteristics of UUT oriented functional control. When placed in the manual control mode the ATE essentially operates as a manual test station. This capability was achieved at relatively low cost, and without degrading performance in the automatic mode of operation.

Manual control was achieved in two ways. The first was to use, wherever possible, standard commercial instruments which provide manual control switches on their individual front panels in addition to the remote programming capability. The second method was applied to the general purpose switching relays, the test point selection crossbar switches, and to those instruments which did not normally have front panel manual control. This method is computer assisted using the digital computer which is an integral part of the test station.

Computer assisted manual control uses functionally labeled panel mounted switches in the same way as a typical manual test station. However, in this case there is no electrical connection between the switches and the test station. The only electrical connection is to the station control computer. The switches are wired in a binary code and connected to a computer input gate. The computer is able to selectively address the switches and read binary codes which identify the respective positions to which they have been turned. Figure 6 illustrates the connection between the computer and the manual switches. Diodes are used for matrix isolation between switches.

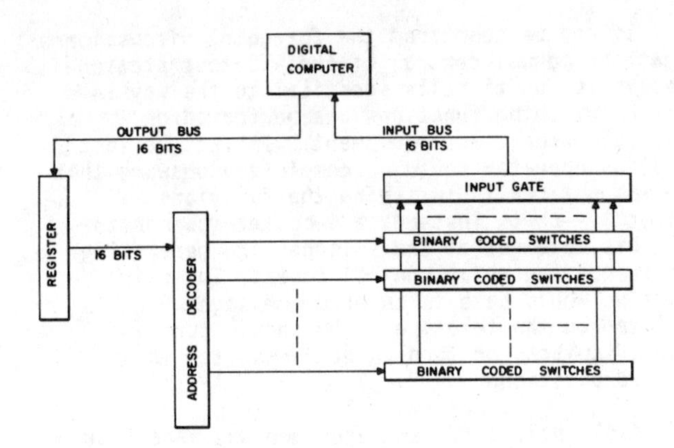

Figure 6. Manual Switch Circuitry

When the ATE is in the manual control mode a special program is read into the computer core memory. This program causes the computer to constantly interrogate the manual switch positions, and fabricate coded control words which program the test station to the corresponding state. These control words are identical to those normally supplied by an automatic test program, but in this case they are generated by the digital computer itself in response to operator input through the manual switches. Special INSERT push button switches prevent the computer from reading the selector switches until the test set operator is satisfied that he has selected the desired condition. It can be seen that this method of manual control requires minimal electrical connection to the ATE, requires practically no changes in design of the automatic portion of the ATE, and cannot affect the performance of the ATE when it is performing automatic tests. Figure 7 shows a computer assisted manual control panel for a programmable time delay generator, and for two programmable threshold detectors.

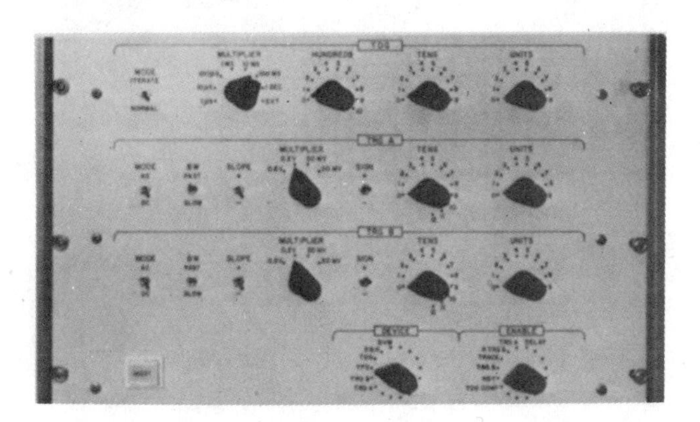

Figure 7. Typical Manual Control Panel

Computer assisted manual control of the general purpose switching relays required a somewhat unique approach. Since the relays are general purpose the UUT oriented function of a specific relay depends upon the particular unit being tested. A relay's function also depends upon the wiring of the patchboard adapter for that device. Therefore, the manual control program for relay control consists of two parts. The first part is a general purpose routine which is the same for all devices tested. The second part is a program table which is unique for each device tested. This program table defines the relationship between a switch position and the desired set or reset state of corresponding relays. The control switches for the general purpose relays are therefore also general purpose. There are 56 two-position switches, 22 three-position switches, and 20 seven-position switches provided on a single panel for relay control which is shown in Figure 8. An over-lay is placed over the switch panel to provide the functional nomenclature of a particular UUT. Each different UUT has a unique over-lay and a unique switch/relay program table. Figure 9 is a portion of one such over-lay.

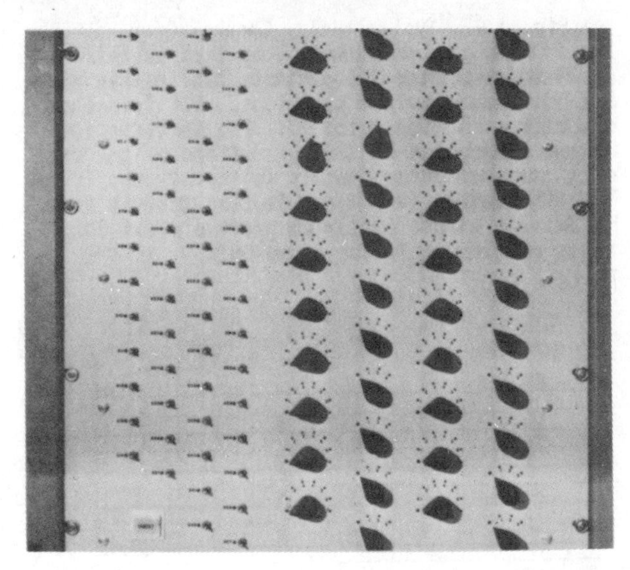

Figure 8. Manual Control for Relays

The switch/relay relationship as determined by the switch/relay program table is completely flexible. Any switch can control up to 16 different relays, and there is no restriction on which relays a particular switch can control. The relays are magnetic latching and the set or reset state of each relay is controlled for each valid position of its controlling switch.

The on-line test program compiler is used to compile the switch/relay program tables. Figure 10 shows the truth table for a typical switch and the corresponding compiler input statements as they would appear for that switch when compiling the switch/relay program table.

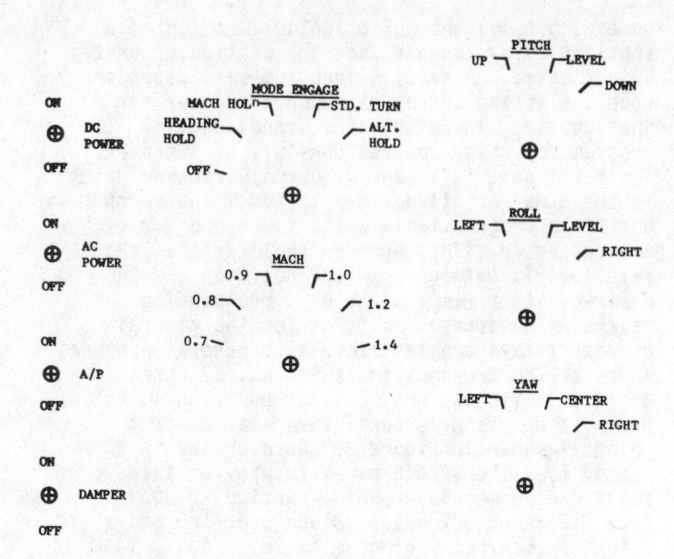

Figure 9. Typical Relay Control Over-Lay

The PATTERN statement indicates that in this application switch S36 controls four relays and has five valid switch positions. S36 is one of the seven-position switches. The specific four relays controlled by S36 are defined by the four RELY statements immediately following the PATTERN statement. It can be seen that it is a relatively simple matter to provide a switch/relay program table for a new UUT.

S36

	Switch Positions				
RELAYS	1	2	3	4	5
1K5-16	R	S	S	R	R
3K4-14	R	R	R	S	R
3K3-12	R	R	S	S	S
4K5-16	R	R	R	S	R

Compiler Input Statements

```
S36  PATTERN  4, 5, RRRRSRRR,SRSRRSSSRRSR
     RELY  1K5-16
     RELY  3K4-14
     RELY  3K3-12
     RELY  4K5-16
```

Figure 10. Typical Switch/Relay Correlation

The switch/relay program table uses a relatively compact format. Each switch requires two or three 16-bit computer words, and each active relay requires a single computer word. Thus, a table which uses all 98 switches to control 200 relays would occupy approximately 450 words. The general purpose routine which fabricates the relay control codes from the table occupies 366 computer words.

It can be seen from the foregoing discussion that the manual control of the ATE test station relays is functionally identical to the way in which switching functions are performed on typical manual test equipment. In fact, a test station operator could be completely unaware that he was not really switching the functions directly--that, instead, a computer was monitoring his movement of the switches and generating corresponding relay control codes. The only clue he would have is an error message which is printed on the teletype if he should turn an unused switch, or turn an active switch to an unused position.

A slightly different approach was taken for control of the test point selection cross-bar circuits. It was felt unnecessary to use an over-lay technique since UUT nomenclature for test points are not generally meaningful anyway. Test points are generally referenced by arbitrary connector pin designations. The panel for computer assisted manual selection of test points is shown in Figure 11. Each of the dark vertical slots contains a slide switch knob which is not visible in the photograph. Each vertical slide switch controls the selection of input to a specific measurement instrument. A measurement instrument can be connected to any one of five separate measurement buses which are indicated by the horizontal lines. If the slide switch for a measurement instrument is placed at the intersection with a measurement bus, the necessary codes are generated to program the cross-bar to that corresponding state. Each measurement bus can be connected to one of 84 test points which is controlled by the switch pairs on the right of the panel. These arbitrary test point numbers are added to UUT schematics for use by the test station operator.

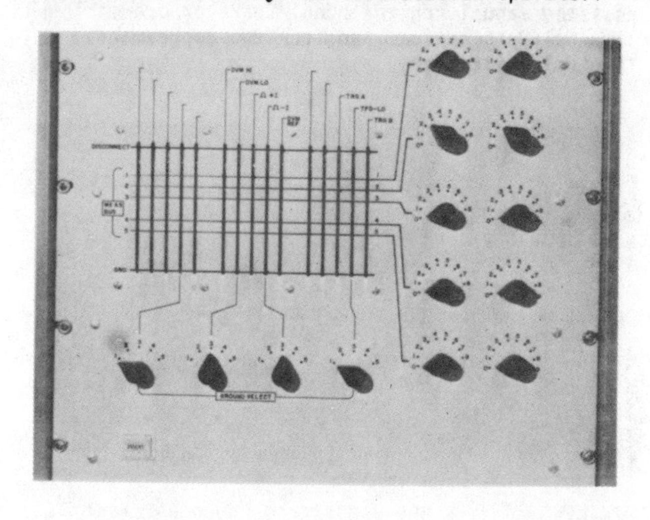

Figure 11. Manual Control for Test Points

The entire computer assisted manual control program requires 1208 computer words exclusive of the switch/relay program table. It occupies that portion of core memory which is normally occupied by test programs during automatic testing.

AUTOMATIC GENERATION OF TEST PROGRAMS

It can be seen that while the computer assisted manual mode gives the appearance of completely manual operation, the same codes are generated as those used in an automatic test program. Therefore, if all test set functions were under computer assisted manual control, it should be possible for the computer to store the resultant sequence of control codes as they are fabricated. Those stored codes would then immediately comprise an automatic test program--thus completely eliminating the classical test programming procedure. The total effect would be one of the operator first performing a test manually with the computer monitoring and storing his every action. From then on the computer could repeat the test automatically and much faster. While this is an over-simplification, it is the essense of work presently being conducted on an experimental basis for which the term Instant Programming has been coined.

One obvious over-simplification in the above description is the fact that several essential functions occur solely within the mind of the operator during manual test. These include such functions as evaluation of results against limits, mathmatical calculations, and branching type decision making. The computer would not generate codes corresponding to these functions since it is not even able to sense their occurrence. It is therefore, necessary to add a capability which allows the operator to insert these functions at the proper time solely for the purpose of generating an Instant Program. This is being done by the use of additional function switches in some cases, and by the use of teletype entry in others.

It is expected that when using the technique the operator will first execute the test without storing the generated code. He will probably work from a standard flow chart After he has eliminated the errors in hardware and test procedure, he will repeat the test for a program generation run.

An interesting feature is the way in which program instructions for the relays and crossbars are generated. It would, of course, be prohibitive to generate a control code for each relay controlled by all the manual control switches each time the Insert switch is depressed. This would generate redundant test program words. However, it would also be expensive and cumbersome to provide a separate INSERT switch for each control switch. The solution was to retain an image in core memory of the position of all relay control switches the last time control codes were generated. In this way the next time the INSERT switch is depressed control codes are generated only for those switches whose positions were changed. The image of all 98 relay control switches occupy only fourteen 16-bit core locations. The cross-bar is handled in a similar fashion. Its image occupies 17 core locations.

Limitations

Unfortunately it is too early at this writing to quote results from Instant Programming applications. However, there are two obvious limitations to the technique which should be mentioned. The first limitation is in the lack of test program documentation. A test program which is generated with a compiler has the advantage of a comprehensive source language listing. This provides an understandable and accurate documentation of the program. With Instant Programming no such documentation is provided since the detail machine language codes are generated directly. Since computer automated test programs usually consist largely of a series of return jumps to standard executive sub-routines followed by specific variable data, it should be a relatively easy matter to process an automatically generated test program through a form of reverse compiler.

A reverse compiler would create English language statements from the test program codes. However, it is prohibitively restrictive on the ATE test capabilities to require that test programs contain only those functions which are performed by standard executive sub-routines. In those cases where in-line machine language coding is generated, the reverse compiler would be unable to distinguish test program instruction words from data words. It is, therefore, anticipated that the optimum procedure would be as follows:

a) A reverse compiler would interpret those test program words which refer to standard executive sub-routines into their English language equivalent. This would include the major part of a typical test program.

b) The reverse compiler would flag other program words, and interpret them both as machine language instructions and as data.

c) The test programmer would then inspect the flagged words and select the proper interpretation. At the same time he would provide the equivalent English language annotation.

The second limitation of Instant Programming concerns the use of logical branching and looping within the test program. These functions are easily provided within the capabilities of Instant Programming, and should be used to a limited degree. But their extensive use in a test program would probably create a complex logical structure which would be difficult for the programmer to visualize and control. In those cases where extensive branching and looping are required, it would be better to consider using the more conventional forward compiling procedures to develop the test program.

Future Expansions

From the foregoing description it can be seen that both the computer assisted manual control technique and the Instant Programming technique require the use of a small real time digital computer. In the computer controlled ATE system described, the necessary computer happens to be part of a standard test station. However, the technique is equally applicable to punched tape programmed test equipment if an ancillary computer is provided solely for the purpose of developing the punched tape test programs. In this application the computer generates 8-bit punched tape codes rather than the 16-bit computer words, but the technique is similar. The only electrical connections to the test set are those simulating the paper tape reader. The computer simulates the slower reading speed of a tape reader with programmed delays between tape codes. The computer also tests the normal tape reader stop and advance signals in order to provide the proper response to the test system. It may be necessary to bring other test system signals back to the computer, such as an indication of a tape search mode following a limit comparison, but these signals should be relatively few.

This application to controlling a punched tape test set serves to emphasize one of the basic principles of the computer assisted manual control. That is, that the manual control switches do not indicate current test system status because of any feedback from the test system. Rather they indicate status through open loop control of the test system by the switches. Of course, if there was a malfunction in the test system such that it would no longer follow the manual control commands, then the switches would no longer correctly indicate system status. However, this could still be the case even if feedback did exist.

The limitations of Instant Programming discussed in the previous section would probably not occur in the application of programming a tape controlled test set. Due to the rigorous and sequential format of punched tape codes, it should be possible for a relatively simple compiler to translate all program codes into their English language equivalents. In addition, the restricted branching and looping capabilities of a tape controlled test set would insure that the logical complexity of a test would never exceed the programmer's ability to visualize and control.

CONCLUSION

It is an unfortunate fact that test programming costs often exceed the hardware costs, and the problem increases with general purpose test systems. In such systems, additional hardware and system capabilities provided for the purpose of reducing programming costs can pay handsome dividends. The expanded capabilities should be oriented toward minimizing test program debugging effort since this phase generally accounts for 50% to 75% of the total programming task.

A major problem facing the programmer is the lack of ability to observe and control the functional state of automatic test equipment while he is debugging test programs. Humans do not work well in the realm of digital codes when manually manipulating test system hardware. Comprehensive manual control, while not necessarily used during automatic testing, can be very useful for facilitating the debug of hardware and automatic test procedures prior to test program debug.

A computer assisted manual control shows great promise as the means for automatically generating test programs. This is accomplished by causing the computer to remember the sequence of functions executed by the operator while he conducts the tests manually. After that the same tests can be repeated automatically and, of course, at much greater speed.

PART V: Software Tools and Techniques

Automatic testing technology owes much to computer technology. Even before general-purpose computers were integrated into ATE systems, the concepts developed by computer designers were implemented in the design of the program controllers. Even though the hardware design contributions of computer technology to ATE are significant, much more has been contributed in software design concepts. The bulk of the software design contributions to date have dealt with software tools or aids to test program preparation rather than in the design of the test programs. This is logical because test design relates more closely to the testing function, that is, the dynamic control of electrical signals, than it does to data processing. Software tools, on the other hand, deal with the processing aspects of test program preparation. Because of this heavy involvement in processing, the computer provides a natural partner for the test engineer who has to integrate the testing functions into a logical flow process executable under stored program control.

Since test program design and validation relates very closely to conventional software design and debugging, it should not be surprising to find that programmers' tools such as problem-oriented languages, assemblers, compilers, and simulators can also be valuable tools for test program production. This part introduces the reader to some software tool terminology and typical applications of such tools in the field of ATE.

The first paper, "Software Design Techniques for Automatic Checkout," defines many of the software terms for the benefit of the reader who is not an experienced programmer. More than that, it offers a structured design concept for the orderly development of software processing modules. Such an approach provides a means for efficient design, management, and configuration control of the software tools. It also facilitates imposing the much needed discipline on software projects, which historically have been difficult to budget and manage.

From the earliest days of ATE application, it was recognized that the only practical approach to developing a large quantity of test programs in a reasonable time is to develop a test language and associated processor that enables the test engineer to prepare his own programs without having to first become an expert programmer. Many test languages evolved along with associated compilers. Each seemed to meet the short-term needs of a given application. It soon became evident, however, that ATE system designs and application requirements changed too rapidly for any fixed language or processor design to keep pace with these needs. Hence, the development of universal compilers was undertaken as a means for rapidly generating new compilers as the old became obsolete. The universal test equipment compiler described in the paper entitled "UTEC—A Universal Test Equipment Compiler" describes such a compiler that has found successful application. Today there are some who argue that compilers are obsolete now that practical interpreters have been developed. Perhaps interpreters are preferable in many applications. Yet there is still practical value to an off-line program preparation tool as well as for on-line interpretation. In some instances use of both tools is the answer. Experience has shown that there is no such thing as the "best solution" for ATE. Each user must become knowledgeable in the available tools and techniques and decide for his own application which solution is best. Universal compilers have proven practical in some instances and therefore are worthy of study and consideration.

Simulators received substantial attention from ATE people in the 1960's. Very few ATE users undertook the development of a simulator but many investigated the practicality of such a software tool. Those who did try simulators were generally disillusioned with them. They were expensive to develop, difficult to maintain and operate, expensive to run, and yielded little help in the area of test program validation. The paper entitled "Conceptual Design for an Automatic Test System Simulator" reports on a concept for a simplified simulation. It attempts to minimize development cost by limiting simulation to gross performance characteristics of the test system with an absolute minimum simulation of the unit under test. While such a simulator cannot be used as a substitute for validation on a real ATE system, it does provide a cheap way to verify test program logic design and a means for eliminating a great many of the bugs which otherwise would waste substantial validation time. It can also automatically generate some support documentation useful to the design engineer during validation of his test program.

Recently, simulation has enjoyed substantial new interest by the ATE community. The new emphasis is on modeling the UUT for purposes of test design rather than for program validation. One paper, "Computer Aided Test Generation for Analog Circuits," describes a system derived from a circuit design model. To date, analog simulation has been only partially successful. The interesting aspect of this paper is that it describes a system that allows for manual test designs to be integrated with computer-generated test designs to become a complete test program. This enables the test designer to use a combination of manual and computer-assisted techniques using the better approach, as he sees fit, for each test to be designed.

The last paper, entitled "Establishing Test Generation Requirements for Digital Networks," describes a simulation system for digital test design. It has been selected as the most used and most promising of systems available. It has been proven successful for modeling the more complicated digital module designs represented by today's avionics modules containing up to 4000 NAND-equivalent circuits. The system has potential for even larger models, but with some expansion to the existing design. Whether the system described fits a particular user needs depends on the application. It is clear, however, that if a practical economical system is to be developed for designing test programs on a large scale, digital simulation must be part of the system. Experience has shown that improving manual design techniques will not provide any substantial cost reduction. In fact, the complexity of current and future UUT's is such that engineering analysis of the UUT and program design without the aid of a computer simulation may not be achieved at all. Certainly the quality of digital test programs prepared manually indicates that the manual approach cannot produce a test program that is adequate except for the simplest UUT designs.

SOFTWARE DESIGN TECHNIQUES FOR AUTOMATIC CHECKOUT

by

Dorothea H. Jirauch
General Dynamics, Convair
San Diego, California

The checkout of sophisticated electronic systems is becoming progressively more complicated. As hardware systems become more and more complex, the checkout procedures for these systems become more complex. Many systems have advanced to the point where it is no longer feasible to do the checkout manually.

A completely automated real-time checkout system on a large high-speed digital computer is becoming an accepted way of life.

Real-time computer technology is advancing rapidly. Progress is rampant in both the computer hardware and software fields.

The software field, itself, is relatively new. Its rate of growth has been phenomenal. The technology changes so rapidly that a sabbatical of a year could antiquate a programmer's basic skills. It is probable that this frenetic pace has produced a deplorable lack of organization in this field.

The purpose of this paper is to discuss some of the important aspects of the design of the software for an automatic checkout system.

Software System Components

The main sections of the software for an automatic checkout system are:

Test Programs

These are the modules that do the actual checkout. The test programs should be flow charted in detail as early as possible. The service modules in the operating system cannot be firmed up until the needs of the test programs are entirely specified. The test programs should be coded in a higher language that activates a set of service modules to accomplish the details. This allows for easy modification to the test programs. They should be coded, if possible, by the test-engineers. These modules need not be re-entrant provided every non re-entrant subroutine used is unique to the test program. The operating system should take the burden of real-time off the shoulders of the test-programmer. However, anyone tying into a real-time system should be aware of some basic principles, for example, disabling the interrupt system can cause chaos. In any structure, certain keypoints bear the bulk of the load. The proper positioning of these stress points is extremely important. They should exist somewhere in the operating system and not within the test programs or on the shoulders of the operator.

Interrupt Processors

The interrupt processors service the interrupts as they occur. If each interrupt has its own unique interrupt processor, the processor does not need to be re-entrant provided that any non re-entrant subroutine is unique to it. The interrupt processors must be concise and very efficiently coded. This implies symbolic level coding. Unless the time required by the interrupt processors is carefully evaluated, the interrupt system can use all of the computer time leaving none for the remaining requirements. An interrupt processor that does nothing but count in a clock cell every 0.1 ms can use 10% of the computer time. Every interrupt processor should be carefully analyzed for efficiency. Its necessity should be evaluated versus its time consumption.

Control Module

The control module is that portion of the operating system that controls the flow of priorities. It is advantageous to use an extremely flexible decision table algorithm. It should be very efficient. Elaborate control of twenty or more real time tasks can be achieved for about 1% of the computer time when properly coded. It should be able to handle both cyclic and non-cyclic tasks. It should have the capability of task addition or deletion during run time. The operator or the test programs should be able to influence its decisions. The "pushdown list" concept can be used for last on, first off storage of a real-time priority system. However, the more flexible "free list" concept is ideal for such capabilities as voluntary release of control from a test program with the ability to return to the suspended test program via the console.

Service Modules

The service modules perform specified tasks for the test programmer or the operator. Such modules should be re-entrantly coded. They free the test-

Reprinted from *IEEE Auto. Support Systems Symp. Rec.*, Nov. 1966.

engineer/programmer from the burden of real-time oriented problems and the many necessary housekeeping tasks. They can not only be linked in the traditional tree arrangement by the test programmer, but can be used as independently functioning mechanisms by the operator. These modules reduce duplication of coding. They need to be debugged only once. This results in efficient use of memory as well as man power. This logically oriented set of modules is open ended, thereby making expansion of the system very veasible. Much emphasis should be placed on good expansion capabilities. These modules reduce the probability of an error in a test program disturbing sections of memory associated with the hardware or with other test programs. They can furnish the interface and conflict protection between:

a. The operator and the control module.

b. The operator and the I/O.

c. The operator and the hardware.

d. The operator and the test programs.

e. The test programs and the control module.

f. The test programs and the I/O.

g. The test programs and the hardware.

These service modules allow the test programmer to influence the decisions of the control module. They can give the test programmer command of any hardware function, complete with verification and tontinuour safety checking, by a single reference. They can furnish the test programmer with the ability to connect or disconnect interrupt service modules and to arm or disarm any interrupt. They can give the operator run-time intervention without interfering with the real-time flow. They can also automatically send time-tagged messages to the operator when any significant action is taken or when any error condition develops.

I/O Modules

The I/O is the portion of the operating system that controls the flow of the input and output. Proper design of the I/O section in a real-time operating system is imperative and requires a very specialized knowledge. The I/O modules in a real-time system must operate in parallel with the computation, stealing only an occasional cycle for communication purposes. Buffering of the information to the I/O devices must be implemented to prevent jamming of the real-time flow by a temporary overload on one or more I/O channels. Sophisticated techniques for disc I/O can enable the system to keep information flowing from the memory to disc or from the disc to memory or tape without slowing down the execution flow of the non-I/O sections. Good paging techniques and the modular construction of the system minimizes memory fragmentation during test program interchange.

Simulator

The simulator is a software module designed to be a temporary substitute for the prime equipment (target hardware to be tested). Such a package allows the prime and any special purpose interface equipment to be dispensed with during the early stages of debugging. It makes trouble shooting of the software possible without complications from the new hardware. There also exists the very real possibility that hardware being developed concurrently with the software might not be available at this time. The simulator required by an automatic checkout system usually needs to be dynamic at run-time. It requires much ingenuity to develop a good simulator. The designer must be an experienced programmer with a strong hardware background. The ideal simulator should require no modifications to the coding of the software modules. Assembling of these modules in the simulator mode should be all that is necessary. This can be done by including a deck of special simulator-oriented definitions. The simulator can also be triggered by a clock interrupt to simulate real time in all except the I/O modules. This should uncover most of the real-time bugs. However, an I/O bound situation could be missed while in the simulator mode. It is difficult to conceive of an automatic checkout problem where a well-designed simulator, available at the proper time, does not save considerable man hours and calendar time.

Personnel Requirements

A large real-time automatic checkout system requires close cooperation between a group composed of prime equipment engineers, test-oriented engineers and system-oriented programmers.

The lead prime equipment engineer should have some test equipment and programming background. Such a background helps him to design the equipment to minimize complexity of the automatic checkout. It also enables him to communicate more readily with the test-oriented engineers and the system-oriented programmers.

The lead test-oriented engineer should have a strong hardware orientation as well as a well-rounded programming background. He supplies the bridge between the prime equipment designer and the software designer. He must be capable of working directly from the equipment diagrams and specifications and conveying this information to the lead programmer. Such a man is rare and hard to find but does exist.

The lead system-oriented programmer should have an engineering-oriented background, experience in real-time operating systems and a strong sense of organization.

The choice of these various lead people is so important that it is wise to let the schedule slip for several months if necessary in order to obtain the right people. A choice based on expediency can result in the schedule slipping many times that amount before the checkout system is completed.

A more desirable situation is to have the entire programming staff under the same management as the equipment design staff. Such an arrangement avoids a lot of problems and simplifies many it does not avoid.

Real-time automatic checkout is no place for the incompetent or the inexperienced. A poorly selected staff can give rise to problems that have no known solution short of throwing away the result of all previous effort and starting over.

Obviously the selection of staff members cannot be made on the basis of 'who is not too tied up to be thrown in the gap'. Such an attitude usually results in turning the gap into a deep chasm.

Since very close cooperation is necessary between the lead test-oriented engineer and the lead system-oriented programmer, compatibility of personalities is most important at this strategic point.

A staff of test programmers, preferably test-engineers, are needed to program the checkout tests while a staff of programmers are needed to program the operating system.

The lead system-oriented programmer should pick his staff most carefully. He should realize at the beginning that all of his own time will be spent in the design and coordination of the software effort. Any lead man that thinks he will have time to do any playing around on the computer, etc. is deluding himself.

Since the design and coordination of the project will require him to spend an appreciable amount of his time away from the vicinity of the software programmers, he should pick an extremely capable person to assist in the design of the software and to be second in command. This will assure smooth running in his absence and protect the project from his possible loss.

His next selection should be a competent I/O man. No other section of a real-time system is more critical than the I/O section. One hears of too many I/O bound real-time systems to doubt this. I/O men with real-time system experience were on the 'scarce list' last time I looked. It is a good idea to start looking for this one early in the project.

Another selection to be made early is a man to handle the software checkout. This man should be responsible for developing a simulator and supervising the entire checkout procedure of the software programs. Debugging a real-time operating system is such a tricky and time consuming factor in an automatic checkout, its outcome cannot be left to chance.

Both the I/O and the simulator man need more lead time than the remaining programmers since the software checkout phase is dependent on much of their work being operational.

Another important team member should be some one capable of redrawing and correcting flow charts, checking up on the state of the documentation, writing abstracts on the software modules, updating the manuals and other like tasks. This requires someone with good experience and capability. Usually such a person lacks only the formal education required for a more advanced position.

The remaining software programmers should be picked for their competence and their ability to produce software modules that have been carefully and precisely defined by the software system designers.

Care should be taken to choose a staff that have not previously had conflicts with each other and who are hard working and cooperative.

Each module should be assigned to a programmer and he should specify an alternate programmer to check his work. This drastically cuts the number of logical and mechanical bugs that get as far as the debugging stage. It also has the very desirable effect of supplying insurance against the loss of any one programmer.

It is a basic mistake to plan a structure in such a way that if one member is removed the remaining structure falls apart. Personnel backup should be always kept in mind.

New or Old Computer?

The selection of the digital computer for the checkout effort requires much thought. The hardware capabilities available on today's market are fascinating. In almost all large automatic checkout projects the latest, most up-to-date equipment that will be available in the next year is put on order. This is done despite the fact that very often a more than adequate, proven computer is readily available.

We now have the components to check out the new unproven hardware item. These components consist of an unproven computer complete with a magnificient new set of vendor-supplied unproven software, an unproven interface between the unproven hardware and the unproven computer and last, but not least, an unproven automatic checkout software system.

When a new computer is selected several months should be allowed in the checkout schedule for late delivery and validation. The new computer should be checked exhaustively with small diagnostic programs after installation and routinely thereafter. Nothing is as frustrating and time consuming as looking at complex software programs for bugs when the trouble is in the hardware.

A new computer is usually equipped with a new assembler, compiler, utility pack, etc. Even though the vendor makes a conscientious effort, the delivery date of his software usually slips. With luck these new software packages have all but a handful of the most insidious bugs removed. Watch out for those that are left! They can be so devious that only an extremely experienced programmer can pinpoint the source of trouble.

Adequate allowance should be made on the checkout schedule for an appreciable amount of computer down-time in the first few months.

Optimism should be kept at a minimum.

A competent programmer and engineer should be assigned the full-time job of getting the project through this trying period as smoothly and as rapidly as possible. Use of the machine should be declared off limits to the remaining staff until it is reasonably sound.

Despite all the difficulties involved in using a new computer, there are times when advancements in computer hardware have been sufficient to justify the decision to use a new one instead of a familiar, proven model.

Software Specifications

General purpose digital equipment requires software. Complexity of such software is sometimes greater than that of the hardware being tested.

This complexity has been increased by the fact that everyone concerned feels it is easier and less expensive to do things with software than with hardware. That this is not always the case is often overlooked.

Another contributing factor to the complexity of the software is that programmers seem to be a rather eager breed, anxious to show their ingenuity by performing miracles. This optimistic attitude often leads to their drastically underestimating the difficulties involved. The hardware engineers, equally optimistic, are so happy to have all their existing problems solved that they throw themselves energetically into the creation of new ones. The requirements for the software should be controlled as rigorously as are those for the hardware.

One of the most important decisions in the design of a software system is a definition of what constitutes a reasonable level of sophistication for the problem as it exists now and in the foreseeable future. The objectives of the undertaking should be clearly spelled out before any design takes place. Constantly changing goals can keep the design staff going around in circles.

Management should assign their most competent personnel to define the requirements of the checkout system. Tight control of these requirements should be exercised by the hardware personnel. Tight control of the implementation of the specifications should be maintained by the software personnel.

The key man in the preliminary stages of the design is the test-engineer who should be capable of communicating both with the hardware designer and the software designer.

In order to make a wise decision on the appropriate level of complexity in the checkout system, the test requirements must be defined in some detail.

The primary objectives are usually not too difficult to spell out but it is often difficult to obtain detailed information from the hardware designers.

A well-qualified test-engineer can glean much information directly from the specifications of the prime equipment but in no case should the prime equipment designer be ignored during this phase. Only the close cooperation of all concerned can give the desired result.

If the prime equipment designer feels time is too short for him to stop to spell out test requirements, the approach of presenting a proposed test to him usually produces the desired reaction. He promptly points out his objections, these can be corrected and the specifications firmed up.

When this approach fails, the software designer should never supply test requirements without proper input from the hardware people. Such a step leads to unnecessary recoding, resulting in additional cost and loss of calendar time, near what should have been the completion of the program. No definition is better than an incorrect or an inadequate one at this stage.

For the good of the entire project a hold should be placed on the definitions and management notified of the delay and the reason for it.

While early specifications of the software system is an essential feature, it is very important that such definitions are not too restrictive. Every design specification should expect modifications and allowances for them. A design, in a dynamic environment, without flexibility rapidly gets out of date.

Software Documentation

Software techniques have been coming into existence fast and furiously. However, a software algorithm does not present the same feeling of solidity that a piece of hardware does.

This has allowed a myth to be fostered upon the field of automated checkout. This myth consists of three basic precepts.

1. It is not possible to approach the design of a large real-time software system in an organized manner.
2. It is not possible to document a system until it is operational.
3. There is no need for documentation after a system is operational.

This amazing lack of an organized approach has been a handicap to much-needed progress. Good design procedures for a real-time software system differ little from those of a hardware system.

It is the direct responsibility of the lead software designer to see that the documentation is comprehensive, correct and timely. It should predate the implementation. No excuse should be recognized for his failure to do this.

A standard procedure should be set up for handling all documentation and rigidly enforced. One person should be assigned the semi-clerical job of checking up on the documentation. He should give every possible assistance to the programmer to ease the documentation load. This relieves the entire programming staff of much necessary, time-consuming boring clerical work.

Documentation should be filed as the problem progresses. Flow charts etc. should be thought of as belonging to the project rather than to an individual programmer.

The documentation should include:

Specifications

Specifications should define the interfaces between the hardware and software. They should also define the basic capabilities of the software. The software specifications should be formulated by a team of knowledgeable problem-oriented engineers and computer specialists. This group should be neither so large as to be unwieldly or so small as to be vulnerable to the loss of any one member. Two each of hardware and software men seems optimum for most applications. The problem-oriented engineers should make frequent and productive contacts with the prime equipment engineers during this formulation.

Dictionary

One of the first documentation efforts initiated should be the compiling of a dictionary of terms used in the program. Communication within a group always presents some problems. When the difficulties are increased by a whole set of new terminology that means one thing to one person and something different to another, much time can be lost. The time alone spent in the discussion of which definition is correct, without counting the loss of time due to misapprehension of a concept, can be significant.

Concepts

Any concept, regardless of how complicated it is, can always be subdivided into a set of less complicated but still completely logical concepts. In turn, any complicated logical concept in this set can be partitioned into logical subsets. This technique can be applied iteratively until the result is a group of logical concepts any one of which is well within the scope of a trained mind. These individual concepts are easily documented. Each concept as it is defined by the system designers should be recorded in detail. The sum total of these concepts must necessarily be the solution to the original problem. This is not an idealistic approach, it is good common sense. By the application of this packaging technique 95% of the software documentation can be layed out before coding starts. These explanations are the beginning of the documentation framework that will support the whole software project.

Structure Charts

Since the actual flow in a real-time program is

dynamic at execution time, a flow chart as such cannot be drawn for the highest level flow. This is replaced by structure charts that show the interaction of various parts of the system. Examples of this are interactions of the operator and the operating system, the interaction of the interrupt system and the control module, the interface between the control module and the controlled routines etc. These structure charts present an organized picture down to the level where flow charting is possible.

Abstracts

A form for abstracts should be laid out early in a project. Every module in the program should have an abstract. The designer has defined the modules by subdividing the concepts until each module is specified. At this time the purpose, the required input and the expected output should be spelled out in the appropriate sections of the abstract. A general description of the method, limitations and pertinent remarks should be entered by the programmer after the flow charting is finished. The status of the module should be updated on the abstract as it progresses from the flow chart stage, to the coding stage, to the debugging stage, to production.

Flow Charts

The flow chart is the most powerful single tool a programmer has. It is more flexible than any programming language. Flow charting should be done in detail by the programmer. He should review it closely with his alternate and have it okayed by the system designer. The programmer who codes without first flow charting feels he is too experienced to need a flow chart, but in reality he is not yet a mature programmer. He always reminds me of the boy on the bicycle who yells "Look Ma, no hands".

Listings

Listings of the coding of each individual module are part of the final documentation. While the coding is being done, comments should be sprinkled in liberally. A method of associating each box in the flow chart with the code corresponding to it should be standard practice in all modules. An identification can be written on each box and then used as the entry symbol for its coding.

Progress Charts

The current status and history of every known software module needed for the successful completion of the project should be kept on a chart. Such information as expected and actual completion dates of the design, flow chart, coding, debugging and the names of the programmer and the alternate should be kept current on the chart. There should be a progress chart for the debugging of every module. These can indicate if it has been checked on the simulator or not, which of the planned testing criterion has been passed, and so on. Listings should be reviewed in detail by the alternate programmer before debugging is started on the module.

Manuals

Nine-tenths of the user manual is automatically written when the specifications, dictionary, concepts, structure charts, abstracts, flow charts and listings are put into one book. The member of the group who has been doing the clerical work of keeping the documentation current should be able to write the remaining one-tenth of this manual with the assistance of the lead test-engineer. This manual should be test oriented and explain in detail the requirements necessary to make any test. It should list what should happen in normal circumstances and what might happen in abnormal ones and what action to take in these circumstances. It should also define the operator's interface with the operating system during testing.

Software Checkout

The debugging of the software for a large scale real-time automatic checkout system requires much thought and effort. It cannot be left to chance.

A group of programmers, test engineers and prime equipment design engineers can rapidly saturate even a high speed computer if left to their own devices. This saturation presents problems that can be avoided by careful planning and control of the debugging procedures.

A competent, experienced programmer should be assigned the full time job of developing and carrying out a complete debugging plan for the software. This assignment should be given as soon as the general nature of the checkout job is understood to give sufficient lead time for the debugging procedures to be operational by the time it is required.

Verification procedures must be thought out fully and carried out carefully. The verification of as many branches and combinations of conditions as practical can be accomplished only by a well organized and methodically executed plan.

The verification procedures must consider the interaction between the various components of the complete checkout system. These components can be grouped into software, computer, peripheral equipment, interface hardware and prime equipment categories. Intermittent errors in such a system are almost impossible to track down. The possibility always exists that the trouble is actually in an already supposedly checked out component.

One of the best techniques to isolate the software from the hardware bugs is to replace the interface and prime equipment with a simulator. This removes much unneeded complications.

If the computer and peripheral devices have also been given a clean bill of health by adequate testing, most programmers will start to look seriously at their own work for the difficulties.

The simulator should be designed, coded and relatively bug free before needed by the software system modules. This implies plenty of lead time in the debugging assignment.

In the simulator mode, all modules should be checked independently if possible, then in logical sets, and finally, all together. Only when the simulator mode debugging procedures have been carried out extensively should debugging involving actual prime equipment and interface equipment be attempted.

By no means should debugging be considered as anywhere near complete at this time. Hopefully, 90% of the programmer goofs are past history, but the remainder of the debugging is going to be rough.

Real time bugs not caught by the simulator are apt to start cropping up.

For the first time the communication link between the hardware designer and the software designer is being tested. This link is being tested by a software package (with possible bugs) on a hardware item (with possible bugs). Any trouble however might be neither in the hardware or in the software, but in a misconception on the part of the software designer about what the hardware designer said or meant to say.

Debugging at this point should proceed with caution to avoid injuring the prime equipment now tied into the system.

To minimize the debugging time in this stage, good relations and mutual respect should exist between the hardware and the software personnel. If everyone assumes the trouble might be his, it is traced more easily. Much frustration can be avoided by cooperation in these interaction areas. Necessarily much time will be lost looking in the wrong places, however if 'people interaction' develops, the debugging time will go up by at least a factor of four.

Allowances should be made for considerable down time when the prime equipment is under repair. Since most software and hardware systems are designed concurrently, hardware design errors as well as normal hardware assembly bugs should be anticipated. The necessary modification of the hardware and software to compensate for design modification can be time consuming as well as costly.

The software system designer must keep the problem of debugging in mind throughout the entire design. Modular design techniques, etc. are a considerable help in debugging. Strict supervision of documentation in both the software and hardware areas is mandatory. It is impossible to develop adequate debugging procedures without organization.

Conclusion

The successful design of the software for an automatic checkout system is comprised of many elements working together to form a whole. Some of the more important things involved are:

1. Good management, capable of making wise and timely decisions.
2. Strong technical leadership in both the hardware and software groups.
3. A hard-working staff with the proper experience.
4. Software designers with good organization and foresight.
5. A well defined set of specifications and ground rules.
6. Detailed definitions of all test programs and system modules.
7. An excellent documentation procedure.
8. Good communication amongst the personnel.
9. Cooperation, team-work and mutual respect between all members of the staff.
10. A strong debugging procedure.
11. Optimism (in the right places).
12. Work.

UTEC-A UNIVERSAL TEST EQUIPMENT COMPILER

by

Roland L. Mattison
and
Robert T. Mitchell

Radio Corporation of America
Burlington, Massachusetts

Presented in this paper is a compiler built for use with languages designed to control automatic test equipment. The compiler operates in two phases: first, in a generation phase, a set of tables is built to describe a source language, an object language, and a hardware configuration; then, in a translation phase, these tables are used to create an object program from a source language test program. The operation of the compiler and the means for specifying the tables are also discussed.

Introduction

When generating compilers for use with automatic test equipment (ATE), a substantial need arises for flexibility in both the source and object languages. Flexibility is desirable for two reasons: (1) The field of ATE construction is rapidly expanding, and (2) the hardware and support software design, development, and debug cycles are often simultaneous. In earlier, more standard compilers, the modifications and/or extensions of either language could easily create chaos for the systems programmer.

In an attempt to facilitate compiler implementation and growth, the Universal Test Equipment Compiler (UTEC) has been developed. To insure a large degree of flexibility, UTEC has been created with no particular ATE system in mind. Instead, the compiler can operate in a generation phase during which the configuration of the ATE system as well as the descriptions of the source language and object language are introduced into the compiler. This information is processed and stored into various tables. Once these tables have been satisfactorily produced, the compiler can operate in a translation phase. During the translation phase, these tables are used to transform source language test programs into object programs configured to the desired ATE system. A generation phase language has been implemented which facilitates the process of describing the various languages and the ATE configuration. This language will be referred to as the meta-language.

A typical ATE system consists of various programmable devices for applying stimuli to, and obtaining measurements from, a unit under test (UUT).[1,3] A requirement peculiar to ATE compilers is the creation of a wire list specifying connections between the ATE and the UUT. An equipment designator has been included in the system to handle the wire list and to insure that the wire list remains fixed despite source program recompilations. This is necessary due to the cost incurred in the production of this wiring.

The wide range of computers currently used in ATE dictates that the output of UTEC be a symbolically addressed code which must then proceed through the second pass of a normal two pass assembler. Since this reduced assembler could be different for each type of ATE, it will be excluded from the following discussion.

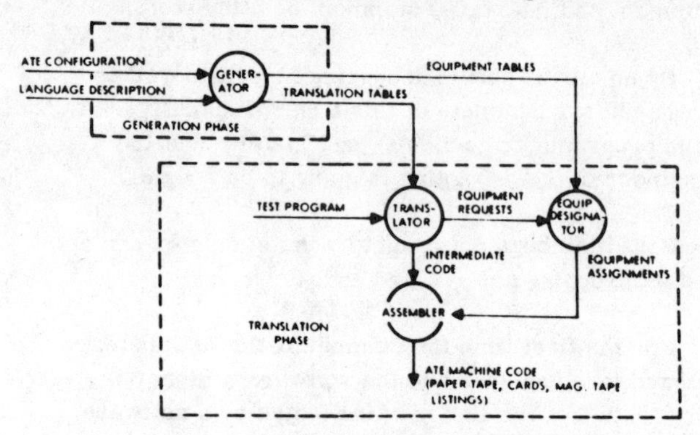

Figure 1. UTEC System Flow

The flow of information through the UTEC system is depicted in Figure 1. The source language description is defined to UTEC using the meta-language and is fed into the generator. From this, the generator produces translation tables for use by the translator. The generator also accepts a specification of ATE hardware configuration and produces equipment tables for the equipment designator. When a source program is input to UTEC for translation, the translator uses the translation tables and outputs an intermediate code ready for assembly. Whenever ATE equipment must be specified by the translator, it inserts a symbolic into the intermediate code, and requests the required equipment from an available equipment pool in the equipment tables. The request is tied to the intermediate code by the symbolic. The equipment designator now processes the equipment

Reprinted from *IEEE Auto. Support Systems Symp. Rec.*, Nov. 1968.

requests and, using the equipment tables, produces equipment assignments for each symbloic in the form of a matching list. The assembler processes the intermediate code from the translator and replaces the equipment symbolics with the actual equipment code as supplied by the equipment designator. The translation phase of UTEC-the translator, equipment designator, and assembler-is controlled by a monitor which takes control of all compiling options and controls the flow of information through the translation phase.

The following discussions describe each section of UTEC, and are intended to give the reader enough information concerning the details of its implementation to provide him with a knowledge of UTEC's capabilities.

The Meta-Language

We now present a language; SYNSEM (SYNtax and SEMantics) for explicitly defining a problem oriented language (POL)[4]. SYNSEM itself is a twofold problem oriented language which (1) specifies the syntax or format of a POL, and (2) specifies the semantics or meaning of the allowed formats of a POL. SYNSEM is therefore divided into two sublanguages, SYN for specifying syntax, and SEM for specifying semantics.

Problem oriented languages currently in use with ATE are tabular in format. The reason for this, and examples of such languages have been previously presented[5,7] and, therefore, will not be considered here. Let is suffice to say that fixed fields are generally adhered to, with one field set aside for the function or verb and the remaining fields for modifiers of various types. Each verb-modifier complex is referred to as a source statement.

The goal of SYN is to allow format syntax type information to be specified for each verb of the POL. This information is encoded into a table by the generator and is used by the translator whenever the verb is used in a source program.

SYN is comprised of various disjoint subsets of any commonly used character set. Three of the subsets are given below:

NUMBERS = |A,B,C,D,E,F,G,H,I,M,N,O|

LETTERS = |P,Q,R,U,V,W,Y|

MAIN UNITS = |K|

The syntax of each source language verb is specified by describing each of the allowed modifiers of that verb To specify a modifier which can only be a numeric quantity, a letter is chosen from NUMBERS and repeated so that the number of times the letter appears equals the maximum number of digits the modifier may contain. Alphabetic modifiers are handled in a similar way by choosing from LETTERS

If desired, a MAIN UNITS modifier may be used with any verb. When the letter K is recognized by the generator, the 4 characters immediately following the K are taken as the MAIN UNITS and entered into a dictionary with the verb. MAIN UNITS are used to further distinguish the verb when more than one source statement uses the same verb.

As an example, consider the following SYN statement to specify the verb CONNECT with the modifier VDC:

CONNECT AAA KVDC BBB PPPP

There may be two numeric (A and B) modifiers and one alphabetic (P) modifier used with this verb. The MAIN UNITS for this particular CONNECT is VDC.

Once SYN has been used to specify a given verb, a SEM "program" is written, later to be executed by the translator, which analyzes the verb-modifier relationship and generates the desired intermediate code for the source statement. The SEM language is composed of a number of semantic instructions, some of which are described below. A maximum of 750 such instructions can be used in any one SEM program. A statement in SEM consists of a semantic instruction followed by its possible modifiers. A three digit label is optional for all statements. A three digit branch is required with some instructions and optional with others. This branch, in conjunction with the label, provides symbolic addressing and conditional branching within a SEM program. The SEM instructions are divided into three categories: (1) Code producing, (2) Modifier handling, and (3) Control.

CODE, CVAR, CALPHA, and CSIGN are examples of four of the code generating instructions. CODE tells the translator to output to the intermediate code the characters which are literally specified with the CODE instruction. For example:

CODE 3 PS1

will cause the three characters "PS1" to appear in the intermediate code. CVAR, CSIGN, and CALPHA each are used with an identifier which the translator references to find the data to be output. These identifiers can be references to verb modifiers or variables established in the SEM language program.

CSIGN NUM1

will generate a + or - depending on the algebraic sign of the number stored in the SEM variable NUM1.

CVAR NUM1 4 2

will cause the numeric value of the variable NUM1 to be coded using four characters with two implied decimal places. If NUM1 = 46.913, the characters 4691 will be output into the intermediate code.

CALPHA W1 3

will cause the three leftmost characters stored in the variable W1 to be coded.

TEST, RANGE, and the four arithmetic operators ADD, SUB, MUL, and DIV are some of the modifier handling SEM instructions. TEST causes the translator to compare a referenced quantity with a group of characters specified following the instruction. RANGE causes a check of a referenced quantity to see if it is numerically between two limits. Execution of either a TEST or RANGE instruction by the translator can cause a branch in program flow to a labeled SEM statement if the comparison fails.

TEST A 3 AMP 40

causes a comparison of the three leftmost characters of the variable A with the three characters AMP.

RANGE B 20.0 30.0 40

causes a comparison of the number stored in B to see if $20.0 \le B \le 30.0$. **If the above comparisons are satisfied, the translator executes the next sequential SEM statement; otherwise, statement 40 will be processed next.**

JUMP, SCWL, ROUTINE and EQUI are each SEM control instructions. JUMP is used with a SEM statement label and causes an unconditional transfer by the translator to the labeled statement. SCWL informs the translator that all of the intermediate code generated for a particular source statement must be saved with a label for future use. The SEM language is provided with a subroutine capability through the ROUTINE instruction. ROUTINE may be followed by a parameter list of from one to seven dummy parameters A CALL instruction, followed by the actual parameters, is used to invoke a SEM subroutine. The EQUI instruction is used to cause the translator to generate a symbolic equipment request for the equipment designator. Its modifiers must be a unique symbolic, which will be placed in the intermediate code by an ECODE instruction, a type number referencing a particular type of equipment, and a set of "connections" to which a specific piece of equipment of the specified type should be wired.

A control card called REQUIRED is used between the two parts of the SYNSEM language and lists all required modifiers in the SYN portion. These modifiers must appear with each usage of the verb.

The following example shows the combined use of the SYN and SEM languages to specify the verb CONNECT modified by VDC.

```
   CONNECT    AAAA  KVDC   JC   JD
   REQUIRED   AC
   NEWN       UN1
   CODE       1     S
   RANGE      A     0      50   10
   EQUI       5     UN1    C    D
   ECODE      UN1
   CVAR       A     4      2    20
10 RANGE      A     51     120  30
   EQUI       6     UN1    C    D
   ECODE      UN1
   CVAR       A     5      2
20 CODE       2     ES          40
30 ERROR      A     OUT OF RANGE
40 END
```

The above program is suitable for input to the generator which would create the necessary table entries for later use by the translator after it sees the CONNECT VDC verb in a source program. For example, suppose the statement.

CONNECT 4.2 VDC J1-4 J3-5

was processed by the translator. The code produced by the previous definition would be

S:00005:0420ES

EQUIPMENT

SYMBOLIC

which would cause the ATE to connect 4.2 volts of direct current between points J1-4 and J3-5. The equipment symbolic number 00005 was produced by the SEM instruction NEWN.

To specify equipment in a particular ATE configuration, two SEM instructions are used: SYMBOL and DATA. All equipment is divided into types, e.g., power supplies, signal generators, voltmeters, and each type is given a number to identify it. One SYMBOL instruction and as many DATA instructions as there are pieces of equipment in a type are used to define that type. The SYMBOL instruction tells how many pieces of equipment in the type and how many connection terminals each has. Each DATA instruction gives an equipment name and the ATE connection terminals for it.

As previously pointed out, when a new POL or a modification to an existing POL is defined to UTEC by means of SYNSEM, the syntax of the language and its semantics are stored into tables by the generator. Since ease of language modification is a requirement, three of these tables have been implemented as linked lists.[6] The format list is used to hold the syntax specification for each verb in the language. The logic list is used to hold each SEM instruction specified in a verb definition. The logic modifier list holds SEM instruction modifiers which are not suitable for entry in the logic list. A dictionary is also used which contains the name of each function defined by SYNSEM as well as various pointers to the lists. A pair of equipment tables, the hardware name table and hardware usage table, are built by the generator to store equipment name codes making up an ATE configuration.

The Generator

The generator division of UTEC accepts the definitions of verb syntax and semantics written in the SYNSEM language, and assembles this information into all the necessary tables and lists. It also has the ability to delete and equate verbs in the lists, and to build the equipment tables. The generator is used whenever a DEFINE, EQUATE, DELETE, or EQUIPMENT control card is encountered and is divided into four corresponding sections as follows:

Define: Following the DEFINE control card, the SYNSEM language is used to define verbs. First the syntax of a verb is given using the SYN language. The verb is placed into the dictionary, the syntax specification is analyzed, and the format list is built. The symbolic character of each argument is entered into an argument table for later reference. At the completion of analyzing the syntax, the argument table contains the one letter symbolic of each argument, in the order in which they will appear in the source statement. The REQUIRED control card containing the one letter symbolic of each required argument follows the SYN syntax specification. The format list is modified to indicate which arguments in the syntax are required with each usage of the verb.

After the REQUIRED control card is processed, the generator must load the SEM program, which gives the semantics of the verb, into the logic and logic modifier lists. When a modifier of a verb is referenced by a SEM instruction, the one character symbolic of the SYN language is used. This character is looked up in the argument table and its integer position number is used for the logic list entry. When a source statement is parsed by the translator, each argument is loaded into a table at the same position as is used for containing its SYN symbolic character in the generator. This allows each SEM instruction to have access to source statement modifiers. Variables may be established in the SEM language by a symbolic name. This symbolic name is placed into the argument table after the symbolic SYN modifier characters, thus establishing a reference location for logic list entries and for storage of variables during translation. Each source statement is translated independently of all others, in that no variables of one SEM program are passed on to another. However, the SEM instructions SX and TX provide storage locations for use throughout an entire source program compilation, giving each definition access to the same location and providing for exchange of information between definitions. If one entry in the logic list is not sufficient to contain all the necessary data for a SEM instruction, a pointer to the logic modifier list is placed in the logic list, and as much space as necessary is used in the logic modifier list. This feature is used whenever large strings of information are used with instructions, i.e., giving entire error messages. A two digit op code for access by the translator, and a branch and link address are always in a standard location in each logic list entry.

Equate: Many verbs in a particular POL developed for use with automatic test equipment are similar in syntax and semantics. For example, the source statement for connecting a stimulus to deliver volts is very similar to the statement for connecting kilo-volts or milli-volts. The equate section of the generator was therefore developed whereby two or more verbs may share the same definition, and therefore the same list area. The name of the verb to be equated is placed in the dictionary and all the information associated with the equated verb is used with the new one, thereby using the same definition. In order that the small differences of the two verbs can be taken into account, a SEM instruction called CHFLG is used in the original definition. This instruction requires two indicators which are stored in the dictionary. A define type definition always sets them to zero, but they may be set to any desired value by the language designer using the EQUATE option. The CHFLG op code can test the value of these indicators and thereby set up branching logic in the SEM language program to control the translation.

Delete: The purpose of this section is to remove previously defined verbs from the dictionary and to restore their list space for future use. This is done by simply naming each verb to be deleted following a DELETE control card.

Equipment: All equipment which the ATE system has available is defined in the SYNSEM language by the SYMBOL and DATA instructions. These instructions follow the EQUIPMENT control card. All the equipment is divided into classes or types, each type having some commonality. For instance, all power supplies having the same characteristics of output and number of terminals may be classed as one type. The SYMBOL instructions gives class information and precedes DATA instructions giving the names and terminal connections of particular equipment in that class. In this section of the generator, the equipment names for each class are stored in a hardware name table, thus building equipment pools from which equipment selections can be made, and work area is allocated in the hardware usage table based on class specifications. Both of these tables are used by the equipment designator.

Translator

The analysis and translation of source programs and the subsequent output of intermediate code is handled by the translator.

When analyzing a source statement, the verb is first checked against the list of defined verbs in the dictionary. When a match occurs the associated dictionary entries are used as references to the format and the logic lists where information specified by the SYNSEM program for this source statement verb has been stored by the generator.

Using the format list as a guide, the translator parses each source statement and creates an argument table. Each modifier of the source statement verb is stored in this table in such a way that it can be referenced by the SEM program as explained in the generator section. The scan finishes when the entire format list for the verb has been considered.

Once the format scan has been completed, the translator turns its attention to the logic list where the algorithm for generating intermediate code has been stored for this verb. Each entry in the logic list is a numeric representation of one of the SEM instructions. A two digit op code is extracted from each entry which identifies the particular SEM instruction requested. Once the instruction is known, the translator is able to completely dissect the logic list and modifier list entries for this instruction and perform the desired operation. All references by the instruction to the verb modifiers or variables are made by simply referencing the argument table position for that modifier, or variable, as the parsing algorithm already has inserted the modifiers in the table. Consider the following SYNSEM specification:

```
CONN        AAA    KVDC     JC     JD
  .
  .
  .
RANGE        A     10.0     20.0   100
```

The generator creates the argument table as shown in Figure 2.

Figure 2. Argument Table of Generator

Position	Contents
1	A
2	K
3	C
4	D

Since the character A is in position 1, this position number was used in the logic list when the RANGE instruction was processed by the generator. When the source statement:

```
CONN   14.6   VDC   J101-42   J16-33
```

is parsed by the translator, the argument table is filled as shown in Figure 3.

Figure 3. Argument Table of Translator

Position	Contents
1	14.6
2	VDC
3	J101-42
4	J16-33

When the translator discovers the RANGE instruction number in the logic list, it decodes a reference to position one in the argument table for the number it is to test. In the example, the number 14.6 is checked to determine if it is between 10.0 and 20.0.

Each logic list entry provides the translator with the position of the next instruction to be considered, or, in the case of conditional instructions, the translator must pick the next instruction from two or three choices after it performs the current instruction.

The translator continues through the logic list until the END op code is discovered, at which time it has completed its analysis and code generation for the source statement under consideration. The next statement is read and the entire process repeats. When the translator reads the END verb, it has completed the intermediate code for the source program.

Equipment Designator

Each time the translator processes a source statement which requires the use of ATE equipment, an entry on a tape is generated by means of the SEM instructions EQUI or PREAS. This tape is called the request tape. The translator itself has no ability to select equipment from the available equipment pool in order to satisfy the needs of the source statement. The SEM language program used to translate these source statements requiring equipment first generates a unique symbolic number which will be used by it to symbolically refer to an equipment name in the intermediate code it produces. It then determines the type of equipment required by the source statement and generates the request tape entry. Each such entry generated tells the type of equipment desired and the symbolic number used to identify it, and tells how the terminals of that equipment should be connected. In the case when a specific piece of equipment must be used in a particular manner, the translator also processes an equipment preassignment by producing a request tape entry which gives the specific name of a device and tells how it is to be connected.

The function of the equipment designator is to read and process the entries on the request tape produced by the translator. It attempts to match an equipment name of the correct type to each symbolic number and produce information describing how each piece of equipment is to be connected. In the assembly of the intermediate code, each symbolic number is replaced by the matching equipment name as provided by the equipment designator.

The equipment designator operates using two tables: the hardware name table and the hardware usage table. The hardware usage table is divided into two sections for each equipment type: the hardware assignment section and the hardware request section.

The hardware assignment section for each equipment type contains one assignment indicator and terminal connection space for each piece of equipment of that type. The indicator gives the availability status of the equipment at all times. The connections made to each piece of equipment are stored in the space allocated when the equipment is used. The hardware request section for each equipment type can contain a number of requests for equipment of that type. A request consists of storing the unique number along with terminal connections desired.

The requests processed by the designator fall into two classes: (1) Those which name specific pieces of equipment, and (2) those which symbolically seek an assignment of any piece of equipment of a specified type. When fulfilling requests, two passes are made over the request tape with the items in classes one and two being handled on passes one and two respectively.

On pass one, the designator simply reads the requests, and in the hardware assignment section, sets the indicator for the named piece of equipment to reflect a used condition and stores the connection references. Before the first pass, all indicators reflect an equipment available status. After pass one, the indicators of the equipment named in pass one are set to indicate a used status.

On pass two, the designator tries to assign one piece of equipment to each symbolic request. In addition, it creates the matching list to be used by the assembler when processing the symbolic references in the intermediate code.

Each request causes a scan of the hardware assignment section for the type of equipment requested. If the connections of the request match those of a piece of equipment already used, the request is matched with that equipment. If the connections of the request do not match those of any already used, a new piece is assigned to match this request. If all the equipment of the type requested has been used, the request is put into the hardware request section and saved. When the entire request tape has been read and processed in pass two, the designator must process the unfulfilled requests remaining in the request section if any exist.

The manner in which these unfulfilled requests are handled attributes to the uniqueness and purpose for the equipment designator. A source program update feature exists, the use of which will be discussed with the monitor. This update will allow a source program to be recompiled without significantly changing the equipment assignments and wiring between ATE and UUT. This is possible because of the equipment designator.

In the first compilation of a source program no unfulfilled requests remaining in the request section can be satisfied. However, at the end of an initial source program compilation, the equipment designator produces equipment assignment source statements called update preassignments which state how each used piece of equipment is connected. Through control of the monitor, these update preassignments are the first source statements processed in the next compilation or update, and produce class one requests which name specific pieces of equipment. Therefore, in update compilations, the first pass of the equipment designator sets the status indicators of the previously used equipment to indicate an update preassignment.

When unfulfilled requests do exist on update compilations, the designator scans the assignment section for all equipment which was update preassigned but not used in pass two of the equipment designator operation. The indicators of these equipments are reset to reflect an available status. An attempt is then made to assign the unfulfilled requests to the equipment made available. If the request still cannot be satisfied, it remains in the hardware request section as an error condition. This procedure insures that equipment will be assigned in the same manner as the previous compilation as far as possible since equipment requests are first matched against previous assignments. Only minimum changes are made to satisfy unfulfilled requests on source program updates.

A wire connection list is produced from the hardware assignment section giving all the equipment used and how it is to be connected. All changes to the equipment usage from the previous compilation are listed. Error statements are produced based on entries remaining in the hardware request section. The equipment designator also produces new update preassignments at the end of each compilation, and passes these on to the monitor for use in later updates.

An additional feature of the equipment designator is its ability to connect equipment of one or more types in series or parallel, and to substitute equipment of another type if the pool of one type of equipment has been exhausted. Both of these features are controlled through class two requests from the translator. Because of hardware characteristics, only certain classes of equipment can serve as substitutes for other classes, and this information can be specified through the meta-language, thus controlling the translator produced requests. How equipment is connected together is also specified at the meta-language level. The meta-language can thus control the translator to interact with the equipment designator in any desired manner, giving a large degree of flexibility in equipment usage.

The Monitor

The monitor is the bookkeeping and control section of UTEC. It guides a source program from its various input forms through compilation and produces various types of output of the object program.

On an initial source program compilation, the source program is input in card form. Each source statement is given a sequence number and processed through the translation phase. A typical ATE test program is subdivided into many discrete tests, each given a name much like statement numbers in digital computer programs. One form of UTEC output controlled by the monitor is a test-by-test listing which is a merged listing of source and object code on a test-by-test basis. That is, each source language test is immediately followed by the object code produced for it. This is an aid during the program validation process. Other outputs from UTEC include an equipment connection and usage list produced by the equipment designation as previously mentioned.

The monitor stores on magnetic tape a copy of the source program with sequence numbers. The update preassignment source statements generated by the equipment designator are stored preceding the actual source statements on this tape.

An update to the source program is accomplished by simply stating which sequence numbers of statements are to be deleted, between which two sequence numbers new statements are to be added, or which statements are to be changed. The monitor produces a new source program for input to the translator using the stored program from the last compilation. In this manner, the update preassignments are processed as described earlier, followed by the source program as it existed on the last previous compilation, with the indicated changes having been made by the monitor.

The monitor also provides for communication between UTEC and the user concerning various options available. UTEC can be instructed to automatically include standard equipment self tests for each piece of equipment it assigns in the object code. The monitor can also provide straight source or object language listings in addition to the test-by-test listing. The user can instruct UTEC, through the monitor, as to the form of the object code output; i.e., paper tape, punched cards, or other.

Conclusion

At this time, UTEC has been completely written and checked out using FORTRAN IV, and a language developed for use with one type of automatic test equipment (LCSS) currently being produced by RCA has been implemented. The implementation of another language for a second type equipment is now being considered.

It is interesting to note that after having defined the language to UTEC using the meta-language, the users could evaluate the quality of the language and its usefulness, and suggest changes and improvements. These changes were easily incorporated into the language almost daily during a shakedown period, thus allowing them to be tested within days after they were conceived. The overall effect was to stimulate ideas for improvement. As a result, a language much more effective than that originally specified was developed.

References

1. Evanzia, B.J., Automatic Test Equipment; A Million Dollar Screwdriver, Electronics, August 23, 1965.
2. Ingerman, Peter Z., A Syntax - Oriented Translator, Academic Press, 1966, ch 1, pp 13-19.
3. Mayper, Victor, Programming for Automated Checkout: Part I, Datamation, April 1965, Vol. 11, No. 4, pp 28-32.
4. Mayper, Victor, Programming for Automated Checkout: Part II, Datamation, May 1965, Vol. 11, No. 5, pp 42-46.
5. Ryle, B.L., The Atoll Checkout Language, Datamation, April 1965. Vol. 11, No. 4, pp 33-35.
6. Wilkes, M.V., Lists and Why They are Useful, Proc. ACM 19th Natl. Conf., August 1964, Phila., Pa.
7. Scheff, B.H., Simple User Oriented Compiler Source Language for Programming Automatic Test Equipment, Communications of the ACM, April 1966. pp 258-266.

CONCEPTUAL DESIGN FOR AN AUTOMATIC TEST SYSTEM SIMULATOR

by

F. Liguori
Software Systems Engineering Laboratory
Emerson Electric Company
St. Louis, Missouri

Abstract

This paper describes the concepts pursued in the design of a "software" simulator for a large-scale, multiputpose, automatic test system. The test system (General Purpose Automatic Test System) has been in field use for several years, and the demand for new operational programs has increased beyond the available machine time to validate the programs. Hence the simulator must provide a pratical means for prevalidating programs intended for the real test system.

This unique application of a simulator has many interesting features including complete software construction, modularized organization, and automatic generation of signal tracings and flow charts. The simulator is operable on a general purpose computer and requires little input data other than the machine language program normally produced for the real test system.

Introduction

Simulators are being successfully applied to the solution of a large variety of problems in many diverse disciplines. In practically all applications the role of the simulator falls into one of two general classes as follows:

1. Environmental simulation - Where an existing device, system or man-machine combination must be tested prior to application in the real environment in which it must ultimately perform.

2. Model simulation - Where a design concept is tested prior to commiting the necessary resources required to build a real system.

There is a third possible class of simulators which could be used to model a fully developed system. First impression of such an application is that while it is possible, it would be impractical if not purposeless. For one might logically argue, why simulate what one already has? Is not the real thing always a more perfect model of itself than any simulator, no matter how sophisticated? Of course any simulator or model must be a less than perfect duplication of the real thing or it would not be a model but an exact replica, fully equal to its real-life counterpart. There are, however certain characteristics of some real things that do not lend themselves to experimental or developmental activities as well as a less exact but more flexible model. The field of operational program development for certain classes of automatic test systems offers a great potential for the practical application of just such a simulator. This paper represents a conceptual design and rationale behind functional requirements for an automatic test system simulator currently being developed at Emerson Electric Company under direction of the author. The system being simulated is the AN/GSM 204(v), commonly called GPATS (an acronym for General Purpose Automatic Test System). This system is shown in Figure 1. The concepts developed in this paper are purposely general so as to be applicable to many Multipurpose Automatic Test Equipments, commonly called ATE.

Reprinted with permission from *Proc. ACM National Conv.*, Aug. 1969.

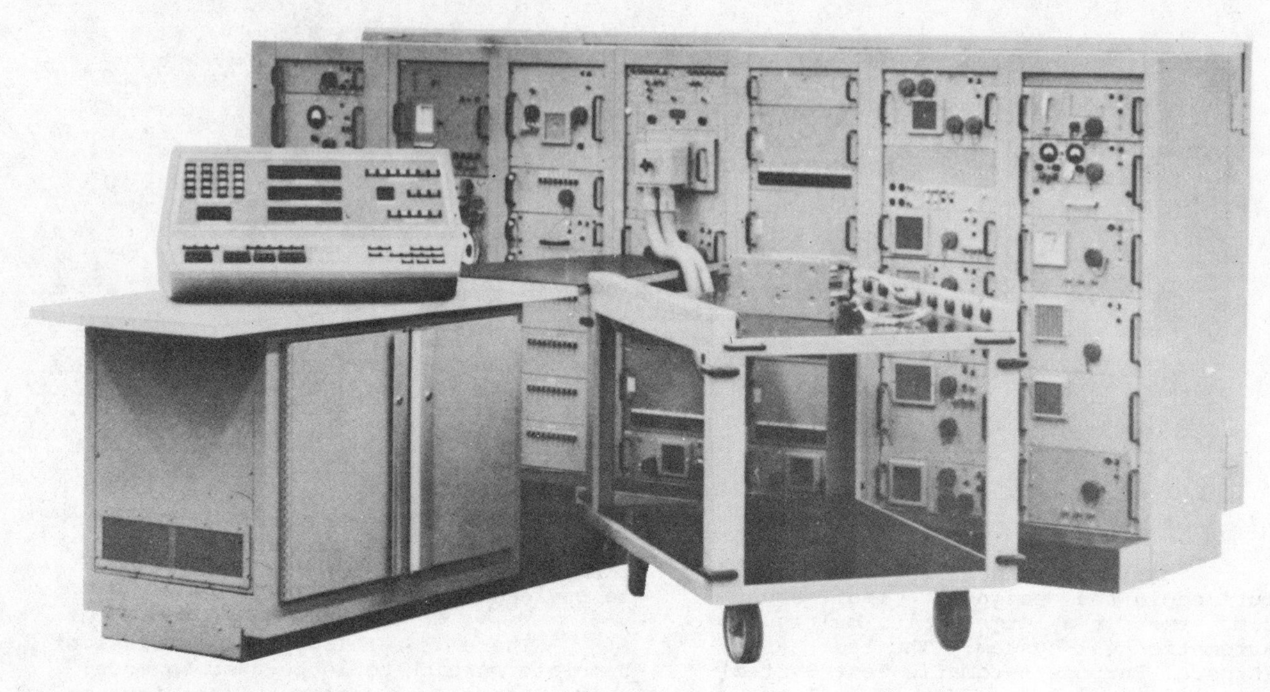

Figure 1

General Purpose Automatic Test System (GPATS)

Simulator Characteristics Determination

Simulator, like models, must admit to being something less than that which they are to simulate. Those characteristics of the "real thing" which are not realized in the model need not represent serious shortcomings, however. By reducing the real thing (the ATE), to a list of functional characteristics, a hierarchy of importance can be ascribed to its characteristics. The relative importance of these characteristics must be based on the purpose of the simulator rather than the ATE. In ATE applications the purpose of the simulator is to check out programs which must eventually play on the ATE. Hence, it is the programmable characteristics of the ATE that must be given top consideration in simulation. In contrast, the ability of the ATE to withstand severe environmental conditions in its real application is irrelevant to the task of debugging programs. Hence,by selectively choosing the pertinent ATE characteristics of the simulated system, simulator complexity can be substantially reduced without materially affecting its intended purpose. This indeed is the secret of good modeling. It is this approach that allows a new aircraft design to be effectively tested in a wind tunnel using a precisely shaped skin of the aircraft model without simulating any of the complex controls internal to the real aircraft.

The other basic requirement of effective simulation is that the simulator be relatively easy to use. This consideration places certain operational restrictions on the simulator design. For example, a restrictive characteristic of an ATE simulator could be that the control statements required to exercise a given UUT program designed to test a particular unit must be substantially less trouble to prepare than the task of exercising the program on a real ATE. A functional block diagram of an ATE with a typical Unit Under Test (UUT) connected is shown in Figure 2.

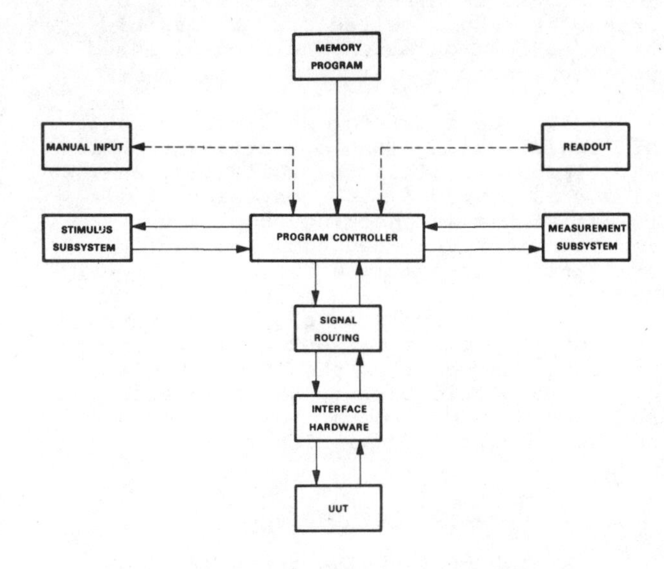

Figure 2

Simplified Block Diagram of an ATE

The Application

To answer the question "Why simulate what one already has?" requires an understanding of what is different about programming multipurpose automatic test systems. In the vast majority of computer applications, both business and scientific, programming the system is essentially a non-recurring process. Once the operating system is developed, and operational routines debugged, the programmer's task is complete. Debugging will taper off and though never really ending in some applications, very little additional creative programming is done except for occasional expansion of the system application. This is not so with Multipurpose Automatic Test Equipment (ATE). Such systems are virtually complete laboratories of complex electronic equipment, power supplies, signal generators and measurement devices tied together by a complex of cabling and switching and all controlled by a stored-program operated device. The system may utilize a specially designed programmer controller (PC) or a conventional digital computer. In either case, it is the program that makes the system go. It controls all of the electronics involved in stimulating a Unit Under Test (UUT), receiving and evaluating the responses and drawing logical conclusions on the performance and/or failures of the UUT.

The typical multipurpose ATE may require several hundred operational programs for testing UUT's before it can meet its minimum requirement for maintenance support of a tactical system such as the avionics complement of a modern military aircraft. In fulfillment of its multipurpose objective, the ATE must support several tactical systems concurrently and provide for future applications as well. It is precisely for this flexibility that programmed ATE systems are proving to be such a valuable maintenance tool in the rapidly changing world of electronics.

Today the limiting factor in evermore successful use of multipurpose ATE is not the technological constraints on system design, but rather the ability to continually produce the tremendous quantities of UUT programs required in reasonable time at reasonable costs. Ultimately the limiting factor in program production is the availability of ATE system operating time to validate (debug) new programs. This problem is particularly acute when the ATE must serve both the production testing needs and the validation time requirements. Even with a system dedicated to the validation effort, experience shows that between 10 and 45 minutes of ATE time is required to validate a test. Here a test is defined as one complete electrical set up, signal application, measurement and evaluation of a UUT parameter. On the average, this requires 10 to 15 operations or programmed words. With the typical UUT requiring 100 tests or more, one rapidly approaches a need for several multi-million dollar ATE systems to turn out validated programs in the quantities anticipated for some applications. The ATE industry is thus confronted with the dilemma of having to continually produce tremendous quantities of operational programs in order to be effective while program production increasingly encroaches on ATE operating time. Simulators must provide the solution to this dilemma if ATE is to reach its full potential.

Requirements of an ATE Simulator

A number of functional needs and operational restrictions can be defined for an ATE simulator. The primary functional requirements are:

1. A program "pre-validated" on the simulator must require substantially less validation time on the ATE than if it had not been exercised on the simulator.

2. Cost of operating the simulator must be substantially less than for a real ATE.

In addition, there are several very desirable characteristics as follows:

1. The simulator should be structured entirely of software. This eliminates maintenance and allows almost instant reproducibility of the simulator, should more than one be required.

2. Virtually all inputs to the simulator should be derived directly from the UUT program. This substantially eliminates simulator user errors and minimizes the demands placed on the user. Thus the simulator provides a real service and not an additional programming burden.

3. The simulator program should be executable on readily available, general purpose data processing systems.

4. The simulator should be modifiable. Thus changes must be easily incorporated in a user-oriented language when improvements and model redefinitions are called for.

5. The simulator should be expandable, as in the ATE system it simulates. Thus, provision must exist to add new models of "Building Blocks" just as can be done to the ATE hardware configuration.

6. Simulator outputs should be in a readily useable form. Logic violations should be directly identified and irregular usages called out as possible error sources on an engineering-oriented hard copy output.

7. The simulator should be capable of "logic tracing" and outputting trouble-shooting aids. Thus the validating engineer is provided with convenient tools to assist him in checking complex testing functions that can not be directly evaluated by the simulator.

8. The simulator should run through the entire program without human intervention. An operator should not be required to participate in the simulation run once the UUT program has been loaded on the computer for simulation.

Simulator Design

A simulator design for achieving the stipulated requirements is shown in the simplified block diagram of Figure 3. The diagram shows the functional design of the simulator, much as if it were a physical system. Actually, all elements shown are simply software models of the required functions all of which are integrated and controlled by the "program executive". The executive draws from the subroutine library as required for repetitious tasks. In the discussion that follows, each function in the diagram is related to requirements of the simulator called out earlier.

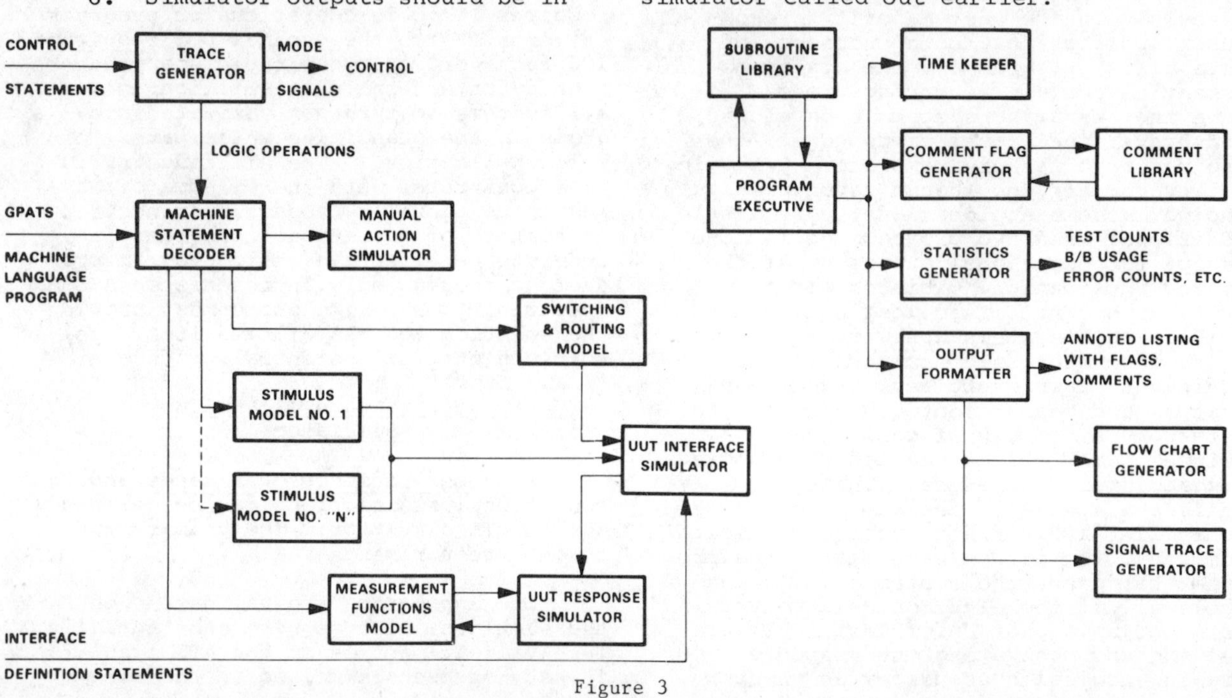

Figure 3

Simulator Block Diagram

The bulk of the simulator program is comprised of a number of building block models. Each building block model represents a corresponding chassis or functional section of the ATE system. Typically a GPATS configuration can vary from fourteen to forty three such building blocks. Typical building blocks include a variable DC voltage supply, high frequency oscillators and a multimeter. The actual building blocks are separately programmable so a model simply simulates the programmable characteristics of a given building block. By representing each building block with a separate software model (sub-program) not only is a closer simulation achieved, but the software simulator also has the advantages of modular design enjoyed by the GPATS. Expansion of the simulator (or re-configuration) to match variations of GPATS systems is thus easily achieved by adding corresponding software building blocks.

In order to simulate execution of commands by building blocks, the simulator contains a machine statement decoder program which behaves as the hardware system "controller". It interprets the commands in the UUT program and calls for the various building block models as required. Rather than actually utilize the applicable building blocks, the program version of the controller evaluates (through legality checks) the called for function. Violations of the programmed rules are stored in the comment flag generator to be printed out at a convenient time.

Some commands call for utilizing more than one building block at a time. In those cases, the switching and routing model is employed to logically tie elements of the system together. Again, an evaluation of the commands is made rather than actually responding to them as is done in the real system.

Measurement commands interrogate the measurement functions model where called for measurements are evaluated against program limits given in the UUT program. Checks are made to ensure the legality of called for measurement functions and that the limits of the metering subsystem are not exceeded.

The interface simulator provides a means for considering the effect of standard interface circuits or elements which might be required in the testing environment. Generally, these effects are neglected because of the individual attention required to simulate each UUT requirement. Such data are not automatically derivable from the UUT program and thus present a real burden on the simulator user. Hence, interface simulation is neglected except when specifically entered by input control statements to the simulator.

The same problem of tailoring would be required for each UUT response. To dynamically simulate each UUT would be a task more complex than human evaluation of the UUT program without the aid of a simulator. Hence, only nominal response values are used to check out UUT program logic. The UUT program contains upper and lower limit values for each measurement evaluation. If for example, a resistance is to be checked for 5 $\pm$ 1 ohm, the upper limit would be stored as 6 ohms and the lower limit as 4 ohms. In executing the program a value within these limits is a "go" which calls for proceeding to the next performance test in the sequence. A UUT response greater than 6 ohms would be evaluated as a "high", calling for a different path in the program to be taken to diagnose the problem with the UUT. Similarly, a value of less than 4 ohms would follow the "low" path. To simulate the UUT program run, each possible path must be traced. Rather than having the user input actual values for UUT responses which would at best be estimates of realistic responses, the responses are automatically generated by the simulator. This is done by adding +1 to the high limit (6 in the example) and adding -1 to the low limit. These sums are then used to simulate UUT response along with either limit value to trace the high, low and go paths respectively. Selection of the trace to be followed is an option left to the simulator user. He may elect to trace only the go path, all paths or specified routing for called out test numbers. Hence the simulator can exercise all paths of the UUT program by deriving simulated UUT responses from the UUT program itself.

As each command is interpreted by the simulated decoder, a predetermined, typical execution time for the called out function is sent to the timekeeper. Each of these is summed to predict UUT program execution time for the trace called out in the selected simulation mode. This provides a very close approximation of the time the program would require for execution on GPATS. Note that the real time of the simulator run can be much shorter than required in the real system. Electrical signal settling times in the real system often call for millisecond delays between events. The simulator has no such settling problem, hence it can operate on an accelerated time scale, yet keep a record of predicted GPATS real time needs.

Information collected from the simulator run is stored on magnetic tape and dumped via a line printer, either on or off line with the data processor used in simulation. The output formatter arranges the data in easily interpreted form with flags and comments associated with violations and suspected problems. From this printout the user evaluates the UUT program run and is able to identify a large percentage of the programming problems without having used the GPATS system for the evaluation. Eventually a clean run will be achieved and the program is then tried out on GPATS. Obviously the more complex testing problems will not have been identified by the simulator run so some on-line GPATS time is needed for final validation. This on-line time, however, is greatly reduced.

An additional feature of the simulator is its ability to automatically generate a flow chart of the UUT program it evaluates. This is simply a graphical reproduction of the trace called out in the selected simulator operating mode. Also, signal tracings for each test can be automatically generated which show signal flow through building blocks. Both diagrams are a great aid in analyzing the more complex test set ups in the UUT program. These tools greatly reduce problem analysis time when validating on GPATS. To manually generate these diagrams would be very time consuming and error prone, yet this is what the engineer must do to resolve some of the more difficult problems during validation.

Finally, the simulator can accumulate some of the knowledge gained on each UUT run and output helpful statistics such as the number of tests using a given GPATS function or a complete matrix of building blocks usage for any program.

In summary, the simulator provides two very important services:

1. It automatically identifies programming or logic errors in a UUT program without need for GPATS.

2. It provides flow charts and signal tracings to assist in validation tasks on GPATS where the more complicated test problems must ultimately be resolved.

These services are provided by the simulator using simply the same machine language UUT program for input that is normally generated in the program production process. A few mode selection statements are the only original coding required for any simulator run. Because the simulator is simply a program model of GPATS, many simulators may be employed in parallel by duplicating the master tape and renting additional data processors for simulator program execution.

Bibliography

R. J. Meyer, "Computer Controlled GPATS", IEEE Proceedings of 1967 Automatic Support Systems Symposium.

R. V. Jacobsen, "Digital Simulation of Large Scale Systems", A.F.I.P.S Conference Proceedings, Volume 28 (1966).

W. M. Syn, "DSL/90-a Digital Simulation Program for Continuous System Modeling", ibid (1966).

R. D. Brennan and R. N. Linebarger, "An Evaluation of Digital Analog Simulator Languages", I.F.I.P., 1965 Proceedings, Volume 2 (1965).

V. C. Rideout and L. Taverini "MAD BLOC, Simulation", Volume 4, No. 1 (January 1965).

John J. Clancy and Mark S. Findeberg, "Digital Simulation Languages: A Critique and a Guide", A.F.I.P.S. Proceedings, Volume 27, Part 1 (1965).

W. Metcalf "Progress Report on GPATS", IEEE Proceedings of 1965, Automatic Support Systems Symposium.

R. D. Brennan and R. N. Linebarger, "A Survey of Digital Simulation: Digital Analog Simulator Programs", Simulation Volume 3, Number 6 (December 1964).

F. Lesh, "Methods of Simulating a Differential Analyzer on a Digital Computer", ACM Journal, Volume 5, Number 3 (1958).

COMPUTER AIDED TEST GENERATION FOR ANALOG CIRCUITS

A. M. GREENSPAN
L. J. RYTTER

AAI CORPORATION
P. O. Box 6767, Baltimore, Maryland

INTRODUCTION

There has been a great deal of work done to automate the process of anlyzing and preparing tests for digital circuitry. There are numerous digital support programs that have been developed and are now actively marketed such as TESGEN, LASAR, ALICE and FAIRSIM. These programs and other digital computer aided design (CAD) programs have received considerable attention because of the increasing variety of jobs that digital circuits are being applied to and because added size, complexity and use of digital circuitry made automation of both the design and analysis task almost mandatory for efficient handling.

Analog circuitry has been ignored to a great extent during the period when digital CAD programs were being developed. There was some work done on analog CAD programs such as ECAP, AEDCAP, TRAC, SCEPTRE, SYSCAP and others. However this work was primarily directed toward the problem of circuit design rather than analysis and test. There was some justification for the lack of attention given to the analog problem. The general consensus was that everything was going to be digital and that analog circuits would decrease to a relatively insignificant few. The few analog circuits that were left could, the story went, easily be analyzed and tested manually. Consequently, no automation would be required. The predicted disappearance of analog circuits has not taken place; nor have the circuits that have been developed proven to be amenable to easy analysis and test.

Many groups in the electronics industry have followed the development of CAD applications programs. None with more interest and expectations than those involved in developing Automatic Test Equipment (ATE) test programs. A number of the digital CAD programs cited above were, in fact, developed by ATE manufacturers for the purpose of helping them to turn out quality test programs in a reasonable amount of time. However none of the analog CAD programs were developed by ATE manufacturers; which explains in part, why they did not find any direct application toward the development of analog ATE programs. The lack of attention given to analog CAD programs by ATE programming groups continued long after it was clear that a tool was needed to help reduce costs, raise the quality and increase the speed with which these programs were prepared. Neglect by ATE programmers of analog CAD is also attributable to the fact that, unlike some of the digital circuits, which were so tedious that manual programming was impractical; analog circuits could generally be programmed in a time and for a price that the market would accept. However this willingness of the market place to accept the cost and time constraints necessary to purchase an analog ATE program has been eroding. It has been necessary for ATE programming to develop ways to improve their analog programming techniques. This need has led AAI Corporation to look to the potential power offered by analog CAD programs to help improve the productivity of our ATE programming staff.

ATE Programming

The task of preparing an ATE program can be divided into five major tasks; Analysis and Program Design, Interface/Adapter Design, Clerical and Support (coding, key punch assembly, procurements, etc.), Validation, Documentation and Demonstration. A brief description of each of these and their relationship will be given, to allow the reader to gain some insight into the ATE programming process and thus be better able to understand how AAI has used analog CAD programs to make the process more efficient.

Analysis and Program Design

This is the heart of the ATE program generation phase. The ATE programmer must take the available documentation (ie, test

Reprinted with permission from *WESCON Conv. Rec.*, Sept. 1973.

specifications, schematic, maintenance manual, etc.) and with good engineering judgment and analysis, mold this data into an ATE program. This involves selecting the performance and diagnostic tests as well as the test limits. This also involves analysis of the UUT failure modes in order to establish the manner in which UUT failures manifest themselves as well as determine the nature of the ATE fault isolation steps which will be necessary to identify the point of failure. The final product of this process is an English language flow chart.

Interface/Adaptor Design

The interface is used to resolve compatibility problems. Currently, all multipurpose ATE systems utilize some sort of general purpose interface to adapt the UUT to ATE and augment the ATE capabilities. The use of interface adapters is so widespread that they are often considered a part of the ATE system, but actually the interface design is as much a part of the software preparation process as the test program itself. The interface interconnection diagram adds further detail to the flow chart by identifying specifically, routing, switches, stimuli, measurement test points, and input/output terminals which may have only been described symbolically in the flow chart.

Program Clerical and Support Activities

Program clerical and support activities consist of the process of converting a completed UUT English Language Test Procedure into a machine coded program ready for validation on the ATE. This includes transformation of the program into the medium used to operate the ATE, i.e., magnetic tape, paper tape, mylar, etc.

Validation

The validation process involves exercising the UUT on the ATE under control of the test program. This is the only means for conclusively proving that the program set is ready for field use. This process includes the inserting of faults into the UUT to verify that the diagnostic portion of the program functions properly.

Documentation and Demonstration

The final step in the program preparation process is demonstration. It is at this time that the test program's ability to meet system support requirements, and function as an integrated set (e.e., tape, operating documentation, and interface hardware) is proven. It is also the time at which the documentation prepared to support the test program, including listings, flow charts, test plans specification, TRD's, etc., are verified against the final product.

The times typically allocated to each major tasks in the process is shown in Figure 1.

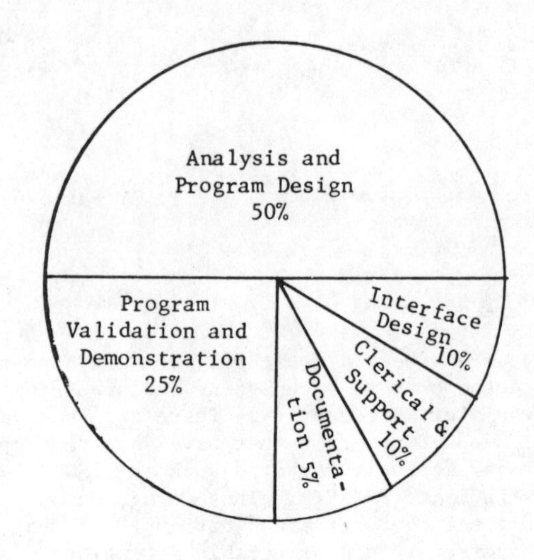

Figure 1

Application of CAD

The five tasks described above can in turn be divided into creative and repetitive catagories. When a task has been identified as repetitive, it can be made amenable to automation by rigidly defining its form and format. Creative tasks require a skilled human. The CAP (Computer Aided Programming) system, as it is currently developed, aims primarily at the Analysis and Test Design phase of ATE program generation. In this area, the creative tasks involve determining the optimum test approach and techniques for a specific circuit. It also requires recognition of unique circumstances and configuration which must be specially treated for automation.

Redundant and well defined tasks involve calculation of circuit parameters for given conditions at relevant circuit nodes, and sorting the results of these calculations into meaningful groups. Another redundant well defined task involves drawing the English language flow chart. Certainly drawing and interconnecting the proper symbols is a task suitable for automation. Approximately 40% of the analysis and test design task can be catagorized as repetitive and well defined therefore amenable to automation.

The time required to validate and demonstrate an ATE program is directly related to the validity of the analysis in the Analysis and Test Design phase. The use of CAD to do the analysis should reduce the amount of validation time; the degree to which this is true is difficult to determine.

Thus it seems reasonable that significant savings are possible through automation of the repetitions and well defined programming tasks involved in preparing analog automatic test equipment programs.

CAP

One method for automating the task of ATE programming involves the application of CAD programs. Intuitively it seems that it should be possible to apply reverse engineering to a given circuit using CAD programs in order to extract the data necessary to write functional and diagnostic tests of circuits. However, there has been little success in attempts to harness the power of these programs to the ATE programming task. One reason for this is the lack of an effective processor which could take the data extracted by the CAD programs and effectively apply it. In addition the type of data extracted, the form and format of this data and the circuit models upon which the CAD programs are based is not compatible with the requirements for automatic testing.

Recent developments at AAI Corporation have solved many of these problems. At present CAD programs are being used to automate many of the tasks necessary for development of functional and diagnostic test programs for analog circuits tested on automatic test equipment. This paper will describe the techniques and procedures which are being used.

CONCEPT

Development of a quality automatic test program for both detection and isolation of faults requires that the test program designer have three basic pieces of information:

1. GO/NO GO limits of tests under consideration.

2. A Fault List

3. And the effects of the faults on the circuit.

With this information, the program designer can select the minimum number of tests which will detect all faults. These tests are by definition the best functional tests. Next, the designer can select a set which will isolate to a satisfactory level in the minimum number of tests. These secondary tests are by definition the best diagnostic tests. (Consideration should also be given to the accessibility of nodes where the tests are to be conducted.)

The AAI CAP System uses Computer Aided Design (CAD) to generate the GO/NO GO limits and to simulate the faults in a circuit. A Fault Sort Algorithm is necessary to convert the CAD program raw data output into a useful form. The Algorithm developed for the CAD system processes the output data from the worst case, faulting and sensitivity runs to determine the effect of faults on the circuit.

ANALOG CAP SYSTEM STRUCTURE

The analog CAP system works in conjunction with a computer aided design program. The CAP system processes input data and the output data for the CAD program in order to convert what was primarily a design tool into a useful ATE program generation tool.

A block diagram of the system is shown below.

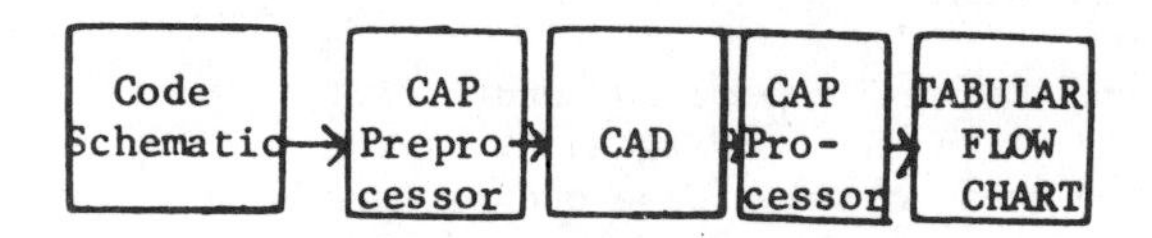

Analog CAP System Diagram

CAPABILITIES OF ANALOG CAP

The analog CAP system is presently designed to fault analyze DC and steady state AC circuits. Capabilities are:

1. CAP can handle analog circuits containing 300 fault parameters and up to 100 tests for detection and diagnosis of faults. (A fault parameter is a fault mode of a component.)

2. The desired level of diagnostics is a program input variable which can be selected by the UUT programmer.

3. CAP takes into account both out-of-tolerance faults and catastrophic faults.

4. CAP optimizes the UUT program by selecting the best tests and discards redundant tests or tests that do not produce any new information.

5. CAP generates the GO/NO GO limits and simulates the faults on the circuit.

6. CAP processes the data to determine if a fault is detectable under the set of tests specified to the system.

7. Test equipment accuracies are taken into account.

MAJOR FEATURES

The Analog CAP system has three major features: DC fault analysis, AC fault analysis and a Fault Sort Algorithm. The CAD program provides the analysis and the CAP System converts the analysis data into a condensed ATE program.

DC Fault Analysis

To conduct a DC analysis, the CAP system uses the DC voltage at the nodes of the circuit as its information for detection and isolation of faults.

The implementation of this analysis requires two passes through the CAD program. One to obtain the worst case information at the nodes of interest and the other to obtain the component failure data. The data is conditioned and stored in files for use by the Fault Sort Algorithm feature of the CAP system.

The UUT may be analyzed under different external operating conditions by making additional sets of worst case and failure runs after changing the input circuit description. All information from previous runs is saved and new information is added to the old.

AC Fault Analysis

The AC fault analysis is similar to the DC fault analysis except that the CAP system uses the amplitude of an AC signal at the nodes of the circuit as its information for detection and isolation.

The UUT may be analyzed under different external operating conditions (i.e. different frequencies, etc.) just as in the DC fault analysis. The saving of information from previous runs is the same as the DC case also.

Fault Sort Algorithm

The Fault Sort Algorithm section of the analog CAP system is the heart of the system. This is the section that takes the raw, bulky computer aided design output data and converts it into a useful form. The Algorithm optimizes the ATE program by selecting the best tests and discarding redundant tests or tests that do not produce useful information. The Algorithm output contains a tabular flow chart, a parameter number/component name cross reference, test limits for each test and a test number/node number cross reference (Figure 2). This output is used to determine if more data is needed to obtain the proper isolation and which components could not be isolated.

USING ANALOG CAP

CAP is used in both the interactive and batch mode. The user submitts commands to the CAP system in the interactive mode and the CAD analysis is done in the batch mode. This is the most cost-effective operating procedure.

Describing a Circuit

A circuit is described to the analog CAP system by creating a circuit file with a four-character name in a form acceptable to the CAP pre-processor and computer aided design program (Figure 3). Information on how to do this is available in a CAD reference manual. All components should be identified by numbers, except components which are external to the circuit under consideration. These external components should be identified with the component type character followed by "L" and a number. This is done so that fault modes of the external components are not considered.

Entering the Analog CAP System

To enter the analog CAP system; first log on to the computer system (In current use is the Univac 1108 computer) via a suitable terminal and get into the control level. In the control level, the computer system responds with "!".

Type in "AAI" and depress the return key. This enters the operator into the CAP system.

Preliminary Data

In order for the CAP system to determine what data, if any, has been saved from previous runs, the operator is asked for his circuit name. The operator should respond with the four-character circuit file and depress the return key.

If a file of the circuit nodes of interest has not been created, the CAP system will query the operator for this information (Figure 4).

TEST	NØDE
1	VN1
2	VN2
3	VN3
4	VN4
5	VN5
6	VN6
7	VN7

Fault Sort Algorithm Output

Figure 2

TEST AND RESULTS	CØMPØNENTS	PARAMETERS
4 0, 5 0	5	1, 4, 5, 16, 17,
4 0, 5 1	4	1, 2, 3, 5, 20,
4 0, 5-1, 7 0	4	1, 2, 3, 5, 10, 13,
4 0, 5-1, 7 1	1	7,
4 0, 5-1, 7-1	4	1, 2, 3, 5,
4 1, 3 0	3	1, 3, 5, 10, 14,
4 1, 3 1, 5 0	2	1, 5,
4 1, 3 1, 5 1	1	3,
4 1, 3 1, 5-1	3	1, 5, 19,
4 1, 3-1	2	11, 18,
4-1, 3 0, 5 0	2	1, 5,
4-1, 3 0, 5 1	3	1, 5, 8, 21,
4-1, 3 0, 5-1	2	3, 13,
4-1, 3 1	0	0,
4-1, 3-1, 5 0	2	1, 5,
4-1, 3-1, 5 1	4	1, 5, 6, 9, 12,
4-1, 3-1, 5-1	2	3, 15,

TESTS NEEDED

3, 4, 5, 7,

NUMBER ØF TESTS NEEDED.... 4

PARAMETER NØ.	PARAMETER NAME	CØMPØNENT NAME
1		R1
2		R2
3		R3
4		R4
5	HFEN	T1
6	SH	C1
7	SH	C2
8	SH	D1
9	SH	D2
10	CS	T1
11	ES	T1
12	ØP	R1
13	ØP	R2
14	ØP	R3
15	ØP	R4
16	ØP	C1
17	ØP	C2
18	ØP	D1
19	ØP	D2
20	CØ	T1
21	EØ	T1

TEST	NØMINAL	MINIMUM	MAXIMUM
1	1.173E-12	1.167E-12	1.179E-12
2	1.173E+00	1.167E+00	1.179E+00
3	5.867E-01	5.839E-01	5.896E-01
4	5.425E-01	5.360E-01	5.492E-01
5	6.415E+00	6.153E+00	6.655E+00
6	8.000E+00	7.920E+00	8.080E+00
7	1.283E-05	1.169E-05	1.397E-05

Fault Sort Algorithm Output
Figure 2 (continued)

The coded input file would be:

```
R1,6-2=6E3(5)
R2,5-6=2E3(5)
R3,4-0=680(5)
RL,7-0=2E3(5)
C1,1-2=1E-7(10)
C2,5-7=1E-7(10)
R4,8-6=1E-3(5)
D1,2-3=1N914
D2,3-0=1N914
T1,2-5-4=2N2222
E2,0-8=6(1)
AE1,0-1=.1(1)
```

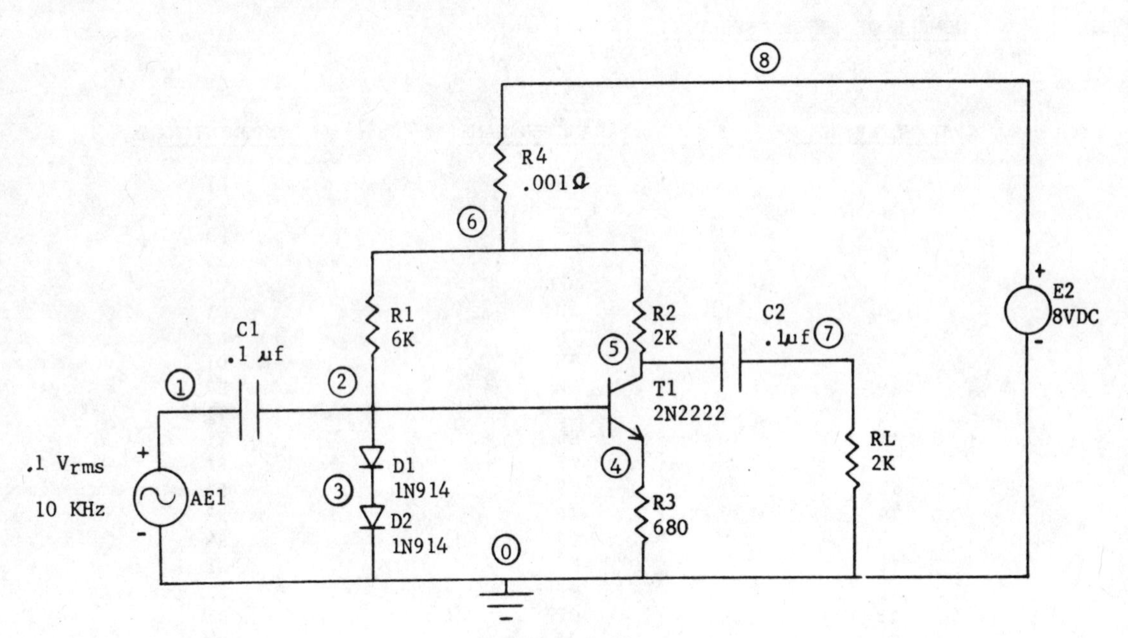

Sample CAD Coding

Figure 3

```
!AAI
CIRCUIT NAME: CADT
NØDE FILE DØES NØT EXIST, CØNSTRUCT NØDE FILE
NØDE: 1
NØDE: 2
NØDE: 3
NØDE: 4
NØDE: 5
NØDE: 6
NØDE: 7
NØDE:
```

Figure 4

Options

Once the preliminary data has been inserted, the CAP system will respond with "OPTION :". At this point, the operator can specify which option he wants. Some options are dependent on the running of other options first. If the needed options have not been run, an error message will be printed out. The possible options are:

(1) DCWC - DC Worst Case (Figure 5)

(2) DCFL - DC Failure (Figure 6)

(3) ACWC - AC Worst Case

(4) ACFL - AC Failure

(5) ALGR - Fault Sort Algorithm (Figure 7)

(6) FLCH - Flow Chart (Not Completed)

(7) LANG - ATE Language (Not Completed)

```
ØPTIØN : DCWC
GIVE TIME LIMIT IN CPU SECØNDS FØR RUN
LIMIT : 50
WANT ØUT ØF TØLERANCE FAILURES ?
ANSWER YES ØR NØ : YES
WANT ØUTPUT ØN LINE PRINTER ?
ANSWER YES ØR NØ : YES
SRU'S:4.7
ASSIGN PRIØRITY TØ BATCH JØB
PRIØRITY : 9
GPS362    TRUN-PNC/LIB$      19799
```

DC Worst Case Option

Figure 5

```
ØPTIØN : DCFL
GIVE TIME LIMIT IN CPU SECØNDS FOR RUN
LIMIT : 50
GIVE FAILED CØMPØNENTS IF ANY
CØMPØNENT :
WANT ØUTPUT ØN LINE PRINTER ?
ANSWER YES ØR NØ : YES
SRU'S:3.8
ASSIGN PRIØRITY TØ BATCH JØB
PRIØRITY : 9
GPS362    TRUN-PNC/LIB$      20059
```

DC Failure Option

Figure 6

```
!AAI
CIRCUIT NAME : CADT
ØPTIØN : ALGR
 STØP
SRU'S:5.5
 STØP
SRU'S:8.3
 STØP
SRU'S:4.8
 STØP
SRU'S:8.1
WANT ALGØRITHM ØUTPUT ØN LINE PRINTER ?
ANSWER YES ØR NØ : YES
GPS362    L$$-PNC    16858
```

Fault Sort Algorithm

Figure 7

CONCLUSION

Experience at AAI has shown that the use of CAP is an effective means for reducing the cost and time required for preparing programs to test analog circuits.

At present the use of CAP must be judiciously handled due to problems which have not, as yet, been overcome. Some examples of these problems include modeling non-standard circuit components properly and obtaining proper data concerning component parameters.

The effect of improper modeling can result in the CAD processor not converging to a solution or incorrect solutions for node voltages. Improper parameter data will also distort the results of the CAD processor.

The result of the problems described above and other minor problems has served to limit the scope of circuits that can be programmed with CAP. However, development of the system is continuing to extend the types of circuits that can be handled and to expand and improve the model library. It is clear that application of the potential power of analog CAD programs supported by appropriate processors is a feasible and practical method for reducing the time necessary to perform the well structured and redundant tasks involved in the preparation of analog ATE programs.

ACKNOWLEDGMENT

The authors would like to thank W. A. Scanga for his helpful comments in during the preparation of this paper.

ESTABLISHING TEST REQUIREMENTS FOR DIGITAL NETWORKS

J. J. Thomas, President
Digitest Corporation, P.O. Box 10611, Dallas, Texas 75207

INTRODUCTION

A variety of automatic test stations are now available, all capable of applying thousands of stimulus patterns and monitoring their response to provide a thorough test for digital networks in a matter of seconds. Unfortunately, the derivation of such stimulus is not such a simple matter, and until recently often required more analysis than went into the design of the network being tested. By sacrificing test quality, however, it was possible to reduce this effort by a factor of about 20 to 1, as was almost inevitably done for the larger networks.

For example, in the past the engineer might derive 500 stimulus patterns which by inspection seem to exercise "most areas of the network to some extent". He might then simulate these stimulus to obtain the responses, completing the test generation in about one month, for a 100IC network (100 integrated circuits ≈ 1000 elementary gates). Since it would take many times that effort to evaluate his test for actual percent detection, and since he himself had very little idea what it was, it generally passed as the 90% to 100% required.

Recently, however, techniques have been developed to evaluate tests for large networks. For example, a D-LASAR subsystem "ISOGEN" generates a Fault Isolation Dictionary and evaluates percent detection on a detailed gate fault basis for networks to 400ICs in less than one hour 1108 or 360/65 computer time. These evaluations have shown that while manually prepared tests for small networks (<30ICs) sometimes rate 90%, those for networks greater than 100ICs seldom rated better than 50% detection. Such evaluation had been out of the question by the various "fail simulators" because they inherently ran 100 to 500 times slower.

Systems such as D-LASAR now allow the test engineer to avoid tedious volume analysis and instead apply his expertise to special-case network problem model work-arounds. In return, the system does the following:

1. Derives stimulus which create and relay sensitized paths to the output for high quality fault detection by analysing the network's complex sequential logic.
2. Accurately simulates the stimulus to provide reliable responses, taking into account initialization (propagation of unknown states, and known relationships between them), timing, race conditions, and asynchronism.
3. Generates the Fault Isolation Dictionary and lists undetected faults by determining which faults (on an elementary gate equivalent basis) cause which response patterns/pins to fail.
4. Provides fault isolation on the station, fully automated.

As a result, fault detection covering better than 98% of FA1/FA0 faults, and statistically 95% of occurring failures (including dynamic failures, intermittents, shorted adjacent wires, multiple failures, majority loading, shorted input junctions, etc.) is obtained, as well as an average actual fault isolation to better than 2 ICs. For a 100IC network the effort is in the range of one man-month (experienced) and two hours of 1108 or 360/65 computer time.

To obtain the same quality in the past would have required either about 2 years of labor, or about 1 year labor plus about 60 hours of computer time, or roughly 20 times the effort, unless as previously indicated 40% detection quality was allowed to pass for 100%, in which case the one month labor again applies. With such a 20/1 schedule dependency on invisible quality, unsatisfactory testing was the natural by-product, and ATE testing was generally not as effective as the technician's manual test ingenuity.

ATE testing effectivity can be assured today only if the following ground-rules are adhered to:

1. Percent detection must be made visible by automated analysis.
2. "Fault" as used in percent detection evaluation must be defined as elementary gate (AND, NAND, OR, NOR) node junctions FA1/FA0 (static failures at 1 and 0). Thus, e.g., the 9-nand functional equivalent of a JK flip-flop shows it to have 32 unique detectable failure modes (all verified actual) as opposed to the 2 to 9 failures normally attributed to it.
3. Races, station skew, asynchronism and initialization factors must be made visible by automated means and corrected in order to obtain station test repeatability, which is otherwise a major problem.

The feasibility of meeting modern test requirements is now quite dependent on the application of automation aids, and conversely, an understanding of the latest aids is essential to the establishment of such requirements. Toward this end the basic performance functions of the D-LASAR system shall be discussed, after first considering the need for high quality testing.

Reprinted with permission from *WESCON Conv. Rec.*, Sept. 1973.

YIELD VS. PERCENT DETECTION
(See derivations, Appendix A)

If, say, digital component ICs off the production line have a fault-free yield of about 85% prior to testing, and if production line testing quality is such that 98% of all faults are detected, then of those passed, $.85^{1-.98}$ or 99.675% are actually fault-free. If 200 such ICs comprise a sub-assembly (SRA) then $.99675^{200}$ or about $e^{-200(1-.99675)}$ or 52.5% of such SRAs will be fault-free prior to SRA testing, not counting SRA packaging faults. If SRA test quality is 95%, then $.525^{1-.95}$ or 96.83% of those passed are actually fault-free. If six sub-assembly SRAs comprise an assembly (WRA), then $.9683^{6}$ or 82.43% of such WRAs will be fault-free prior to WRA testing. Finally if WRA testing detects 70% of all faults, then of those which pass, $.8243^{1-.7}$ or 94.38% are actually good, and only 1 in 18 WRAs exposed to use are faulty. In practice this quality is difficult to attain for large systems.

On the other hand, if the more typical test qualities of 95% for ICs, 50% for SRAs and 30% for WRAs are applied, then after testing, only 3.3% of the WRAs passed would in fact be fault-free. Such ATE test quality is obviously not competetive with the manual technician who is left to correct 29 out of 30 WRAs, each containing an average of 3.5 faults! Even if the SRA test quality were 90%, 50% of the WRAs would be put into use faulty. With a 95% SRA test the WRA casualty would still be 1 out of 3, though a 97% IC test would reduce this to a tolerable 1 out of 5.

Since automatic test generation aids do not yet directly handle networks over 400ICs, it is rarely feasible to obtain quality WRA-level tests, and is thus essential that 95% detection or better be obtained on the sub-assembly SRA level. But 95% detection of occurring failures requires about a 98% detection of all detectable ideal FA1/FA0 failures (undetectable failures are those which can never affect the output). Assuming 15% of occurring failures are non-ideal, and that tests which detect 98% of ideal failures detect 90% of the non-ideal failures, then an average detection of 97% follows. If we then allow that an estimated 2% of occurring failures are detectable only by dynamic testing (not normally practiced on ATE) then the over-all figure of 95% results.

FORMER TEST GENERATION AIDS

At one time, the digital simulator was the "solve-all", used to perform the following functions:

1. Verify digital design
2. Obtain response to given test stimulus
3. Simulate failures so as to evaluate which ones affected responses, and were thus "detected". Also by recording which response patterns/pins were affected by which failures, a Fault Isolation Dictionary resulted.

However, item 3 above was feasible only for networks smaller than 30ICs. For example, a 200 IC network has a NAND gate equivalent of typically 2000, and requires that about 2000 stimulus test patterns be each simulated against its 5600 unique actual FA1/FA0 failure modes. At .1 seconds per simulation $\frac{2000 \times 5600 \times .1}{3600}$ or 310 hours 1108 computer time is required. For 30ICs the time is reduced to $(\frac{30}{200})^3 \times 310$, or 1 hour. (This limitation was not overcome until the brute-force concept of fail simulation gave way to that of propagating criticality information from latch to latch thru the network in a single pass thru the stimulus, as D-LASAR's ISOGEN does.)

The simulators usually operated on User-prepared logic equations, or else made calls to component subroutines. Handling of initialization, races, and asynchronism was in most cases very limited, and accuracy and test repeatability suffered.

Since these simulators did not operate on a gate-equivalent basis, they had no realistic means of identifying actual component failure modes, and thus in most cases recognized only IC interface node failures, so that the existence of 50% of the unique hardware failure modes was generally not even acknowledged. Of course, with fewer failures, the fail simulators ran faster, and various rationalizations were made to support this approach, a reasonable compromise at that time.

Stimulus which detected 100% of the acknowledged failure modes, did by accident detect about half of the unacknowledged modes, realizing an actual 75% detection (when 100% claimed). However, isolation was almost nil for unacknowledged failures since no corresponding response pattern/pin failure matches were to be found.

The principal task, however, is not the simulation of existing stimulus, but rather the derivation of stimulus which will cause all failure modes to subsequently alter one or more response patterns/pins and thus be detected. If the network were purely combinational (no memory) then the application of all 2^n possible patterns (n=number of input pins) would create all possible network states and thus detect all consequential faults.

However, since most networks are sequential, sensitized paths must be relayed critically from memory latch to latch until they reach the output, each relay requiring another stimulus pattern. Also, prior patterns are required to establish memory states at a known and desired state invariant to their initial state. Some networks require the derivation of thousands of self-initializing sets of stimulus. A single stimulus set may contain 1000 stimulus patterns, and yet detect only a few additional failure modes.

Some earlier attempts were made to "derive" stimulus by simulating trial-and-error stimulus, saving only useful patterns. The success of such systems was very limited, since if there are m ICs, then there are likely about m latches; and self-initializing stimulus sets are likely to be roughly $2^{\sqrt{m}}$ patterns long. But since each pattern in a set may be chosen from any of 2^n (n=number of input pins) the situation is analogous to that of a combination lock having 2^n numbers on its face, and requiring $2^{\sqrt{m}}$ turns. To try all combinations would require $(2^n)^{2^{\sqrt{m}}}$ turns or patterns. For 30ICs, 10 inputs, this amounts to $(2^{10})^{2^{30}} \approx 10^{135}$ patterns. Obviously, analysis of the lock mechanism, or network, is preferable to simulating trial and error patterns.

The next question is, "should the analysis be done by a forward trace or reverse trace". Detection of, say, a Nand-Node junction failed at a 1, requires that sufficient stimulus be provided to normally set it to a 0, and that additional stimulus be provided to relay that 0 state occurrence critically from latch to latch until it hits an output pin. If stimulus were being derived by a forward trace, a corrective reverse trace search would be required each time a criticality linkage is broken, and would thus result in far more reverse traces than the basic reverse trace approach itself.

It seems desirable, then, that the stimulus generator use basically a reverse trace mechanism, and that it generate self-initializing stimulus sets.

Effective stimulus generators are few and in general were not available previously. The basic hurdles which had to be overcome were computer run time and the ability to analyse large complex sequential networks.

FAILURE MODE IDENTIFICATION

It is imperative that the actual failure modes of a component be acknowledged, so that tests may be developed which are assured of detecting and isolating those failures when they occur. It has been shown exhaustively by Hayes[1] and others that the logic of a component or network can be modeled in a NAND (or NOR, or AND-OR-NOT, etc.) equivalent form, and that, assuming a correct functional equivalent was used, the NAND-node junction failures FA1/FA0 do in fact correctly represent the basic failure modes of that component or network.

For example, the NAND diagram of figure 1 is a reduced, minimal functional representation of the more common master-slave JK flip-flop.

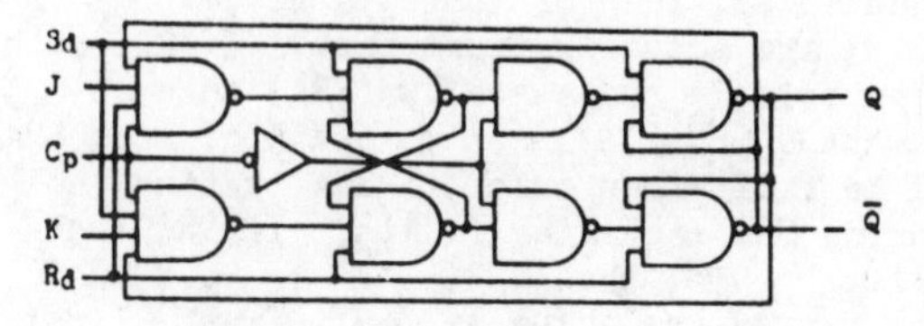

Figure 1

It has 25 nand input junctions to FA1 (inputs FA0 are equivalent to nand outputs FA1), and 9 nand outputs to FA1 and FA0, for a total of 43 failures. But 4 output nodes don't fan-out and are thus equivalent to the input junction they go to, and one nand has only a single input which is thus equivalent to its output, leaving 34 unique failure modes. But 2 (R_D junction to J gate and S_D junction to K gate) exist only to improve timing and their detection would require closer time control than the test station can provide. Thus this JK has 32 unique, detectable failure modes.

A set of tests which at first would seem to exercise the component's truth table functions, but which fails to detect a detectable junction in the component's elementary gate equivalent, is invariably shown to have lacked generality in its attempted truth table verification, a very common occurrence.

It was at one time thought, since all failures will affect an IC interface, that detection of that interface should be sufficient. Of course, if this reasoning were correct, then mere detection of network output node failures would be sufficient, since all failures affect the output eventually. Unfortunately, however, stimulus which are sufficient to set network output pins to a state opposite of their supposed failure (thus detecting that failure) are far from sufficient to relay internal failures critically to that output.

In the final analysis, the elementary gate equivalent approach for failure mode identification and detection is found to be both necessary and sufficient. The NAND equivalent network not only provides the crucially needed failure modes effortlessly, but has other distinct advantages:

1. It provides the automation programs with a simple data base to be used in stimulus generation, simulation, and fault dictionary generation.

2. Sensitized paths, timing, races, asynchronism, latches, etc., all are recognized/handled by the programs in a fast simplified manner, with all the degrees of freedom which the network has. (Component subroutines could never be expected to perform these functions in any feasible run-time or programing effort.)

D-LASAR SPECIFICATIONS

In order to obtain a brief overview of the D-LASAR system its principal features are outlined below. Each subsystem described is executable separately, having all core ($106K_{10}$) at its disposal during execution.

INPUT SUBSYSTEM

1. User-Prepared Network Model Deck: High level modeling, e.g., 405JK,A10/24,238,942,L1,3/407,19/ represents a JK flip-flop, component #405, replaceable package (or location) A10, and the input node numbers for J, K, C_p, S_D, R_D are 24, 238,942,L1 (logic 1), and 3. Q and $\bar{Q}$ are node numbers 407 and 19.

2. Component Library: User may model any component in terms of Nands and previously modeled, or standard library components, and add such models to his component library.

3. Data Base: By referring to the library of component nand equivalents, the INPUT subsystem converts the User's Network Model Deck to a total Nand network, which is recorded in the form of a Node Pointer table which for any node points to the location in a From/To table containing information as to the "from" and "to" nodes of that node. Recording the network logic in this fashion facilitates high-speed random-access forward/reverse tracing, for stimulus generation, simulation, and fault dictionary generation.

4. Input Run Time: About 1 minute per 100 ICs.

5. Model Deck Preparation and Verification: About 1 week per 100 ICs.

STIMULUS GENERATOR (STIMGN) SUBSYSTEM

1. Network Size: to 200 IC directly, larger networks handled in segments.

2. Method: STIMGN starts at a single network output pin and establishes states so as to create a "critical string" back from it (see figure 2). Requirements for secondary nodes (i.e., not critical but affecting criticality) are satisfied last. The resulting node state configuration is then evaluated to see if it contains any latched latches (memory elements). If it does, then the input (stimulus) nodes of the configuration are sufficient to establish the assumed output node (response) states only if the assumed latch states are established via prior stimulus. Thus, working reverse in time,

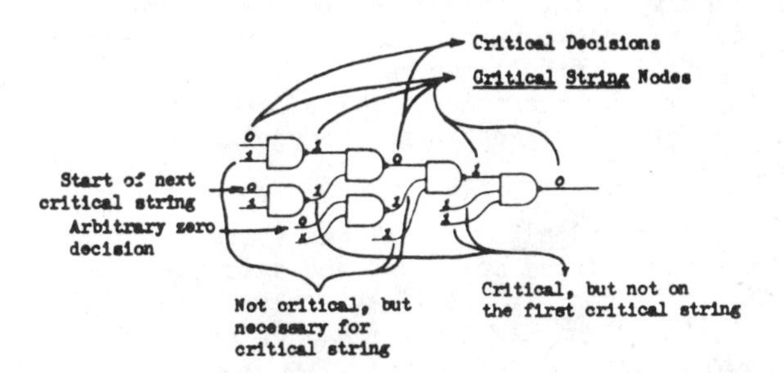

Figure 2

these latch states are used (as the output pin was) as sources to be set via critical strings on subsequent configurations. When a configuration is finally attained which contains no latched latches, the input bits extracted from this last configuration comprise the first stimulus pattern and obviously produce their effect independent of the network's initial state. The stimulus applied in reverse order of the derived configurations thus comprise a self-initializing set. Additional sets are derived, taking only undetected routes, until all failure modes which can be detected via the first output pin/state, have been. This traceback process is repeated for additional output pins/states until the required percent detection is obtained.

3. Complexity Limits: None. Workarounds required for certain cases of asynchronism, indeterminate initialization, large counters, illegal memory states, and redundancy.

4. Fault Detection Capability of Derived Stimulus: Typically 90% to 96% on the first pass thru ISOGEN, with 98% guaranteeable after 2 to 3 passes.

5. Typical Run Times (1108 or 360/65): $\frac{1}{2}$ to $1\frac{1}{2}$ hours per 100 ICs, depending on User experience and network complexity.

6. User Effort: One to three weeks, depending on network complexity. Two months User training required.

STIMULUS OVERLAY (OVRLAY) SUBSYSTEM:

1. Function: Overlays those portions of self-initializing stimulus sets whose patterns are the same, or differ only on bits for which one or the other is a "don't-care".

2. Reduction: Average reduction in the number of stimulus patterns is 2 to 1, and ranges from 1/1 to 6/1.

3. Effort and Run Time: Nil

SIMULATION (SIMUL) SUBSYSTEM

1. Timing: All Nands whose inputs changed have their outputs calculated before any such states are incorporated, so that all signals propagate in unison, one Nand delay at a time. In this way asynchronism, spikes, etc., are handled properly, and an accurate simulation results.

2. Race Analysis: All Nands are given a worst-case ± delay tolerance and a worst-case analysis is done to see if spikes or close signal timing might leave the state of some latch unknown. If so, the User is given a "Race" warning message and affected nodes are set to unknown. The addition of "buffer" stimulus patterns is usually required to eliminate the close timing. Also, the User specifies a worst-case time skew, to correspond to that which exists between stimulus bits from the applicable test station. Such skew radically affects worst-case race analysis, especially if greater than 40 ns.

3. Unknowns: SIMUL is a 3-state simulator, 0/1/X. The total network is initially set to state "unknown" (X) which propagate by their own set of rules as subsequent stimulus begin to "sweep" them out of the network during initialization. Unknowns are later created when a race occurs, and may infest previously known areas. It is vital that network outputs which are X not be represented as a 0 or 1, since test repeatability otherwise becomes an intolerable problem due to initialization variables, and network and test station time variation affects on races.

4. Asynchronism: When a latch is changed by a transient spike whose worst-case minimal width is greater than one Nand delay, the latch is said to have been set "asynchronously" without a race, and the User is warned of the asynchronism.

5. Indeterminate Initialization: In order to show, for example, that a master-slave JK became initialized when J/K was, say, held at 1/0, and clocked 0,1,0, known relationships between unknowns must be maintained. Thus if we assume the master latch was $a/\bar{a}$ then with clock low, slave becomes $a/\bar{a}$. Then when the clock goes high, the lower Nand of the master latch is seen to have an $a,\bar{a},1$ on its inputs which always produces a known 1 on the Nand's output, since either a or $\bar{a}$ must be a 0. Then the master is a known 0, and the final low clock sets the slave to 0.

If two variables ($a/\bar{a}$ and $b/\bar{b}$) have to be propagated (i.e., two latch sources) to create the known, it is referred to as "second order indeterminate initialization", etc. SIMUL handles first-order, and some second-order indeterminancy at this time.

6. Pattern Search Initialization: When a variable number of stimulus patterns must be applied until a sought response is obtained, in order to initialize a network area, that area is said to require a "Pattern Search" conditional loop. Such initial test sequence must be supplied manually, and its initialization effect incorporated by specifying for SIMUL the relevant initial latch states. SIMUL's "Partial Initialization" capability is used for this purpose as well as to reduce higher order indeterminant initialization to first or zero order.

7. Fail Mode: The User may request that a specified failure (single or multiple) be simulated and the resultant altered response patterns/pins be compared to ISOGEN's Fault Dictionary to see what degree of isolation would result. This is a useful means of verifying the Fault Dictionary effectiveness as opposed to the introduction of actual faults on the station. This means, as well as the use of an old Fail-All mode, has shown that ISOGEN's Fault Dictionary (generated hundreds of times faster than Fail-All simulation would) provides an average isolation to better than 2 ICs.

8. Source of Stimulus: While stimulus are normally taken from a file generated by STIMGN, they may be introduced from cards (existing tests, say, to be evaluated) or from a file generated by PATGEN, a routine which conveniently generates a repetitive clock sequence for large counters, etc.

9. Network Size: SIMUL currently handles networks to 400 ICs, and is readily expandable.

10. Run Time: Ranges from .02 to .3 seconds per stimulus pattern depending upon network size, frequency of occurrence of races, and the rapidity with which unknowns are swept out, the latter requiring the complex indeterminate initialization processing.

11. User Effort: Nil, except for race elimination, to be performed automatically in the near future.

ISOGEN

1. Function: Derive a Fault Isolation Dictionary and notify User of undetected faults.

2. Method: Given SIMUL's total node state configuration resulting from the simulation of the next stimulus pattern, trace critical paths back from each latch to update the history of

nand node junctions critical to it. Do the same for network output pins, and save the table of criticals (failure mode detects) for each output pattern/pin, if that table either contains new detects or contributes to fault isolation. 98% of such tables usually don't contribute, resulting in a file reduction of 50/1.

The effect which a failure has on its own criticality in later configurations (due to altered memory states) must be accounted for. Also criticality continuance/discontinuance due to fan-out/fan-in must be properly accounted for, even when the redundant paths contain sequential logic.

3. Network Size: To 400 ICs currently.

4. Run Time: About 1 second per stimulus pattern for 200 ICs and 2 seconds for 400 ICs, several hundred times faster than the fail-all approach.

5. User Effort: Nil.

ISO

1. Function: Given the failed response patterns/pins either on the test station or provided from SIMUL's "fail" mode, compare that response footprint with ISOGEN's Fault Dictionary and print out the junction failures and respective replaceable packages, in order of fail probability.

2. General: ISO is provided on the test generation computer, and may be readily installed on the test station computer, given 1500 words of available or overlayed core.

3. Run Time: The number of words in the Fault Dictionary is approximately 400 times the number of ICs. Thus a 200 IC network requires storage for $80K_{10}$ words. ISO processes this in one pass thru the on-station tape, at mag-tape speed, or something on the order of 15 seconds (faster for dedicated disk).

4. User Effort: Replace the indicated 2 most probable ICs and re-run the test. The test will then pass 90% of the time, requiring an additional replacement/re-run 10% of the time.

D-LASAR IMPROVEMENTS UNDERWAY

STIMGN: For large networks (over 100 ICs), about 1 out of 6 requires that "ILLEGALS" be specified for run time efficiency. An ILLEGAL is a memory state combination which can exist statically in a non-contradictory manner, but for which no sequence of stimulus exists capable of creating it. While STIMGN derives the ILLEGAL cores easily in most cases, a pre-processor is being added to eliminate the current occassional requirement for manual ILLEGAL specification.

Also, since STIMGN derives stimulus by tracing back from single output pins, then for that stimulus set, STIMGN recognizes detection relative only to that one response pin and only in the last pattern of that set. Whereas in actuality when the total effects of the stimulus are recognized at all output patterns/pins (e.g. 20 response patterns, 40 pins gives 800 measurements, compared to 1) major detection increases are realized. Thus in later versions, ISOGEN will be run after each stimulus set generated, prior to generation of the next set, resulting in a major reduction in the number of stimulus. Currently this process is controlled by the User, who makes short STIMGN runs with an output pin selected in a region as yet undetected, alternating such runs with SIMUL/ISOGEN.

ISOGEN: Asynchronous latch setting currently requires manual clean-up in the Fault Dictionary. ISOGEN's handling of asynchronism is being generalized to eliminate this effort.

SIMUL: The capability is being added to SIMUL to add buffer stimulus to automatically eliminate race conditions, eliminating User effort here.

SUMMARY

For the test engineer to have generated tests in the past which approach the quality now obtained with automation aids, he would have had to go thru the following very tedious procedure for, say, a 100 IC network:

1. Identify its 2800 unique failure modes (a tremendous task without a Nand equivalent library).

2. For the first of the 2800 failure modes, derive stimulus patterns sufficient to set it to the not-failed state, independent of the initial state of memory.

3. Derive another pattern to create a "critical string" from that failure to set a latch (the latch must have been previously set to the opposite state). If the failure cannot be made critical to that latch, alter prior stimulus so as to create memory states such that it can, or else find another latch to which it can be made critical.

4. The latch just set critically by the un-failed state of the selected failure mode must now be made critical to some other latch in a similar manner, etc., until the information is relayed critically to some output pin. This process sometimes takes thousands of patterns. The engineer may be down to the last latch-to-latch criticality relay, and find that no critical linkage is possible, and thus have to back up, going back and forth modifying memory states in an attempt to find a critical route. Thus

the detection of a single failure mode may be essentially unfeasible by manual means.

5. For the set of stimulus just derived, determine their total effect at all points in time and determine what other failure modes have thereby been relayed critically to what latches and outputs. Timing, races, asynchronism, propagation of unknowns, indeterminate initialization, etc., must all be accounted for and handled.

6. Repeat 3, 4, and 5 for the next as yet undetected failure mode, realizing the effects which that mode has on the initial memory state going into this step, i.e., from its effect in prior stimulus applications. Repeat the above until all 2800 failures are detected.

7. From the voluminous data as to what failure modes had become critical (purposely or accidentally) to what output patterns/pins (about 4 million entries), throw away those which do not contribute to the detection or isolation obtained up to that point, thus reducing the volume to about 40,000 entries.

8. Re-format and cross-reference (to replaceable packages) the above file, and re-sort into the form of response patterns/pins per failure mode, and group failure modes into their respective undistinguishable fault classes. This is the Fault Dictionary.

9. Prepare on-station fault detection and isolation routines which compare station response with expected responses and compare the indicated failures with the Fault Dictionary to isolate the fault.

No engineer can be expected to spend the years required to perform the above tedious repetitional, essentially impossible tasks ... The engineer supported only by a fail-all simulator cannot generate satisfactory test programs for networks over 30 ICs.

The engineer applying D-LASAR, however, has quite a different task --- while the volume work is essentially automated for him, he must analyse any unusual design peculiarities in the network (which has to be done in any event) and provide D-LASAR with the work-arounds which it requires (according to generally established but sometimes complex rules). Thus a high-level, low volume analysis is required, but only once, to modify the model deck. Then D-LASAR uses this information repetitionally.

Several months of specialized training is required to become proficient in the application of the automation aids, but the engineer who succeeds is capable of carrying many times his weight, and of generating test programs which do the job.

APPENDIX A

DERIVATION OF YIELD VS. PERCENT DETECTION FORMULAE

If p_i is the probability of failure of mode i, then $1-p_i$ is the probability of its passing. Also if there are n such modes in the unit under test consideration, all equally likely, then $(1-p)^n$ (approximately e^{-np} for small p, large n) is the probability that all will pass and that unit will be in the fault free, or "yield" group. Obviously $p_i(1-p)^{n-1}$ is the probability that mode i only will fail, and since there are n such modes, $np(1-p)^{n-1}$ is the probability that the unit will have exactly one failure. Similarly $p_i p_j (1-p)^{n-2}$ is the probability that modes i and j, and not the others, will fail, and since there are $\frac{n(n-1)}{2!}$ ways to take 2 at a time out of n, then $\frac{n(n-1)}{2!}p^2(1-p)^{n-2}$ is the probability that exactly 2 failures will occur.

If d is the percent detection then it is the probability that the single failure case will be detected. For the double failure case d is probability of detection of the first failure, leaving a detection region of dn-1 out of n-1 for the second, so $d\frac{dn-1}{n-1}$ is the probability of detecting both. Multiplying this times the probability of exactly 2 failures occurring, we obtain $\frac{dn(dn-1)}{2!}p^2(1-p)^{n-2}$ as the contribution to yield for the 2 failure case after correcting detected failures. The total contribution to yield for the 0, 1, 2 etc. failure cases is thus $(1-p)^n\left[1+nd\left(\frac{p}{1-p}\right)+\frac{nd(nd-1)}{2!}\left(\frac{p}{1-p}\right)^2+ ----\right]=$ $(1-p)^n\left(1+\frac{p}{1-p}\right)^{nd}=\left[(1-p)^n\right]^{(1-d)}=y_a$, the total yield after d% detection testing. But $(1-p)^n$ is the yield y_b before testing. Thus $y_a=y_b^{1-d}$, the desired expression. Notice that y_a denotes the most probable fraction of the units that are actually good, out of those which finally passed the test. Thus the formula holds whether units with detected faults are "corrected" and passed, or simply disposed of, since a corrected unit has no better probability of being fault-free than one which passed initially.

1. J.P. Hayes "NAND Model for Fault Diagnosis in Combinational Logic Networks" IEEE Trans. Comput., vol c-20, Dec. 1971, pp. 1496.

PART VI: Configuration Control and Management

Effective deployment of any system requires good configuration control and management of the facility. When the system is dependent not only on hardware but also on extensive software system and application programs, the need for configuration control and management is multiplied manyfold. Typically, the purchase price of an ATE system is less than half the cost of its application software (test programs). For many multipurpose ATE systems, test programming costs can be as much as ten times system procurement costs.

The absolute dependence on software, which represents the greatest cost factor in ATE, make software configuration control and management the prime concern of the ATE facility manager. Obviously, good software configuration control is built upon a solid hardware configuration. But the problems of hardware control are much better understood by technical managers so the papers selected for this section stress software configuration and management. Software includes the ATE operating system, test programming aids, and the test programs. All must be controlled and intelligently managed. Once the operating system and programming aids are developed and debugged, they can be considered a fixed-design part of the ATE system. The final variable, of course, is the set of test programs designed to handle each of the units to be tested by the system. In most successful applications of ATE, new test programs are continually being developed thereby making this element of the facility the most costly in the long run. Hence, particular attention must be given to the techniques for preparing and managing test programs.

The first paper, "Software Management through Specification Control," compares the attributes of hardware and software. It shows that while they differ in form, they are very similar in that both are the product of an engineering labor force. Hence, the technical specification, which is the key in any hardware design control system, is equally useful in controlling software development. The advantages of using specifications as the primary control tools in software production are enumerated. Until such a disciplined approach to software development is followed, software projects will continue to operate essentially uncontrolled and unpredictibly.

The paper "Configuration Management of Software" takes a close look at the test program production process and develops a configuration management system that relates directly to the production process. Following the concept of the previous paper, a series of control specifications are defined with each specification relating to each major function in the production process. A second set of specifications also relating to the production functions provide guidelines to the individuals involved in the particular production function. This "factory" approach to test program production, of course, is only appropriate in large-scale efforts where many specialists can be economically utilized to assist the test engineer. It should be remembered, however, that no matter how many specialists are used to help produce the test program set, one individual (the test engineer) must retain total responsibility for the integrity of the finished test program set (program, interconnection hardware, and instructions). Even in the smallest shops, the technique of specification control is valid. In instances where one man must perform all functions, a much less elaborate system of specifications is required. But without some formal documental system of controlling test program production, the quality will be poor and costs high.

One of the great advantages of ATE is that it provides a means for unprecedented quality control of the testing function. Its repeatability and ability to generate documented test results make ATE an ideal quality control tool. There are many forms of documentation that can be designed to become a by-product of the testing function. Industry has only scratched the surface of the ATE documentation potential. The paper "Management of Data from Automatic Test Systems" provides a valuable insight into the types of documentation obtainable from ATE systems and offers some guidelines to effective management of such data.

"Estimating ATE Software Costs for a Support Systems Analysis Study" is one of the very few papers published anywhere on the subject of ATE software costs. It takes a logical analytic approach to developing the cost factors involved in test program production. Like any reasonably successful approach, it is based on two important ingredients:

1) empirical data developed from many years of test program development;
2) a formal documented approach to cost estimating.

There is no perfect formula for software cost estimating, but one thing is clear, until those responsible for program production begin to document their techniques and then use actual experience to reevaluate the cost estimates as the project progresses, programming costs will always be a mystery.

The paper entitled "Technical Management Techniques for Large Scale Automatic Test Systems Engineering" addresses the problem of producing large quantities of test programs for a sophisticated ATE. The paper points out that there is a world of difference between effective production of a few programs by a small group of experienced test designers and massive programming efforts involving hundreds of engineers. This should not be surprising, yet few organizations involved in large-scale applications of ATE have taken a hard look at program production problems on a large scale. Until such a managed approach is adopted, costs and schedules for test program development will remain uncontrolled.

SOFTWARE MANAGEMENT THROUGH SPECIFICATION CONTROL

By F. Liguori
Sen. Proj. Member, Technical Staff
RCA, Burlington, Massachusetts

About the Author

Mr. Liguori received his BSEE degree from Tufts University in 1957 and an MBA from Hofstra College in 1960. He received his formal computer education at the graduate school of Adelphi College. Since joining RCA in 1962, he has been involved in test designing and programming for various general-purpose automatic test systems, including the Multipurpose Test Equipment (MTE) and the Land Combat Support System (LCSS) for the U. S. Army Missile Command and the Depot-Installed Maintenance Automatic Test Equipment (DIMATE) for the U. S. Army Electronics Command, and a newly installed commercial automatic test system. For the past three years he has been a principle contributor to Independent Research and Development studies on advanced programming techniques for the RCA Automatic Test Equipment product line.

Contrasting software with hardware is a convenient way of describing software to non-programmers. This emphasis on differences has led to a common misconception in industry that the two products are inherently different and thus, their development must be handled differently. Close analysis of both tasks reveals often overlooked similarities. In fact, the two products are so similar in content that the commonly used formal techniques for managing hardware developments apply equally well to software development.

This article offers the rationale behind the thesis that hardware and software development tasks are in fact quite similar. It then discusses the applicability of specification control to software development, emphasizing how each advantage of specification control applies equally well or more so to software development tasks.

The conclusion is that software management is not inherently different from hardware management. The same tools of control may be (and in fact have been) successfully used for software development tasks. It is, therefore appropriate, and necessary for project managers to provide more definitive direction of software development tasks through detailed specifications without requiring personal proficiency in software design.

Reprinted with permission from *Software Age*, May 1968.

■ Examination of documents commonly used in industry to control system design and engineering reveals a lack of understanding of the nature of software and methods of software development control. This is particularly evident in Requests for Quotations (RFQ's) involving computerized systems requiring both hardware and software development. Typically, such RFQ's consist of 50 pages or more of very detailed technical requirements. They include not only the necessary performance characteristics, but often even dictate the basic method of implementation of the system to be designed. Of these 50 or more pages, often as few as two or three pages and certainly less than ten percent of the document is devoted to software specification. Yet, software for the system may cost as much or more than the hardware. This over emphasis on hardware is indicative of an unique pre-occupation with hardware characteristics and a lack of understanding of software techniques and requirements. Software characteristics, which by and large dictate operational characteristics of the completed system, are left vague and undefined.

Work statements and design specifications, which are outgrowths of the RFQ, further reflect the vagueness with which software is controlled. The result is that, unless the contractor selected happens to be unusually competent in software development, this aspect of the system is left to be worked out during the hardware development phase when it is impractical to make the optimum hardware/software configuration tradeoffs.

The Automatic Test Equipment product line at RCA has traditionally involved a complex interplay of hardware and software. Experience has proven the value of detailed software work statements and design specifications. This paper develops the rationale behind the similarity between hardware and software development and management tasks and discusses the advantages realized by close control of software tasks through definitive specifications.

Similarity of Software and Hardware Systems

Hardware and software are often contrasted to explain the special problems of software with respect to the familiar base of hardware development. And, indeed there are many aspects of software development that make it different from hardware. Hardware is developed in discrete stages (paper design, breadboard, engineering prototype, etc.) that progressively constrain system changes. Software stage boundaries do not seem to progressively constrain software system changes. The obvious reason is that software appears inherently more flexible (therefore changeable) because it is structured with "paper" rather than fabricated hardware. Ironically, it is this "advantage" of software that is really its nemesis. Nothing can hamper final designs more than susceptibility to change, and nothing is more fatal to schedules than design changes.

A second characteristic of software that makes it more elusive to schedules is its camouflaged nature. Unlike hardware developments which go through visible stages of development, preliminary software and debugged software are of the same general appearance. Even when it is apparent that a task is still in the process of completion, it is much more difficult to determine percentage of completion of software as compared to hardware developments in progress.

Despite the obvious differences in hardware and software it is the thesis of this paper that software and hardware systems have greater kinship than disparity. Furthermore, mismanagement of software systems development stems from overemphasis of differences and failure to recognize and capitalize on similarity. Once this thesis is accepted, it follows that the methods of hardware systems development can be successfully applied to software systems development. This causes two seemingly divergent disciplines to converge in most system applications. Hence, it allows well developed techniques to be applied in a field much less understood by management people[1] who control the integrated system development at the critical point—where software and hardware have a common manager.

The Basis of Similarity

To see the commonality between software systems and hardware systems one must examine not the form of the finished product but rather the resources required to develop each system. In either development, basic resources are time, manpower, and materials, all of which have a common denominator in dollars. These resources are common to all human endeavor so it is necessary to further investigate the types and quantities involved as well as their interrelationships to prove or disprove substantial correspondence between hardware and software development tasks.

When dealing with a system under development, whether it be hardware, software or a combination, schedules typically involve a year or more. Usually, there is no substantial difference in period of performance for hardware and software system development when comparing tasks of approximately equal dollar volume.

In any system development, manpower (particularly engineering labor) is by far largest single cost. Cost analyses of past research and development contracts show that manpower typically accounts for 75 percent of total cost. Variations are geared to the ratio between purchased and fabricated items in the system. Where all subsystems are specially developed, the labor approaches 90 percent of the task, whereas systems configured largely from off-the-shelf units yield much lower labor percentages. If one could determine the percentage of purchased item costs, these items too would be seen to be largely labor burdened except where very high production rates exist. In software develop-

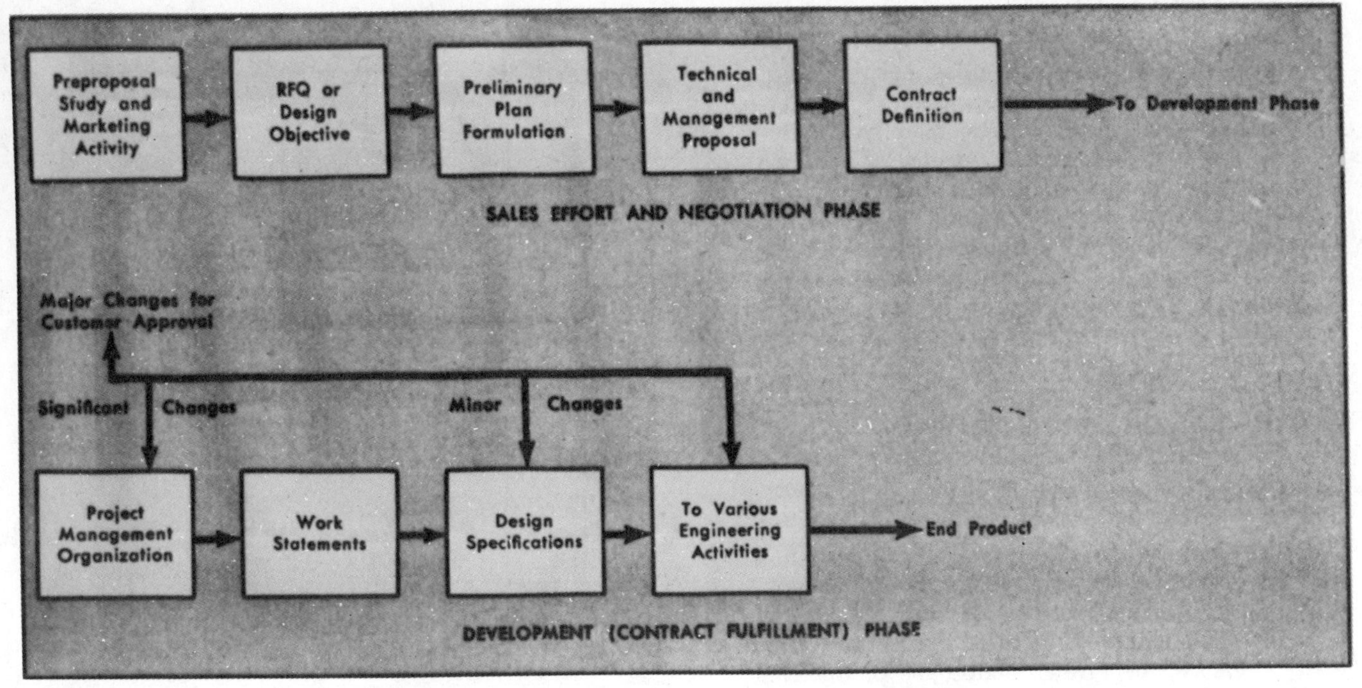

Figure 1. Documentary Control of the Engineering Design Process

ment, also, labor generally runs 75 percent and it can run as high as 90 percent of the task. Even in control system programming where certain adaptor hardware elements are attributed to "software" costs, about 80 percent of the cost is expended in labor, largely engineering.

From this it is apparent that, in either hardware or software systems development, engineering (or other professional) labor comprises the bulk of the cost. Furthermore, the differences between software and hardware system development labor rates are negligible. The professional engineer or scientist plays the major role in both.

Material costs are naturally higher for hardware system development, being in the order of 25 percent, as compared to about 5 to 10 percent for software materials. In software efforts, however, there are generally equipment rental costs involved to debug programs on a computer. These typically are 10–15 percent of total costs. Rentals are often treated as "material" costs, at least in purchasing procedures. With rentals so classified, the ratio of materials to total costs also becomes quite comparable for hardware and software systems.

In conclusion, all factors of production are close to equal for either hardware or software developments involving equal dollar volume. It is also apparent that the largest single cost by far is professional labor. Hence, the most precious ingredient of any system is the "intelligence" expended in its design and construction. Since any effort can be reduced to management of factors of production, management of software and hardware development tasks can be handled essentially in the same manner.

Method of Control

The instruments of control in any system development are managers and procedures. The managers of hardware systems and software systems are almost always technically or scientifically disciplined professionals, at least at the project management level. Salary levels and prestige are approximately the same for projects of equal dollar volume since they each involve approximately the same number of personnel in the same salary ranges. Hence, the only variable between hardware and software management is the set of procedures.

The procedures for hardware management are by far more standardized and apparently more effective than for software.[2] In fact, there appears to be no standard approach to large scale software developments. Although most software managers have learned to rely heavily on documentation for control, the manner of task specification, quality control, product standardization and change control is quite loose in many organizations. With an absence of formal procedures, success or failure in software tasks depends much more heavily on close personal supervision and the quality of such supervision.

While supervisory talent is a required commodity in any endeavor, heavy dependence on such a rare commodity is risky. It is not without justification that American Industry places as little dependence on personalities as possible, preferring to rely heavily on policy and procedure manuals. Software management should follow this example and shift some of the control from personal management to procedural management.

Management Control Tools

PERT and similar management control procedures are commonly used today. Many of the larger concerns are utilizing computers to implement these control procedures, thereby greatly facilitating correlation and dissemination of control data. Software as well as hardware elements of a system are included as milestones in the overall schedules. This has allowed broad control of tasks but the guidance for meeting individual milestones must come from more detailed technical direction. In hardware development, the process is well defined by formal documentation. A typical development process under document control is shown in Figure 1.

This process is a familiar one and need not be explained in detail. Two points are worth noting, however.

(1) Normal process flow is from customer requirement to end product with changes as a counter force. The more significant the change, the greater the impact and, hence, the further against the productive stream it flows.

(2) The major point of contact between the management functions and the engineering effort is the work statement which references the design specification. The specification, therefore, provides the major technical direction from objective to accomplishment.

Advantages of Specification Control Techniques

Borrowing from the hardware management discipline, the specification should be the basic control device in software development efforts. A brief review of advantages offered by specification control of hardware is worthwhile, for each advantage applies equally or more so in software management. A specification, when properly and timely executed, provides:

1. A clear, indisputable definition of the task requirements in terms of design objectives.
2. A legal base for negotiation.
3. A readily referenced source of guidance.
4. A logical means of subdividing tasks and interface handling.
5. A standard of measurement for progress review.
6. A standard of measurement for finished product performance.
7. A source of direction independent of personalities.
8. A means of standardization control.
9. A vehicle with which to build new designs based on past projects.
10. A means for maximum span of control by management.
11. A handy tool for orientation of personnel assigned to the project after formal orientation.
12. A basis of reference and control of inevitable changes.

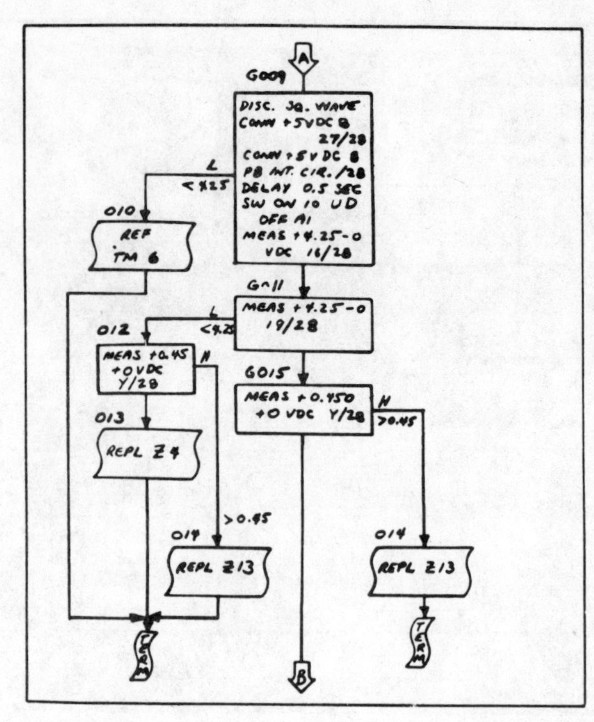

Partial Flow Chart was used to help generate part of a computer listing.

The advantages listed for specification control over less formal, personal control of hardware designs are fairly self evident. Rather than offering trivial explanations of these advantages, it is more worthwhile considering how some of the advantages listed have even more significant impacts on software design management.

Requirements for task definitions and a legal base for negotiation obviously apply to any endeavor and software takes no exception to these basic needs.

An item often lacking in software efforts is a readily referenced source of design guidance. In hardware efforts, system, subsystem and often standard circuit and module specifications set down both the performance objectives of the unit in question and even the methods of circuit implementation. In software efforts, no such tools are provided. Generally, a single work statement, vaguely worded, must serve as sole guidance to all tasks and subtasks. For a programmer doing a small segment or routine, this "big picture" objective offers little guidance. He must depend on his task coordinator to tell him what the precision requirements, interface with the main program, execution speed requirements and memory allocation must be in addition to the specific problem to be solved. Experience has shown it to be ineffective to rely on verbal direction given by subtask coordinators. With such an approach discontinuity generally arises among the program elements. Eventually such discontinuities are resolved and the program debugged, but an unnecessarily large part of the effort would have to be devoted to cut-and-dry techniques. Worse still, some subordinate functions of the program may be entirely

UTEC CODING FORM

TEST ENGINEER F. LIGUORI

DATE 4 MARCH 1968

UUT 9135

RCA BURLINGTON

	TEST NO.	G	FUNCTION						LBL	UPDATE
			.							
			.							
	007		PRINT	REPLACE C1						
			GØ TØ	TERM						
	008		PRINT	REF TM 28.						
			GØ TØ	TERM						
	009	6	DISC	SQ						
			CØNN	+5.0	VDC	27	(GRD)			
			CØNN	+5.0	VDC	(23D)	(GRD)			
			CØNN	+5.0	VDC	(21-14)	(GRD)			
			DELAY	500	MS					
			SWITCH	ØN	10	U	D			
				ØFF	A1					
			MEASURE	+9.25	VDC	-0	16	(GRD)		
1			PRINT	%$** ENTRY POINT #009 **						AC0290
			GØ TØ	011	-	010				
	010		PRINT	REF TM 6.						
			GØ TØ	TERM						
	011	6	MEAS	+4.25	VDC	-0	19			
			GØ TØ	015	-	012				
			.							
			.							
			.							

The source language coding is used for generating the RCA Universal Test Equipment Compiler (UTEC) language program.

SOFTWARE MANAGEMENT

overlooked in the design process. These must then be "patched" into the program, often at decreased program execution efficiency in order to minimize the interaction on the completed sections of the program. By reducing each subprogram task to a separate, though subordinate specification, all requirements may be defined in detail at the outset.

Task subdivisions in hardware design tasks usually fall into neatly defined physical unit boundaries. This is seldom the case in software efforts. Software can be broken down only into a small number of separate programs. One may be for system pre-acceptance testing, another for operational control and a third for maintenance. Beyond such major divisions, there is little natural subdivision. When no definitive specifications are imposed on the task coordinate, the problem is often handled by assigning a major task to a task force of several programmers, all of which have overlapping responsibility. In such circumstances the more ambitious member of the team (not necessarily the task coordinator or even the most capable programmer) must bear the brunt of the task. By substituting indigenous leadership for documented directives a good deal of inefficiency often results. Some work, incompatible with the final approach, is necessarily scrapped while other parts are reworked. Specification control provides an explicit means for subtask definition, ensuring compatibility by facing the problem at the outset rather than when well along the design path.

Without a detailed work statement and subordinate specifications, attempting to ascertain percentage of completion of a software task is generally a guessing game. This problem arises from the lack of visual evidence of accomplishment in partially completed programs. Generally there is far more paper work observable part way down the road to completion than when the program is finished. This is due to the overlapping coverage of subtask programmers, extensive rework and alternate coding generated to cover alternative approaches. All of which point to a lack of definitive direction in the detailed design stage. In hardware tasks, pieces fit together more naturally. Hence, the end of one phase and start of another is more definite, thereby leading to a simpler, more accurate assessment of percentage of completion. If software were generated in conformance to a network of subordinate specifications, fulfillment of such specifications would provide an accurate measure of accomplishment.

Finished product performance is judged against an overall requirement specification and is thus fairly easily ascertained. At that level, software performance can be determined just as readily as hardware performance if not more so. While it is true that some software bugs may exist for years before detection, these are minor deviations to desired performance. Once debugged and operational, programs simply do not fail; so once properly demonstrated, they perform without error forever.

The need for de-personalization of tasks in scientific and engineering endeavors is self-evident. Programming should also be approached just as objectively as hardware designs and specifications provide objectivity, if nothing more.

Standardization in hardware is a recognized necessity for many reasons, not the least of which is the interplay of subunits. Software tasks involving many individuals have similar problems of interplay. The fact that this interplay is less evident makes it more of a problem. Standardization would go a long way toward facilitating this interplay. Specifications, with their use of applicable document references, provide a convenient means of imposing basic standards on the overall requirements of a peculiar task.

Any design well controlled by documentation provides a sound basis for related development efforts to come. To appreciate the design considerations of a program, a good deal more than a finished program is required. With specifications covering each facet of software development, later projects can see exactly the logic, objectives and accomplishments of a past software effort.

In any effort, a manager's span of control is effectively increased by relegating definable direction to written procedures and reserving his personal attention to problem areas as they develop. Most of the software details are directly related to the ultimate task requirement and thus may be documented at the start of a project. Specifications provide such a vehicle of documented control.

Specifications, as definitive documents, provide historical information on past projects. This was discussed as it pertains to follow-on work. Specifications also provide a tool for orientation and training of personnel. This is an equally valuable tool for software as it is in hardware development.

The final consideration for specification control deals with design changes. This has been deferred to last not because it is least in importance but because it is perhaps most worthy of consideration and comment. Nothing can be more detrimental to schedules, costs and even the success of a task than uncontrolled changes. This has been recognized in hardware design efforts. Formal and often elaborate systems for design changes exist at every phase of system design. It becomes increasingly difficult and expensive to make changes the further along the process a product has progressed. Yet, no matter how undesirable, changes are inevitable. They must be contended with but they cannot be allowed to get out of control. Hence, in software, as in hardware designs, only mandatory changes should be permitted and budgets must be realigned periodically to adjust to work load effects of approved changes.

Somehow widespread miscomprehension exists as to the effects of changes on software efforts. Because programs are essentially "paper designs" there is a tendency to feel that nothing is firm. Perhaps it is simpler to change a few coded

```
************************************************************************
007   PRINT   REPLACE C1.                                        279
      GO TO   TERM                                               280
                        07[illegible]      P0750
                                  REPLACE C1.0
                                  *
                                  J001
************************************************************************
008   PRINT   REF TP 20.                                         281
      GO TO   TERM                                               282
                        074        P0760
                                  REF TP 20.0
                                  *
                                  J001
************************************************************************
009 6 DISC    20                                                 283
      CONN    +5.0    VDC     27      (GND)                      284
      CONN    +5.0    VDC     (230)   (GND)                      285
      CONN    +5.0    VDC     (21-24) (GND)                      286
      DELAY   500     MS                                         287
      SWITCH  ON      10      0       D                          288
              OFF     A1                                         289
      MEASURE +4.25   VDC     -C      16      (GND)              290
  1   PRINT   **** ENTRY POINT ,009 ***                  AC0290  291

 UUT 0120-  REV --  -RUN 3   -UUT NAME TAPE SEARCH CONTROL.
      GO TO   013     -       010                                292
                        075        $ PGNR *
                                   $ PS10050 PS1A *
                                   $ PS20050 PS2A *
                                   $ PS40050 PS4A *
                                   DL05007E
                                   $ RSA021 RSAA RSB1034 RSBA RSCA *
                                   $ TPA3737 TPB6060 RDC1 TPAA TPBA *
                                   B
                                   L+004253
                                   P000 ENTRY POINT 375 ****
                                   *
                                   *077
                                   V VDC 3 0750*
                                   Z076
************************************************************************
010   PRINT   REF TH 6.                                          293
      GO TO   TERM                                               294
                        C76        P0766
                                   REF TH 6.0
                                   *
                                   J001
************************************************************************
011 6 MEAS    +4.29   VDC     -C      19                         295
      GO TO   015     -       012                                296
```

A typical sheet from a computer listing of a test program has been prepared for the Land Combat Support System. Source language (left side of sheet) is that of the RCA Universal Test Equipment Compiler and object language (right side of sheet) is machine language for the RCA Land Combat Support System (LCSS).

statements than it is to replace elements in a unit of hardware. But if the program modified costs ten thousand dollars of labor to develop, a five percent change can reasonably cost 500 dollars. Changed programs represent destroyed intelligence which in today's professional labor market is far from cheap. The effect of change on software can be just as costly as hardware changes even though physical evidence of disruption is missing. Stringent methods of change control must be applied to software efforts if schedules and budgets are to be maintained. With specification control of software tasks, changes must be subject to formal control. Availability of this control alone would repay all of the effort expended in setting up a family of specifications to direct and control a complex programming effort.

In summary, this paper has attempted to show the similarity between hardware and software design tasks. This leads to the conclusion that specification control, the method proven effective in hardware management, applies also to software tasks. In fact, in many ways this means of control is more valuable to software control. The existing problems of software development discussed herein are not intended to be critical of programmers. On the contrary, it is somewhat amazing that many programmers in industry are able to achieve goals at all when all of the inherent disadvantages of dealing with elusive software are considered. If anyone is to be chided at all, it is the systems manager who does not give sufficient attention to software at the outset of a system design specification task. The importance of software in computerized systems is self evident. It is time software management methods reflected their due measure of importance throughout each phase of software development. ■

REFERENCES

(1) B. L. Ryle, *"An Engineering Approach to Configuration Management"*, Proceedings of the 1966 Automatic Support System Symposium, St. Louis, Missouri, Section of IEEE.

(2) Project SETE, N.Y.U., *"Problems and Pitfalls in Automatic Test Computer Programming"*, May 1965.

(3) Bellcomm, Inc., *"Management Procedures In Computer Programming for Apollo"*, Interim Report for NASA, TR-64-222-1, 30 November 1964.

CONFIGURATION MANAGEMENT OF SOFTWARE

By F. Liguori *

Summary:
This paper discusses configuration management of software as it pertains to Automatic Test Systems. It defines configuration management, its need and role in the process of producing large quantities of operational test programs for test systems. The discussion is particularly pertinent to the application of automatic test equipment (ATE) in the maintenance depot for testing large quantities of Units Under Test (UUT's). However, the basic approach of using specification documents as the baseline tool for configuration management is pertinent to a large variety of projects involving the production of many computer programs.

1. Introduction

Although Automatic Test Systems represent a new technology barely a decade old, it is rapidly maturing. One sign of this maturity is the switch in emphasis at technical symposia from Systems Hardware Design to Software Design Problems. The next hurdle to pass is that of effective software configuration management. Configuration management is a particularly difficult problem because it deals with a very real need that evades predictable solution. Just as with practical test program development techniques, software configuration management is a function learned by involvement. The rules are largely emperically derived by having made numerous mistakes and ultimately arriving at a working solution. One reason that progress has been slow in good configuration management is that unlike hardware development, it offers very little glamour to attract the better systems engineers. In a way, however, the very neglect of the matter coupled with its importance to overall success of Automatic Test Equipment (ATE) makes configuration management the most challenging problem.

In its application to Software for Automatic Test Systems, configuration management has the same purpose as for any system to be designed and utilized. Its object is to provide an orderly means for developing and deploying a new system in a manner that is consistent with the purpose for its creation. Effective configuration management for an automated test system must deal with both hardware and software for only through proper melding of these essential elements can the system fulfill its maintenance role.

Hardware configuration management, while equally as important as software management, is somewhat less of a problem because engineers and managers have been solving that problem as a regular diet. Software configuration management, while not new, is relatively unfamiliar to engineering managers because practically all of them have advanced through hardware disciplines.

This paper attempts to bridge the gap between a hardware-oriented understanding of configuration management and its proper role for software.

* Manager of the Software Systems Engineering Laboratory, Electronics and Space Division, Emerson Electric Company, St. Louis, Missouri, USA

Reprinted with permission from *IERE Joint Conf. on Auto. Test Systems Rec.*, April 1970.

Emphasis is placed on the control aspect of configuration management using specifications as the control vehicle.

Before discussing details of software control schemes, it is worthwhile noting that in almost all respects software is essentially an engineering product just as is hardware. It should not be surprising, therefore, to find that in most instances the same techniques used for hardware configuration control can be used effectively for software. Capatilizing on this premise, several advantages accrue:

1. The hardware oriented manager needs not learn control schemes alien to his discipline.

2. The same control system and most of the control tools can be used for both hardware and software.

3. Compatibility is readily achieved between hardware and software because they depend on the same control tools.

2. Configuration Management Defined

A better understanding of just what configuration management is will go a long way towards its achievement. Configuration management is simply management jargon for common-sense control of a complicated system whose success depends on many variables. The most reliable method of controlling the variables is through documentation, such as specifications, drawings, technical work statements, and engineering data of various types. The biggest problem with configuration control is due to the rapid obsolescence of these documents in the dynamic environment of system design and field trials.

Obviously, configuration control cannot eliminate changes. It must recognize that changes are essential to successful engineering endeavors and be geared to handle them. However, configuration management must remain the watchdog of the customer/user to assure that the system meets his objectives. In its most basic definition, configuration management is the process of judiciously controlling changes during the evolutionary phases of system development. Configuration management must provide a living baseline from which all elements of the system evolve into an up-to-date, effective product that fulfills the purpose for system development.

3. The Need for Configuration Management

To appreciate the need for configuration management of software requires some understanding of the nature of software and its generation process. There are two basic types of software involved in automatic test systems:

1. Operational software - the test programs executed on the automatic test system to perform tests on a unit under test.

2. Software aids - programs used to assist the test designer in the production of operational software.

Both types of software - the product and the tool - must be configuration managed. Both are in evolutionary stages as the automated test system is developed and continue to evolve even after the hardware has stabilized. The operational programs have the greater need for close configuration management for several reasons:

1. Operational programs became an integral part of the deployed system.

2. New operational programs are continually being prepared whereas the software aids development is essentially non-recurring.

3. Many more people must become involved in operational program development than the few skilled programmers used in software aid development.

4. Configuration Management of Operational Test Programs

Configuration management as it pertains to operational test program generation is stressed here because of the vital role and quantities of test programs needed for effective deployment of the test system.

The four phases of test program evolution through which configuration control must be maintained are:

1. Test data generation

2. Program preparation

3. Validation and documentation

4. Deployment

Program development spans the first three phases of program evolution. During development, configuration management is needed for many reasons, including:

1. Making the program compatible with the test specification, the UUT and interface hardware, the Automatic Test System capability and operator/user.

2. Controlling and coordinating program development during the various phases of program evolution.

3. Generating an up-to-date documentation package consistent with interacting system characteristics.

4. Providing a full understanding of the program generation process for realistic costing and scheduling.

5. Establishing a realistic baseline for process improvement.

During the deployment phase of program evolution, configuration management becomes a more formal process with control passing from the program production organization to the user. In the case of the military user, formal engineering change procedures must be followed. These are well defined by the military for hardware changes and are becoming better defined for software. Review of the change procedures, such as covered by the typical Air Force document ANA445, reveals that for the most part, software changes can be handled using the same procedure as for hardware. This is as it should be because test programs are as much functional parts of the operating system as the logic gates and power supplies. Because configuration management after deployment is vested in the hands of the user who has his own rigorously defined procedures and because program changes are greatly reduced during this phase, this discussion concentrates on configuration management during the first three phases of program evolution.

4.1 Control and Guidance Specifications Related to Test Program Production

As previously stated, the best way to control variables is through controlling documentation,but effective documentation must do more than control - it must also guide. The process of producing large quantities of operational test programs has evolved into a sequence of fairly well defined steps or functions. These functions are surprisingly similar among various manufacturers, at least within a given class of test systems, such as those dedicated to depot maintenance of military electronics equipment. Figure 1 illustrates the typical test program generation process. The raw materials for the process are engineering data, engineering talent and the hardware required for manufacturing interface equipment to tie the unit under test (UUT) to the test system. The tools required include physical facilities, electronic data processing equipment (with software aids) the test system and control specifications.

Volume production techniques dictate that specialists be developed to handle each function in the production process so that "assembly line" techniques may be employed. However, this approach conflicts with developing a coherent, consistent program. Configuration management must tie the factors of production together so as to produce the coherent, consistent programs. This can be achieved by two means:

1. Assigning one test engineer complete responsibility and authority over the various factors of production as they relate to his test program.

2. Defining the responsibility and role of each contributor involved in producing the test program.

Adhereing to the precept of using documentation as the baseline vehicle for process control, a family of controlling (and guiding) specifications can be developed covering each function in the program generation process. Assigning specification titles and numbers to each function, the program generation process previously defined can now be expressed in terms of governing documents. Figure 2 calls out a family of specifications with each corresponding to a function in the test program production process of Figure 1.

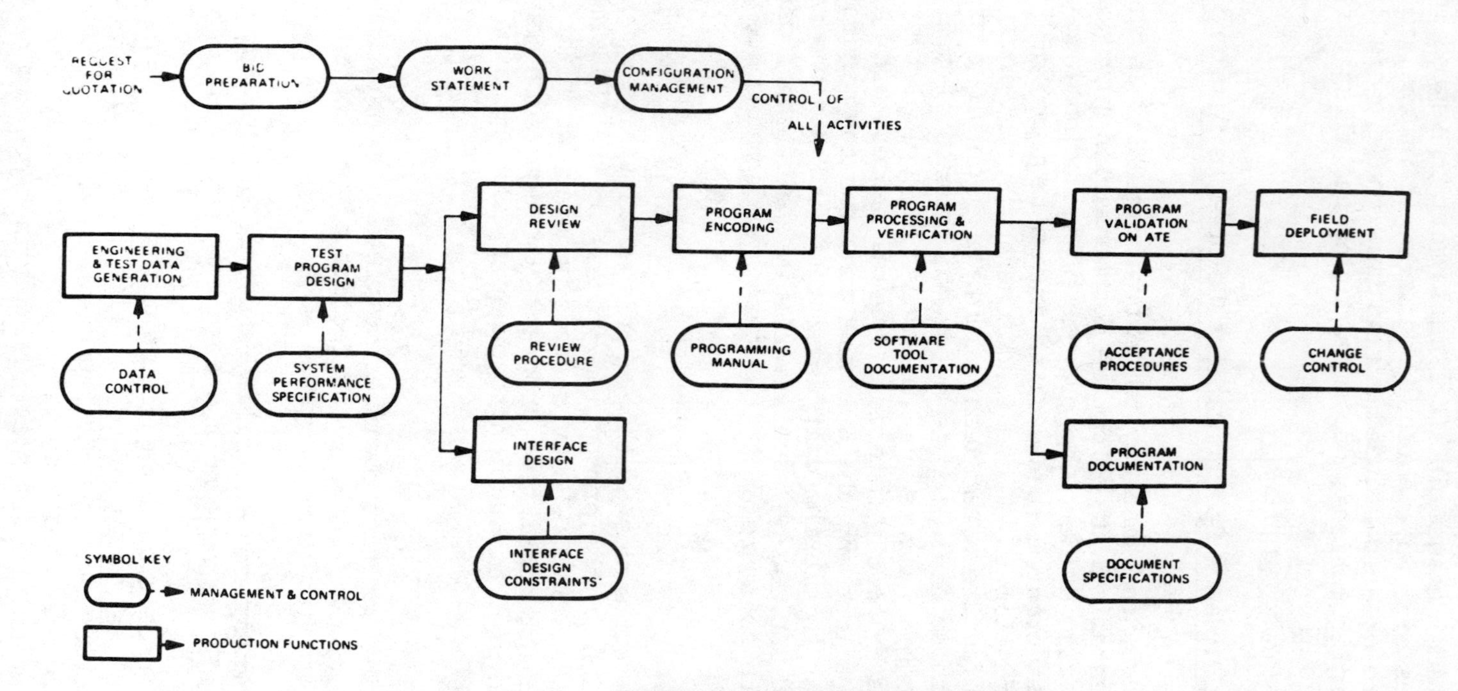

FIGURE 1 PROGRAM PRODUCTION CONTROL PROCESS

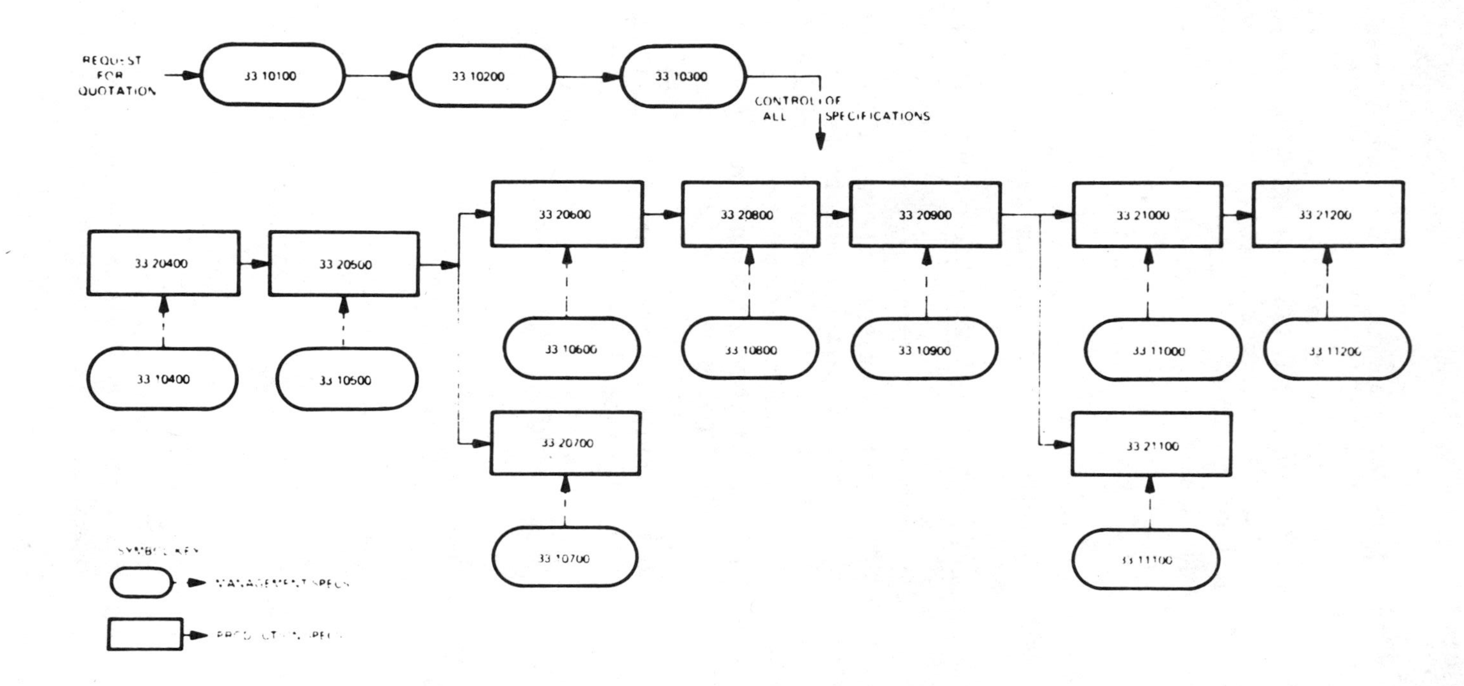

FIGURE 2 PROGRAM PRODUCTION AND CONTROL SPECIFICATION OUTLINE

The applicable specifications divide into two groups:

1. Those related to program production control.

2. Those relating to doing the program production tasks.

This of course, is the classical subdivision of any well organized project between the managers (or controllers) and the workers (or doers). Since the assigned engineering has complete responsibility for the test program, he must be concerned with all of the control specifications. He must also be knowledgeable in each area of specialty covered by the "doer" specifications. Incidently, the engineer also fills the role of the worker in several of the "doer" functions involving test design.

The non-engineering tasks are handled by skilled workers who concentrate on a single specialty such as keypunching, electrical assembly, etc.

4.2 Listings of Typical Test Program Production Specifications

The detailed content of each specification will vary somewhat among organizations and specific requirements of the customer's work statements. To illustrate a typical set of specifications, those called out in Figure 2 are defined very briefly in Tables 1 and 2. Specifications provide the configuration control vehicle and technical guidance for the performance of all tasks associated with test program production. Developing such a system of specifications and keeping them relevant and up-to-date offers the best baseline for configuration management of software. With these tools, people with average technical competence can successfully produce effective test programs with reasonable efficiency. Without such control, even exceptionally qualified people stand an excellent chance of failing in this task.

TABLE 1
Software Management and Control Specifications,
Family Tree and Brief Description

Spec. No. Title and Description

33-1000 Software Management and Control Specifications
Indentured listing of specifications within the test program management family.

33-10100 Software Bid Preparation
Provides guidelines for estimating cost and schedule for the design, production and validation of operational test programs. Includes samples of estimating forms, their usage and emperically derived factors for determining engineering hours, data processing costs and other factors of program production.

33-10200 Software Work Statement Preparation
Provides guidelines and examples for the preparation of work statements used to guide line organizations in the various program production tasks oriented toward fulfilling program production requirements in accordance with the applicable contract.

33-10300 Software Configuration Management

Provides guidelines for effective production control of test programs as generally applicable to the ATE product line. Provides guidance for using the common technical direction contained in software specifications from which particular work statements can be synthesized. This specification correlates the usage of the full family of software specifications.

33-10400 Test Data Control

Provides guidelines for classification, filing and control of engineering documents used as input to the test design and programming process.

33-10500 ATE System Performance Specification

Defines the capability of the test system from a test designers viewpoint. It constrains the test designer to using the system within its boundry of proven performance.

33-10600 Design Review Requirements

Establishes requirements and guidelines for conducting a technical design review of new test plans (flow diagrams or English language test procedures).

33-10700 Interface Design Constraints

Stipulates limitations of complexity allowable in the design of the interface between the ATE and UUT. A sub-specification of this basic specification must be developed to meet the special needs of each contractor in this regard.

33-10800 Programming Manual

This document provides both control and guidance for encoding the previously prepared test plan in an appropriate problem language acceptable to the program processing function. It serves as a "doer" guide as well as a control tool.

33-10900 Software Tool Documentation

This encompasses a family of documents describing the design and dictating the use of various software tools used in test program processing. It includes program users guide and design descriptions for compiling and verification programs such as an ATE system simulator.

33-11000 Validation and Acceptance Requirements

The basic specification calls out methods of proving performance of any completed test programs. Additional specifications in this subgroup call out specific demonstration and acceptance procedures applicable to the customer requirements.

33-11100 Documentation Specifications

This sub-group of specifications dictates both minimum documentation requirements and special customer requirements that must be met for each deliverable test program.

33-11200 Change Control

This specification (generally developed jointly with or by the customer/user) constrains the modification of test programs and their documentation that have been formally accepted and deployed.

Table 2
Software Production Guides and Specifications,
Family Tree and Brief Description

Spec. No.	Title and Description

33-20000 Software Production Specification Index
Indentured listing of specifications within the test program production family.

33-20400 Test Generation and Evaluation
Provides guidelines for the engineer to generate pertinent test data from raw engineering documentation and bench testing and/or evaluating suitability of supplied test data for generating test programs.

33-20500 Test Program Design
Provides guidelines for preparing flow charts and problem-oriented, English language test descriptions which ultimately must be translated into programs to control the ATE.

33-20600 Design Review Preparation
Guides test engineer in his preparation for a test concept design review. Stipulates requirements for type and distribution of the engineering data package used as the vehicle for design review.

33-20700 Interface Design Guide
This is a comprehensive manual used to guide the test designer in his definition of hardware interface for connecting his UUT to the ATE. It also guides the interface unit designer (a different engineer, perhaps) in the design and fabrication of the required interface, using standard techniques and equipment in so far as practicable.

33-20800 Program Generation Guide
This augments the programming manual by very explicitly explaining the use of various program coding forms and describing the responsibilities of the program coder.

33-20900 Program Processing Guide
This describes the procedure for submitting an encoded program to keypunching, assembling the source deck, submission for data processing and other activities involved in translating the program to machine language and making certain logic checks before release for validation.

33-21000 Program Validation Guide
This provides the test engineer with guidelines for effective validation of the test program on the ATE and defines his obligations in preparation for formal acceptance demonstration of his program to the customer.

33-21100 Documentation Preparation
Stipulates the requirements the test engineer must fulfill in preparing design description and user documentation for his test program. Related specifications cover editing and formatting requirements for the person (and utility program) that converts the test engineer documentation to the customer.

33-21200 Program Change Procedure
Provides instructions to the program coder or other individual who implements authorized changes to test programs and associated documentation.

BIBLIOGRAPHY

(1) Bellcom, Inc., "Management Procedures In Computer Programming for Apollo", Interim Report for NASA, TR-64-222-1, 30 November 1964.

(2) Project SETE, N.Y.U., "Problems and Pitfalls in Automatic Test Computer Programming", May 1965.

(3) B.L. Ryle, "An Engineering Approach to Configuration Management", IEEE Proceedings of the 1966 Automatic Support Systems Symposium.

(4) M. V. Ratynski, "The Air Force Computer Program Acquisition Concept", AFIPS Conference Proceedings, Volume 30, April 1967.

(5) L. V. Searle and G. Neal, "Configuration Management of Computer Programs by the Air Force", ibid.

(6) B. H. Liebowitz, "The Technical Specification - Key to Management Control of Computer Programming", ibid.

(7) F. Liguori, "Software Management Through Specification Control", Software Age, May 1968

(8) F. Liguori, "Software Automation in Automatic Support Systems", IEEE Proceedings of the 1969 Automatic Support Systems Symposium.

(9) "Engineering Changes to Weapons, Systems, Equipments and Facilities" U. S. Air Force-Navy Aeronautical Bulletin, ANA No. 445, 12 July, 1963.

MANAGEMENT OF DATA FROM AUTOMATIC TEST SYSTEMS

Paul. W. Accampo

AUTOMATIC MEASUREMENT DIVISION
HEWLETT PACKARD COMPANY

ABSTRACT

A major advantage of using a computer to control an automatic tester is that data generated is returned in a compatible format, eliminating the need to gather and convert it from hand-written records. Most users of testers are not utilizing available computer power to process this data, because their language formats may not allow for convenient handling, their hands are full just trying to get the system into operation, or program changes are so frequent that there is no format consistency.

This paper reviews a few uses to which data can be put, analyzes the kinds of information available in a test program, and suggests a format for collecting data for transfer to a mass storage device such as a disk memory or magnetic tape to form a data base.

A wide variety of statistics can be gathered and generated from a data base, usually in the form of listings or plots. All requirements are met by data contained in and generated by test programs, correlated in different ways.

INTRODUCTION

In many applications for which automatic testers have been procured, the prime reason has been to speed up testing. Indeed, in some applications, the ability to take data at a high rate is the only means by which testing can be performed. The evolution of system controllers from forms of paper tape control to computers has enabled the testing environment to become automated in the true sense of the word, as opposed to simply mechanizing the testing process to achieve a higher throughput.

Automation has been defined as a technology involving communication and control of an environment[1]; the availability of the computer in the tester allows great quantities of data in a readily-processed format to be obtained and used for control of the functioning of the tester, the flow of items through it, the quality of its workload, and also the efficiency of its overall environment.

For several years, organizations at Hewlett-Packard's Automatic Measurement Division have been working toward the establishment of standard test programming techniques. Various methods of reducing required coding and simplifying measurement techniques have been developed using the BASIC language, and are taught in the Division's 9500 System User Training Course. As part of this effort, efficient, easy-to-program techniques for organizing and removing test data were developed.

The key to utilization of the computer's capability is the manner in which the essential information in a test program and about the test process is obtained and placed in a usable format. The following discussion will attempt to define forms of management information derivable from test programs, and indicate some benefits therefrom. With a standard method for organizing data among all test programs, a software package known as an "executive" can be developed, relieving test personnel from the requirement of understanding how a computer processes data. Stored on peripheral equipment such as magnetic tape transports and disc drives, the "raw" test program data is later accessed by programs which correlate information to provide a wide variety of management reports.

The main hindrance to obtaining this information, however, is not technical; rather it is one of conflicting priorities. The first goal of an ATE implementation effort is to get the system into production, and it is sometimes a difficult task to design and finance the required standardization. Some of the problems are:

a. inability of the test language to handle data.
b. inadequate definitions and standards.
c. problems of implementing manual procedures.
d. inadequate programming background.
e. lack of initial motivation to establish data processing capability.
f. increased program complexity.

The point to remember is that, "Automation is not merely the process of doing automatically that which was once done manually".[2] One of the main differences between automation and simple mechanization is the data derived from a process, and the uses to which it is put.

CHARACTERIZATION OF MANAGEMENT INFORMATION

The information that we are attempting to obtain will allow a manager to make decisions affecting three areas–the test system itself, the workload it processes, and other test-related activities.

Tester efficiency is an important criterion. If throughput rate is at all critical, a manager needs to know how fast programs run, time required for operator actions, the number of items processed in a given time, and time spent programming the tester. Some functions, such as time to perform maintenance, are not easily handled automatically; however this does not preclude obtaining information manually, and adding it to automatically gathered data to produce a composite report. Decisions can now include system maintenance intervals, revisions of slow-running programs, running of additional shifts, and improvement of manual handling techniques. Data from the tester provides the means to identify and allocate resources in areas where improvement is needed.

Reprinted from *IEEE Auto. Support Systems Symp. Rec.*, Nov. 1972.

Throughput doesn't buy anything if testing is invalid. If a UUT* does not perform statistically as expected, the test standards or the unit itself may require modification. The ability to accumulate measured values over a period of time and display them in the form of plots or statistical results enables a manager or engineer to change test procedures on the basis of actual UUT behavior. Very tedious to obtain manually, this information allows the user to improve the ability of the tester to separate good and bad units. The same data can be evaluated as a function of time to observe trends.

A somewhat more complex use of measurement data is to correlate measurements in accordance with which ones failed and how they failed (high or low). Each unique correlation represents a failure type which can be given a number and filed. This assists in fault isolation, and provides information on failure frequency type. If a population of units is being maintained, statistics such as mean-time-to-failure can be determined. Information of this type allows management of the system workload. As another example, knowledge of a statistically (hence economically) optimum time to calibrate instruments, derived from known drift trends, enables calibration frequency to actually be reduced while better maintaining the population. Additionally, it may be possible to "smooth" fluctuations in incoming workload through more accurate scheduling.

The tester interacts closely with other activities in its environment. Test times vary with faults encountered, as do requirements for repair parts and repair times. In one case[3], the interrelationships of these activities have been modeled to simulate the effects of the tester in the environment. Automatically generated parameters can conceivably increase the accuracy and usefulness of such a model, as a step toward optimization of an entire operation.

OBTAINING MANAGEMENT INFORMATION FROM TEST PROGRAMS

Most management information is obtained by correlating occurrences of similar events. For example, to determine if one system operator is faster than another, one can locate a program that both have run several times, sort elapsed run times for each, and average them. This example, if it is performed under computer control, assumes that a test program ID, elapsed time, and operator ID are all available. In most cases, however, it is not clear what future information requirements will be, and it is therefore important that all data which describes a test or is created by it is put into data files.

To insure that sufficient data elements are identified prior to writing test programs, it is useful to group them according to their source. Table 1 contains a partial list; however each ATE manager should review his situation and determine additional parameters he may require.

Whether all of this information can or should be stored in a computerized format is questionable, since mini-computers used to control testers lack large amounts of memory. Primarily this limits the type of information processed to that which is numerical in form. Elements such as item name, measurement location, and initial conditions of the unit do not change between executions of the program. Their best form of storage is probably in separately written descriptions which include appropriate diagrams. Users needing rapid access to pictorials and lengthy descriptions have added computer-controlled microfilm readers which, on instructions of a test program, locate and display a page displaying text and/or graphical information. Instructions to locate microfilm pages are then easily stored in a numerical format.

*Unit under test

TABLE 1

TEST SYSTEM OR OPERATOR
- DATE
- TIME
- ELAPSED TIME
- OPERATOR ID
- SYSTEM ID
- LOCATION IN STORAGE
- SIZE (NO. OF WORDS)
- TEST PROGRAM NAME OR NO.

UNIT UNDER TEST
- NAME OF ITEM
- SERIAL NUMBER
- ITEM IT IS PART OF
- TEST RESULTS (MEASURED VALUES)

TEST PROGRAM
- NOMINAL (EXPECTED) VALUES
- TOLERANCES
- PARAMETER MEASURED (VOLTS/OHMS)
- LOCATION OF MEASUREMENT (TP-1, COLLECTOR OF Q9)
- INITIAL CONDITIONS OF UUT (DC POWER APPLIED, PIN 1 GROUNDED)

Nominal values and tolerances in the test program normally do not vary unless it is revised; they are needed for failure analysis, and thus should be maintained with each program. They need only be stored once. The remaining information is almost all numerical, or does not contain significant numbers of alpha characters.

As the test program executes, we desire to collect the information and store it on a mass memory device such as a magnetic tape unit or disc memory, so it can later be accessed. The software mechanism which allows data to be organized and transferred is the array. An array is a designated portion of core memory reserved for data, consisting of elements, each of which is given a number. The array as a whole is given a name. In the BASIC language, the name is a letter of the alphabet. To reserve storage, a "dimension" statement, which specifies the name and array size, is used. Thus,

```
100 DIM A(100)
```

reserves storage for a 100-element A-array. Other types of statements refer to individual elements. Thus the statement

```
105 LET A(6)=4
```

places the number 4 in the storage area reserved for the sixth element of the A-array. Arrays have two powerful attributes;

the first is the ability to obtain data randomly by specifying the element number:

```
120 PRINT A(6)
```

The second is the ability to transfer an entire array from core memory to a bulk memory. This is accomplished by a command of the type*

```
130 DRITE (9000,A(1),1,100)
```

The command writes the 100 elements of the A-array on to a defined storage area on disc memory, called a file, which is given the number 9000. The writing process starts at the first variable of file 9000, and ends at the 100th variable of the file. The software operating system establishes files in advance upon request of the operator who assigns them a name or number, and a length, in an HP system, of up to 16383 elements. If new data were obtained, and it were desired to write a second 100 elements into file 9000, the statement would be

```
150 DRITE (9000,A(1),101,100)
```

Standard arrays are defined to contain Table 1 data. Most of the data required is placed in the D-array, each element of which has a specific abbreviation defined in Table 2. This array contains all of the information originated by the test system and operator, and the serial number from the UUT. The D-array has been standardized at HP as being 30 elements long for all programs.

TABLE 2
D-ARRAY SAMPLE ALLOCATIONS

D(1)	– SIZE OF DATA BLOCK IN A PROGRAM
D(2)	– PROGRAM NO.
D(3)	– NO. OF MEASUREMENTS IN PROGRAM
D(4)-(5)	– UUT SERIAL NO.
D(6)	– FILE NUMBER ON DISK
D(7)	– VARIABLE NUMBER IN FILE WHERE DATA BEGINS
D(8)	– SEQUENCE NUMBER IN FILE (=TOTAL NO. OF PROGRAMS IN FILE)
D(9)	– DISC CARTRIDGE NUMBER
D(10)	– OPERATOR ID CODE
D(11)	– YEAR
D(12)	– MONTH
D(13)	– DAY
D(14)	– HOUR (BEGUN)
D(15)	– MINUTE (BEGUN)
D(16)	– SECOND (BEGUN)
D(17)	– ELAPSED TIME
D(18)	– SPARE
D(19)	– PASS/FAIL FLAG
D(20)-(30)	– SPARE

A test program consists of a sequence of measurements: values of parameters at points on the UUT. Although parameters vary, e.g., volts, ohms, distortion, % regulation, all have the common characteristics of a nominal (expected) value, an allowable tolerance on that value (usually expressed as a percentage), and a measured value. Thus data which defines all measurements can be stored in three arrays:

M–measured values
V–expected (nominal) values
T–tolerance (% or fraction of V)

The M, V, and T arrays can be variable in length depending upon how many measurements are made by the program. Thus if an amplifier's gain is changed over its frequency ranges in 35 measurements, total storage required is

```
10 DIM D (30), M (35), V (35), T (35)
```

or 135 array elements. If measurements are numbered from 1 to 35, the values of the second would be: measured–M(2), expected–V(2), and tolerance–T(2). The essential data of the test program is thus in numerical, computer compatible form, and can be processed in many ways by many different programs.

A point to stress here is that a measurement may constitute many instrument readings, all of which are processed to yield one value. An example might be linearity of a sweep voltage. Many readings can be compared to yield deviation from a straight line, a single number. Since no single point alone determines pass-fail criteria, individual points are not stored, only the computed result. This avoids storage of large amounts of data which, over many programs, could saturate the data-handling capabilities of a mini-computer system.

The M, V, and T arrays characterize program measurements. From them, each measurement can be analyzed for a pass or fail condition, and direction and magnitude of a failure determined. Relating M-array values of a single measurement over many units provides information on statistical behavior of the measured parameter.

Most of the D-array elements change for each UUT, as do all of those in the M-array. These are thus transferred to mass memory prior to concluding the run. The V and T arrays do not normally change, and may be separately stored; however provision should be made for updating them if the program changes. Techniques for storing arrays will be discussed momentarily.

Note that these arrays are identical for all programs, by virtue of their generalized definitions. Their standardization is critical to the ability to write search routines which operate on all programs.

DESCRIPTION AND USE OF AN EXECUTIVE

The term "executive" is applied to a software master routine which calls subroutines as needed to perform a function. Every test program consists of relatively standard functional elements executed in a specific order. For example,

CLEAR SYSTEM
INITIALIZE UUT
PERFORM MEASUREMENTS
PRINT RESULTS
STORE DATA

Some of these functions are system-oriented, and are thus identical for all test programs. If the programs are written using standard arrays, printing data, except for descriptive titles,

*A slight simplification of some statements has been made for clarity of the explanation.

consists mainly of printing contents of arrays. Storing data is transferring contents of arrays. The routines which perform these functions can be implemented as pre-written "boilerplate" subroutines and used for all programs. The test programmer now need only be taught how data is to be organized into arrays; the executive transfers data at the proper time.

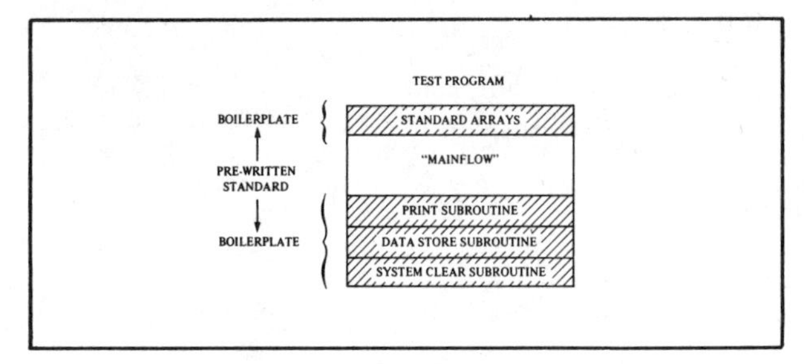

Figure 1. Boilerplate Portions of a Test Program

The use of standard boilerplate routines has another obvious advantage: test programmers do not duplicate each other's efforts, thereby reducing amount of code required to generate a test program.

The Executive referred to is not part of the system software. It is written in BASIC, and operates in the environment of TODS, HP's Test-Oriented Disc System. One might question whether or not these routines should be written into system software, since they are standard to all programs. Requirements for data and display of information vary greatly from one application to another. Writing the executive software in the test language allows for ease of change and adaptation to many situations. A task of writing comprehensive Executive and set of standard subroutines is not a simple matter, and a continual evolution of it can be expected.

DATA BASE ORGANIZATION

A data base has been defined as a centralized collection of all data stored for one or more applications[4]. Each time a test program is run, additional information is added to the base. When management information is desired, a program is written to search the data base and perform the required processing.

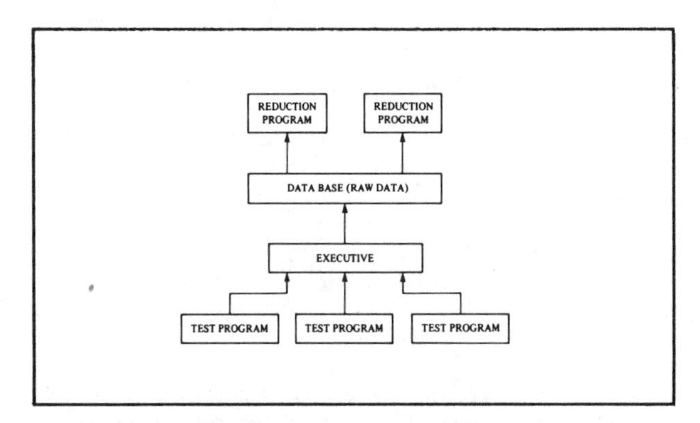

Figure 2. Information Transfer to and from a Data Base

The data base itself exists on the medium of disc or magnetic tape or both. The ideal situation is to have both available since, although they perform identical storage functions, there is a significant tradeoff between speed and volume of data. The moving-head disc is a pseudo-random-access device which can access a specific data word among two million within 100 milliseconds. The magnetic tape transport may take several minutes to locate the same item, although total storage capacity of a magnetic tape reel can be of the order of ten million words. Once a file is found, the transport has rapid access to very large amounts of information. Files are storage areas with beginning and ending marks which allow software to locate them. When established in advance by the system user, the software operating system sets aside disk or tape real estate and remembers its location.

Probably the simplest form of organization for testing purposes is a long cumulative file. The subroutine which writes the data is the same for all test programs, and puts the data down sequentially beginning with the D-array.

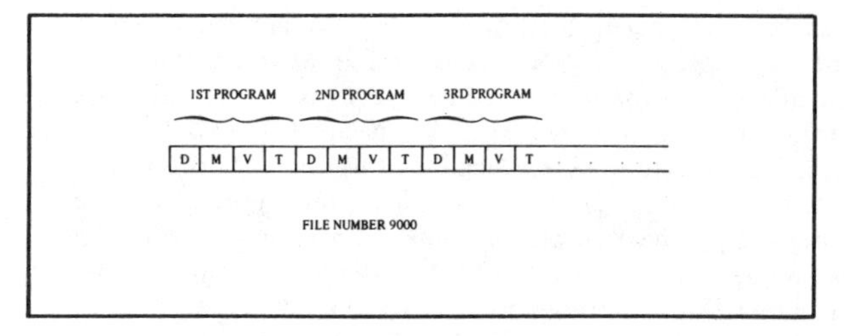

Figure 3. Cumulative File

If there is more data than one file can hold, several files can be established, and the software can automatically advance to the next available one.

The design of a data base should be carefully considered, since search speeds are significantly influenced by its organization. The glaring defect of the cumulative structure shown is the redundancy of the V and T arrays, which triples the total file length. If we are willing to assign several files for each test program, we can eliminate redundancy at the expense of increased complexity of write and search routines.

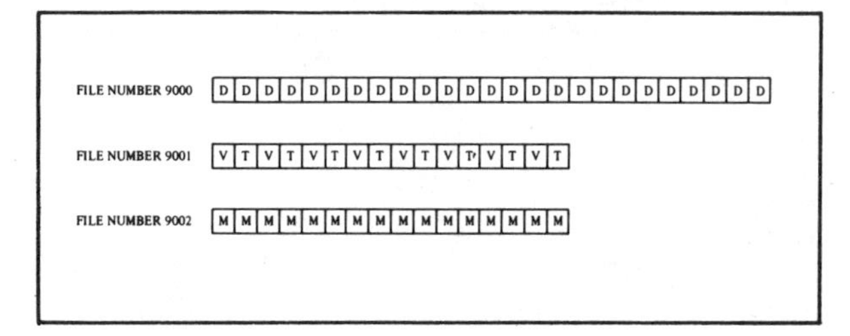

Figure 4. Distributed File Configuration

In this case, a special routine writes V and T arrays once, and the standard write routine puts data into two files, one for D, and one for M. The D-arrays in file 9000 are the only ones which must be sequentially searched. Additional D-elements are assigned to specify the location of V and T in file 9001, and M in file 9002. If most searches require access to the M, V, and T arrays, searching is slowed by the fact that three files must be found (M, V/T, and return to D). If the files were on a single reel of mag tape, the search process could take hours.

Assigning one file for each test program allows the ability to write V and T only once, and search fairly rapidly.

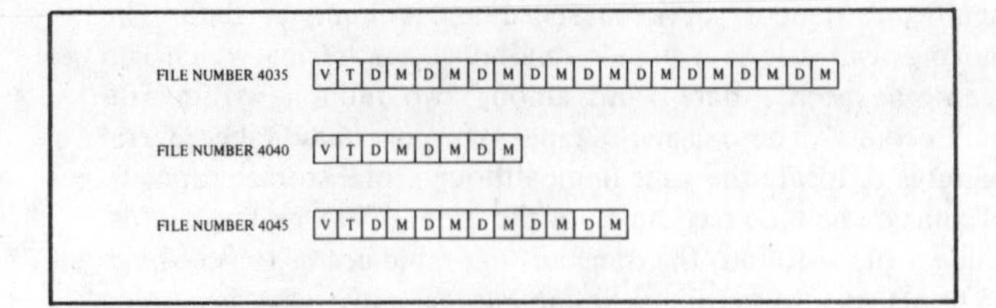

Figure 5. Cumulative Files by Test

A disadvantage here is the number of files which must be maintained. If many programs are stored, searching involves locating and reading many different files; location can be a time-consuming process.

DATA REDUCTION

A simple management report, a list of all test numbers run on a particular day, might be generated as follows: Each D-array is loaded back into core and the elements containing date information are compared with the specified date. If a match occurs, the D-element specifying the program number is printed as a listing output. A more complex search process can be illustrated by the following example: It is desired to compute the average values and standard deviation of an amplifier gain measurement to determine variance over a population. Assume files are arranged in a cumulative fashion and the measurement is No. 14 in program 4090. The flowchart in Figure 6 indicates the actions of the search program. A variable is increased by one each time a valid program is found, to provide the divisor that determines the average and standard deviation. The X-array is a means of temporarily storing all values collected from the M-array prior to performing computations.

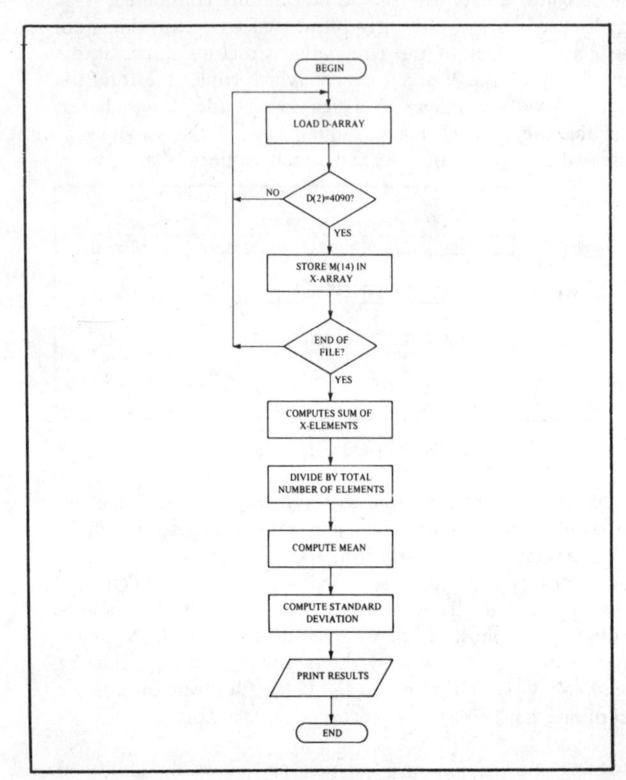

Figure 6. Typical Search Program Logic

As previously noted, a mini-computer is not the optimum system for the type of data reduction being described. The large mainframe computer has larger core memory, and a sophisticated operating system which allows simultaneous processing and data transfer, and more numerous peripherals. A search or sorting algorithm which takes only a few minutes on an IBM 360 may take an hour on a mini-computer operating with a single disc. This does not mean that a great deal cannot be accomplished with available equipment. If a tester is not used during a particular shift, hours are available during which even a slow computer can generate numerous reports. Reduction programs can be written in such a way that data needed for a report request is entered in advance, and requests "stacked". The operator can then enter requests for reports he wants that night and the system will automatically process reports. The major inhibiting factor to a completely automatic process is the need to change disc cartridges or magnetic tape reels.

It is worthwhile to consider the quantities of data which can be stored and the speed with which items can be accessed. Assume we have a test program consisting of 100 measurements. Using the D, M, V and T arrays, data occupies 330 array elements, or in a 16-bit computer, 660 words. Assuming that a typical disk has 500,000 words available for data storage (a conservative number), the test can be run 800 times before the disc is filled. If the test is run once every 30 minutes, one shift per day, a disk cartridge would be filled about every ten weeks. A similar situation in which magnetic tape was used would fill a reel every 75 weeks. We can thus see that the cost of storage hardware, disc cartridges and mag tape reels is not prohibitive. If the initial cost ($10K to 15K) of a tape transport can be justified, there are a number of advantages: 1) Search times are reduced and tape capacity increased because data is transferred in large blocks from tape to disc prior to being processed. 2) Reels of mag tape are cheaper (≈ $15) than disc cartridges (≈ $100). 3) Magnetic tape format is standardized, allowing data transfer from one computer system to another. Thus, if a large computer is available, the main function of the tester can be collection and transfer of data. Even if the initial cost of the magnetic tape transport is prohibitive, sufficient disk cartridge area may be available to establish a data base which can later be transferred to tape through an upgrading of the system.

The data reduction previously illustrated is relatively simple, in that it involves only one pass at the files. If the amount of data extracted from the file exceeds available core, it must be temporarily stored on an additional file. If a disc is full, it can be searched in approximately 30 seconds using previous assumptions and the cumulative file organization. A reel of magnetic tape can be searched in approximately six minutes. These numbers neglect computational time required by the computer, which is assumed to be much less than that to access a peripheral; nevertheless, a reasonable guess of the time required to generate a management information report using a disc, mag tape, or both, from several months of accumulated data, is of the order of 2 to 10 minutes.

If the specific hardware and software configuration is known, one should make a reasonably thorough investigation of the time to search or process files and the amount of data under consideration. Don't save volumes of data which require considerable storage and search times to obtain information of marginal value. Determine this tradeoff before writing reduction programs.

Information derived in this manner can be presented in several forms. A simple one may be a report, in listing form, of a previously run program or accumulated data. One of the most powerful abilities of the computer is to perform mathematical computations on the data. Mean values, standard deviations, extreme values, failure rates and percentages are derivable with only a few lines of programming, as are more complex formulas associated with statistical significance tests. Plots are also relatively easy to generate, using a teletype, line printer, or, for more resolution, a graphic plotter. Trends can be analyzed, and approximated with algorithms that fit formulas to existing data (assuming it behaves reasonably), enabling one to extrapolate and predict future performance.

IMPLEMENTATION ON MULTI-STATION SYSTEMS

Previously implied is the fact that automation is a technology that applies to more than one process. Testing in a facility interacts with many other operations. Control of the entire plant requires data to be fed from each of these to a central facility, where it can be analyzed and control information derived. Since the central facility possesses a relatively large storage capability, the need for disc and magnetic tape at the tester is eliminated. The central system operates in a multiple-mode, sending programs and data to satellite stations, while simultaneously running reduction programs on the data base. As an item moves through a station, data (which in some cases may be manually entered through a terminal) is transmitted to a central station, and added to the data base, enabling a manager, who needs information on the whereabouts and status of an item being processed or who needs to know how well the plant is performing, to interrogate the central through a terminal. The central calls up reduction programs that search the data base and display appropriate information.

SOME PROBLEM AREAS

The test language used must be capable of handling arrays in a fashion that contains all desired data. If it is the only language used on the system, it must have an output format flexible enough to accommodate the various report forms desired. The adaptation of BASIC to automatic testing proved to be a fortunate choice in both of these areas. Moreover, the system software, TODS, uses an identical array format for both BASIC and FORTRAN, enabling the data generated in the previously-defined BASIC arrays to be processed by FORTRAN programs, which run faster and use less core.

Even if the system is capable of handling data, a key obstacle is that the priority of establishing a data base is always lower than that of getting the system into production. The increased program sophistication and the lack of experience of personnel operating a new system typically causes programs to be written in any format that works. Only later does a desire for data utilization and program standardization come about. Courses given at HP have attempted to alleviate the problem by giving examples of executive and boilerplate design, yet each user needs an executive that solves his problems. The best course of action may be to develop an executive in parallel with the initial group of unsophisticated test programs. Programs are later converted to use the executive format on a prototype basis, in order to exercise the features of the executive and make final modifications. Finally, all future programs are written in the refined format. The system remains in production during the entire development process.

Another impediment is resistance to standardization. The principal arguments are 1) "The standard format cannot solve my problem," and 2) "I cannot control my people that closely without removing creativity from their jobs." In these instances, two actions can be taken. The first is to show that a standard format will solve the problem, or, if it won't, to improve it. Our experience indicates that there are very few test situations that the format presented here cannot accept. The second is to establish job objectives which are oriented toward testing, not programming. The problems encountered in attempting to make fast, accurate measurements are quite challenging; however, it should be recognized that test implementation is highly repetitive, and usually a technician-level position. Engineering supervision is usually required to insure that valid programs are written.

In certain specialized testing applications, particularly those in which test programmers are not computer or system oriented, a "language barrier" exists. One of these is in the field of instrument calibration, in which terms such as "% of full scale", "% linearity", "±1 digit", and "parts per million", are commonly used to specify instrument performance. Translation from these specs to the M, V, T array definitions presents some difficulty, particularly since these units must be maintained in the test report, if the document is to be meaningful. If the M, V, and T definitions are not maintained, however, each test program must have a unique reduction program in order to generate statistics. A partial solution appears to be to store the information into arrays, and if the resulting format is not standard, to write at a later date, a routine which changes the format into a standard one. The key requirement to put all required data into arrays remains, however.

Some test programmers have had difficulty in writing complex, looping measurement sequences and keeping track of which measurement goes in which array format. As experience is gained, they develop programming techniques which assist in solving these problems. It appears that once a standard format is developed, an effective training program is necessary for it to be adhered to, and this should be devised in-house.

The perennial nemesis of ATE–the fact that most initial implementations are mechanizations of manual test procedures–touches on the data gathering problem. What, for instance, is tolerance when a specification reads "<50 mv"? Frequently this type of a specification has been given to make manual operations easier, or because a nominal value wasn't known. The ability to perform computations and to make measurements under the controlled conditions of an automatic tester helps define a nominal value and tolerance for these measurements, and thereby provides additional information on UUT performance.

SUMMARY AND CONCLUSION

Development of a data base from data supplied by automatic test programs is a means of deriving management information from which decisions to control a process are made. Although an automatic tester is usually initially justified by its faster testing capability, modern testers are controlled by mini-computers using sophisticated operating systems. With either a disk memory, magnetic tape memory, or optimally, both, the computer can be used off-shift to perform fairly sophisticated reduction tasks and provide a wide variety of management information.

Since this activity normally starts out as a lower priority task than that of putting the tester into production, it must not create a burden on test programmers. Fortunately, as sophistication of programming techniques is achieved, an executive which does exactly this and otherwise reduces and standardizes the programming task can be developed.

Careful organization of the data base is important, since search times are affected and are generally not fast on a mini-computer-based system. Yet if reduction programs are run on a shift during which the tester is not otherwise used, the cost of obtaining desired information is low. In a more sophisticated environment, the data base is transferred to a central processor which provides storage for programs and data from several stations, or simply terminals used as input devices. Information can then be obtained as testing proceeds.

Management information can be used to optimize the operation of the tester through more efficient testing, or through gathering of statistical information which can indicate how often a maintained population needs to be tested or calibrated, and what types of tests are most effective. Knowledge of rate and types of workload at many points within a plant enables co-ordination of all processes, and a true implementation of plant automation.

There are two major challenges facing the ATE manager desiring to make use of his test data. First, he must standardize program formatting so that all programs operate with the same set of arrays, and maintain format consistency throughout all programs. Secondly, he must accomplish the design of a standard format and data reduction programs while the system is in production, since production is its prime justification. The major tools he has to meet these challenges are a clear set of program objectives that specify the importance of standard arrays and reasons for their use, and a thorough effort to train in their use.

REFERENCES

1. Automation, Its Impact on Business and People, Buckingham, Mentor, 1961.

2. "How to Establish an Automated Testing Facility. Problems and Solutions"; W.J. Kahn, Industrial & Scientific Management, Inc.; 1971, p. 13.

3. "Evaluation of Automatic Test Equipment Through Monte Carlo Simulation"; L.W. Wagner, J.W. Edwards, Procedures of the 1970 Automatic Support Systems Symposium, IEEE Catalog 70C52-AES, p.p. 177–184.

4. "Basic Concepts in Data Base Management Systems" Datamation, July 1972, p.p. 42–47.

ESTIMATING ATE SOFTWARE COSTS FOR A SUPPORT ANALYSIS STUDY

Phillip M. Knapp
Assistant to the Executive Vice President
AAI Corporation, Cockeysville, Maryland

Estimating software costs for the development of test programs for an automatic test system is perhaps one of the most difficult tasks in a preparation of a cost analysis for support of electronic assemblies. If no means are available for estimating this significant cost factor, it is impossible to provide a realistic support cost analysis or trade-off analysis where alternate support possibilities exist. One of the purposes of the Support Analysis Program, described in this paper, is to provide a means for automatically estimating this software development cost. The program also includes a means for developing a support trade-off analysis between special support equipment, general purpose test equipment, and the automatic test system.

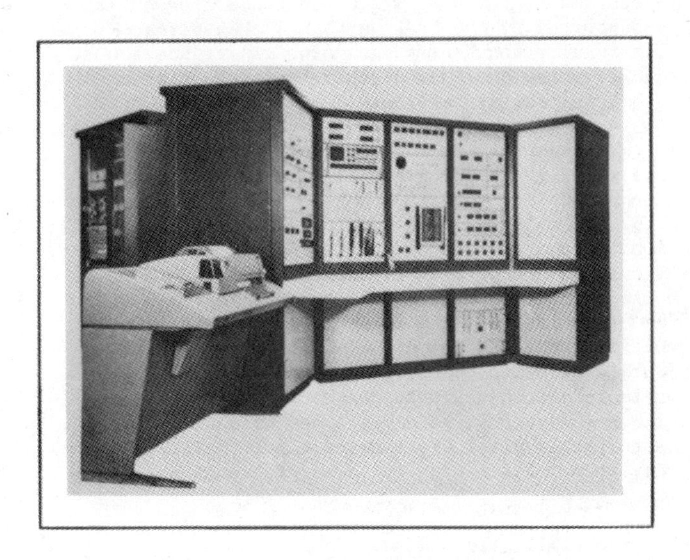

Introduction

The Support Analysis Program was designed to run on the 5500 Automatic Test System presently in use for general purpose avionic support at the seven Naval Air Rework Facilities (NARFs). This automatic test system was designed as a shop aid to reduce avionics maintenance costs by supporting high failure rate items and UUT's (unit under test) with high manual to automatic test time ratios. The problem, therefore, exists at the NARF's to evaluate the desirability of supporting a particular UUT on a 5500 as opposed to continued support on SSE or procurement of new SSE.

At the suggestion of the NARF's, the Support Analysis Program was prepared which would permit a rapid trade-off analysis to be made, evaluating alternate support possibilities including special support equipment, general purpose test equipment and the automatic test system. This Support Analysis Program is presently being utilized and with continued cooperation of the Naval Air Systems Command, data is being fed back to AAI to improve the data base represented by the equations included in the program.

The critical ingredient with regard to ATE is the cost of preparing the test programs. Unfortunantly, no means exists which provide an exact analysis of an exact model for the development of programming time and cost. Many experienced programmers are capable of providing fairly consistent estimates of software development costs, however, few are willing to define the estimating procedure since it involves many subjective as opposed to objective estimating processes. In an attempt to provide an automatic means for developing these estimates, AAI has collected all of the data available for units programmed on the AN/MPM-55 and the 5500 ATS system which include the actual programming experience, machine utilization during program debugging, interface hardware costs, and various factors descriptive of the unit to be tested. This data was then processed using linear regression modeling techniques to develop equations which would use these descriptive factors and produce programming time and costs. These factors include such items as the number of active element groups, number of active connector pins and test points, number of modules (SRA's) in the UUT, number of IC's, logic depth, etc. These equations permit an estimate to be made of software costs for functional, limited diagnostics, and piece part diagnostic programming time. End items (LRU or WRA) and modules (SRA) are considered separately with alternate equations which reflect the depth of test procedures available as documentation for the UUT. Digital, analog, servo, RF, microwave, and pneumatic categories for the UUT or combinations of the above can be evaluated using the support analysis program.

Reprinted from *IEEE Auto. Support Systems Symp. Rec.*, Nov. 1970.

One will find that experts dealing with different automatic test systems may have considerable divergence in their software estimates. It should be emphasized this program is intended to reflect experience with the 5500 automatic test system in the NARF environment. This environment can be quite different than with other automatic test systems, in particular where programming is supported by the ATE manufacturer as opposed to the NARF situation where programming is supported by the NARF personnel themselves. As a result, the NARF programming experience shows results which are considerably less than with most automatic test systems. This can be attributed to the fact that programming is pursued to a depth which is found to be most cost effective as determined by the people utilizing the equipment as opposed to a situation where programs are prepared against a defined programming specification. For a relatively complex module or avionics black box, it is impossible to write a perfect program. If this is realized, then the ATS is likely to be programmed to what it should do and not what it is technically capable of doing. The Support Analysis Program will produce an estimate of the programming time and serves as an experienced bench mark as to the level of programming effort consistent of all previous NARF programming experience.

Program Utilization

The program is designed to run on the Interdata 4 Computer incorporated in the 5500 system. When used, the program utilizes the full 24K bytes of core memory included in the Master 5500 Test Station. Only the computer and its core memory are used when running the program and the test station itself, drum memory or associated controller are not utilized.

The program is loaded through the high speed tape reader and is writen in interactive FORTRAN with data being inputed by the operator and printed out to the operator through the ASR-35 Automatic Typewriter. After the program has been completed, and a recommended support plan printed out, additional computed data values for all variables may be obtained utilizing special subroutines in the program. Approximately 5 minutes are required to input data and produce a printout of a recommended support plan.

Support Analysis Program

The Support Analysis Program shown below in block diagram form asks a series of questions to the operator defining the alternate support possibilities and requesting information which characterizes the UUT. These characterizing factors are used in equations to develop programming costs and machine run time on the 5500 ATS. After all questions and data requirements have been satisfactorily answered, the program then computes the alternate support costs for each of the 3 categories, i.e. SSE, GPTE, ATE, then provides a trade-off analysis, recomputes the support costs and provides a second trade-off analysis to resolve a recommended support plan for the three levels of support. These levels are functional testing, limited diagnostic testing, and piece part testing. The program then produces an automatic printout of the recommended support plan defining cost to test at each of the three levels, method of recommended tests, test cost for each level of testing, programming time, programming costs, utilization of SSE, GPTE, or ATE consistent with the recommended plan, followed by an estimated support cost per UUT in dollars providing data to determine if the unit should be repaired or considered a throw away item.

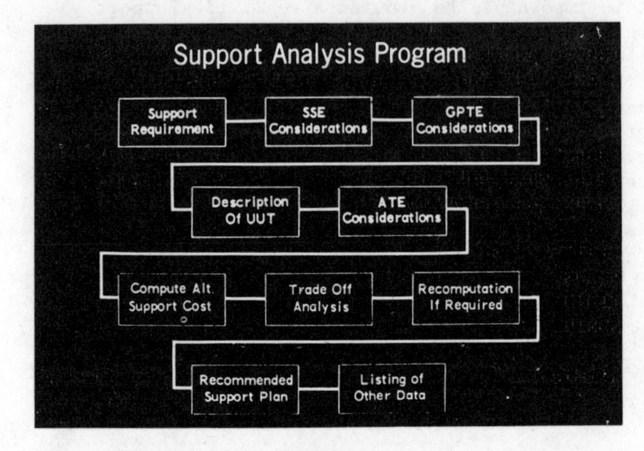

Figure 1

Constants contained in the program relate to average NARF rates, amortization period, cost of ATE, and average hours of availability of equipment per year. These constants are listed below.

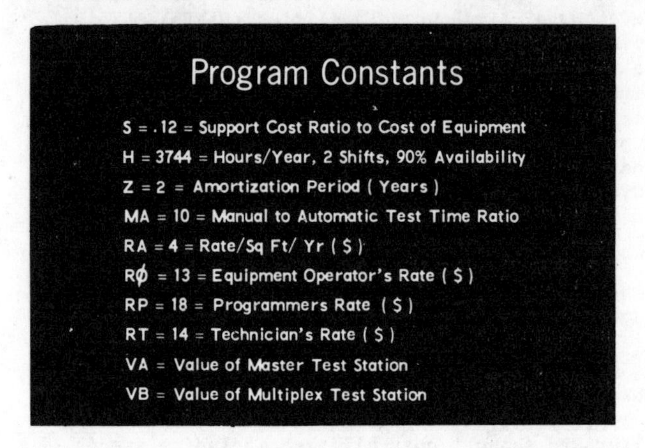

Figure 2

Typical Program Run

The following is a run of a typical analysis with questions asked to the operator via the TTY printed in all caps. Following each question is an explanation and definition of terms utilized.

SUPPORT ANALYSIS PROGRAM 6/1/70 (REVISION DATE)

NAME OR ID OF UUT

= AMPLIFIER A11A

The operator is then requested to answer all questions yes with a 1 or no with a 0. In this example the answers are printed out following each question.

The program may bipass the trade-off analysis and be used to develop ATE costs and programming costs only.

IS PLAN A REQUIREMENT OR PLANT IMPROVEMENT? PLANT IMPROVEMENT

If plan is a plant improvement, a two year amortization period is used for programming costs and hardware costs. If a requirement, the duration of the support requirement in years is requested and this figure used as an amortization period.

NUMBER OF UUT'S PER YEAR? 300

Operator defines the number of candidate UUT's to be supported per year.

DO YOU KNOW RATIO OF FAULTY TO TOTAL UUT'S? YES

RATIO? .6

This ratio is used to define the amount of retesting required for the UUT. Passing a functional test requires no retesting, failing a functional test requires both limited and piece part diagnostics for this quantity of modules which must be subsequently tested for final acceptance. The ratio is defined as variable Y and utilized in a development of Array A.

IS UUT REPAIRABLE? YES

Nonrepairable UUT's are tested at the functional level only.

IS UUT AN END ITEM (LRU OR WRA)? NO

Data is developed for end items for functional and limited diagnostic testing only. Repairable SRA's (Shop Repairable Assemblies) are evaluated through piece part diagnostics.

CAN UUT BE SUPPORTED WITH SSE? YES

Programmable branch to GPTE if not supportable with SSE.

SUPPLIED? NO

If supplied, only the support cost per year is levied against the UUT's and not the procurement costs of the SSE.

ON HAND? YES

If the SSE is on hand, only the support costs are levied against the UUT and not the procurement costs. If the SSE is neither supplied or on hand, it is assumed commercially available and the costs of procurement of SSE are requested. Further questions are asked to define development costs if SSE is not fully developed.

SSE VALUE, $? 60,000

The defined value is multiplied times .12 to develop the annual support cost to be levied against the UUT. If the SSE is to be procured, the procurement costs are amortized and levied against the UUT.

NUMBER OF UUT'S, OTHER TYPES, SUPPORTED BY THIS SSE STATION PER YEAR? 1,000

If other modules can be supported by this SSE station, the annual support costs are divided between the ATE candidate and other modules supported by this station.

SSE CAPABILITY
1 = FUNCTIONAL ONLY
2 = LIMITED DIAGNOSTIC
3 = PIECE PART DIAGNOSTICS

2 (LIMITED DIAGNOSTICS)

If the UUT is a module and is repairable, the operator is forced to define the means by which piece part diagnostics are performed. In this case, piece part diagnostics will be performed on general purpose test equipment.

DO YOU KNOW TEST TIMES? YES

If the operator does not know test times, the manual to automatic test time ratio is applied to computed automatic test times and this value is utilized in development of SSE test costs at the various levels.

FUNCTIONAL TEST TIME, HOURS? .5

LIMITED DIAGNOSTIC TEST TIME HOURS? 1

The operator has then defined the first two levels of testing as .5 and 1 hour respectively.

CAN UUT BE SUPPORTED WITH GPTE? YES

IS GPTE NOW USED? YES

If GPTE is now used, the operator is requested to provide the piece part diagnostic test time. If not now used and test times are not known, the manual to automatic test time ratio will be multiplied by the computed ATE test time and used in determining GPTE test costs.

NUMBER OF UUT'S, OTHER TYPES, SUPPORTED BY THIS GPTE STATION PER YEAR? 500

In a fashion similar to SSE, the annual support costs and procurement costs if any, are proportioned between the ATE candidate and other modules supported by this GPTE station.

GPTE VALUE, $? 10,000

If the GPTE is now used, the defined value is multiplied times .12 to develop the annual support cost per year. If the GPTE is not now used, the value is considered to be the procurement value and this figure is levied against the UUT's.

PIECE PART DIAGNOSTIC TEST TIME, HOURS? 3

IS UUT AN SRA? YES

An SRA (Shop Repairable Assembly) is assumed to be an electronic module. A no answer would terminate the program as may be the case with an instrument or other specialized device not suitable for general purpose electronic ATE.

ARE DETAILED MAINTENANCE PROCEDURES AVAILABLE? NO

Detailed maintenance procedures are considered to be step by step defined tests, presenting stimuli setup, measurement setup, measurement limits, and sequence of tests to define a Go UUT for functional testing. If these are available the operator is requested to define the number of tests required and this data is used in estimating programming time. If only general performance data of the UUT is available, this is not considered to be step by step maintenance procedures. In no case, is it assumed that documentation exists for diagnostics to the piece part level for SRA's.

IS UUT A COMBINATION? NO

If the UUT is a combination, the operator is requested to define which of the following six categories the UUT consists of, and also to define the percentage in each category based on an active element group count. In this case he is requested to provide only a definition of the UUT type since it is not defined as a combination.

UUT TYPE
1 = DIGITAL
2 = ANALOG
3 = SERVO
4 = RF OR VIDEO
5 = MICROWAVE
6 = PNEUMATIC

THE UUT IS DEFINED AS RF

NUMBER OF ACTIVE ELEMENT GROUPS IN UUT? 12

The number of active element groups in the UUT is considered in a somewhat broader sense in this program than normally defined. A count is made on the UUT not only of active components wherein diagnostics will be programmed, but also any additional adjustable elements such as trimmer capacitors, tunable conductors, potentiometers, etc. These are added since programming information must be prepared for calibration and adjustment of these components.

NUMBER OF ACTIVE CONNECTOR PINS PLUS TEST POINTS? 8

This is simply a count of nonredundant connector pins and test points in the UUT which would be connected via interface cables to the ATE. Probe points are not considered in this count.

UUT COMPLEXITY(1-10) ? 3

The operator is requested to make a subjective estimate of the complexity of the ATE candidate UUT based on his evaluation of its complexity against his experience with other modules or end items in this general category. Tests conducted by the AAI Corporation in 1967 involving 25 technicians and engineers evaluating 50 different modules and end items showed a gaussion distribution of complexity estimates. This factor was then incorporated in the development of programmed estimating equations utilizing linear regression modeling techniques.

IS ATS ON HAND? YES

Similar to the SSE and GPTE questions, if the ATS is on hand, only the annual support costs are levied against the UUT. If it is not on hand, then the cost of the ATE is amortized and levied against the UUT.

PRESENT UTILIZATION? .5

If ATS is on hand, the present utilization is requested as a function of total availability time. If the workload presented results in an increase in utilization over 100%, then a recommendation to procure additional automatic test station is made and the support analysis is recomputed based on the additional procurement costs.

IS ATS A MASTER STATION? YES

The 5500 Master Test Station is capable of controlling up to 8 multiplexed test stations. The reduced cost of these multiplexed test stations is factored into the analysis if the ATS on hand is not a master station.

DATA COMPLETE

At this point in the program, all data has been defined and all questions have been answered. Approximately 5 seconds is required to complete the computation. In the case illustrated in the above example, the following recommended support plan will be typed out. Data computed to six places.

RECOMMENDED SUPPORT PLAN

USE ATE FUNCTIONAL TEST

* TEST TIME = 7.6267 MINUTES

* TEST COST PER UUT FOR FUNCTIONAL TEST = $4.92538

USE ATE LIMITED DIAGNOSTIC TEST

* TEST TIME = 8.5333 MINUTES

* TEST COST/UUT LIMITED DIAGNOSTICS = $2.06659

USE ATE PIECE PART DIAGNOSTIC TESTING

* TEST TIME = 24.9708 MINUTES

* TEST COST PER UUT FOR PIECE PART DIAGNOSTICS = $8.0425

PROGRAM ATE PIECE PART DIAGNOSTICS

* PROGRAMMING TIME = 9.7269 HOURS

* PROGRAMMING COST = $1995.08

UTILIZATION OF ATE = 5.43412/YEAR

1 ATE STATION LOADING

SUPPORT COST PER UUT = $15.0344

PROGRAM COMPLETE

The operator may at this point request a listing of all computed data in either a tabular form of final costs and test times for each alternate support possibility, or he may type the word VARIABLES which will result in a complete listing of all variable names and computed values.

Trade-Off Analysis

2 Arrays are used to develop the trade-off analysis. Array A is a 2 dimensional 3 by 3 array having the 3 levels of support as columns, i.e. functional test cost, limited diagnostic test cost, and piece part diagnostic test cost. The 3 rows of Array A are the three categories of support, i.e., ATE, GPTE, SSE. After the computation of test costs for the 3 levels of testing in each of the three categories, elements of Array A are then loaded with adjusted values for these costs.

After Array A has been loaded, Array B is then defined. Array B is a single column Array of 27 rows. Each element in Array B is the sum for the three levels of support in combination with each of the three categories of support. Subsequent analysis of Array B for the lowest value produces the most economical support combination and this then is the recommended support plan which is printed out on the TTY. By loading Array A initially, in all elements, with the maximum number possible in the Interdata Computer and subsequently replacing this with computed costs, undefined support possibilities are therefore driven to the maximum value and in the analysis of Array B are eliminated resulting in only legitimate support possibilities being candidates for the recommended support plan.

Factors loaded into the functional test cost for Array A are adjusted by multiplying times 1 plus the ratio of good to bad UUT's (Y) as estimated by the operator. This will increase the functional test costs by a factor 1 + Y since all units which are failed and must be repaired will have to be functionally tested again increasing this cost. Computed test costs for limited diagnostics and piece part diagnostics are loaded into these elements of Array A and adjusted by multiplying times the factor Y, reducing these test costs, since only the defined ratio of good to bad UUT's must be tested for limited diagnostics and piece part diagnostics.

In the first loading of Array A, programming costs are loaded into the piece part diagnostic level for ATE. If the first analysis of Array B results in GPTE or SSE for piece part diagnostics, then Array A is reloaded with the programming cost for functional testing loaded into this level for ATE. If the second analysis of Array B results in an SSE or GPTE recommendation for limited diagnostics, then Array A is reloaded adding the programming cost for functional testing to the ATE category in Array A. Array B is then analyzed for a third time and this final analysis is the recommended support plan. This is done since the piece part diagnostic program will include both functional and limited diagnostic capability. The limited diagnostic program will include also the functional test capability. By this means the recommended support plan can present to the operator the level to which programming should be accomplished.

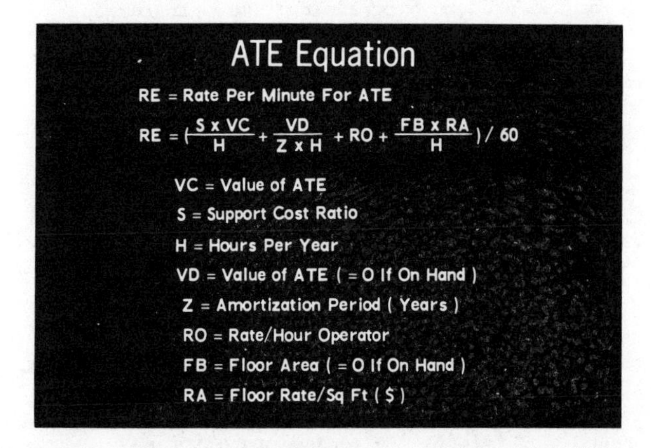

Figure 3

ATE Rate

The ATE Equation, figure 3, shows the cost of the ATE per minute and includes the annual support cost, procurement cost if any, rate of the operator, and the cost of floor area if a new ATE installation is required. This value (RE) is used in developing ATE test cost.

Utilization of ATE is computed by multiplying the number of modules being considered as a candidate per year (N) times the test time in each category recommended for support and dividing by the number of minutes per year of availability of the ATE. This factor is added to the present utilization if defined by the operator and results in a final utilization of the ATE. If this number exceeds 1, an additional multiplex test station would be required to handle the workload and all computations and a new trade-off analysis is made factoring in the cost of this procurement. As a result of this second complete analysis, if the recommendation is still ATE, then the printout will include a recommendation for procurement of an additional multiplex test station. In a similar fashion, utilization of SSE or GPTE is also computed on the same basis and computations repeated as many times as is required to produce a final recommended support plan.

Computation of the ATE Test Cost

Estimating software costs for the development of test programs for an automatic test system is one of the most difficult tasks in the preparation of a cost analysis for support of electronics assemblies. In an attempt to provide an automatic means for developing these estimates, AAI in cooperation with the Navy, has collected all the data available for units programmed on the AN/MPM-55 and the 5500 ATS which include the actual programming experience, machine utilization during program debugging, interface hardware costs, and various factors descriptive of the UUT to be tested. This data was then processed using linear regression modeling techniques to develop equations which would use these descriptive factors and produce programming time, programming costs, and machine run time. These factors include such items as number of active element groups, number of active connector pins and test points, number of modules (SRA's) in the UUT, number of IC's, logic depth, etc.

The estimating process is therefore one of an approximate analysis of an exact model as opposed to an exact analysis of an approximate model. If the particular unit to be evaluated falls well within the envelope defined by the 400 or more units that were analyzed to produce the estimating equations, the resulting data produced may be accurate to within a few percent. If, however, the unit to be evaluated borders on the limit of the envelope, or extends beyond the envelope, the estimating error may be 50% or more. If the support analysis program is used for a large number of units from an electronics system, then the resulting support analysis for the overall system can be considered accurate within a few percent.

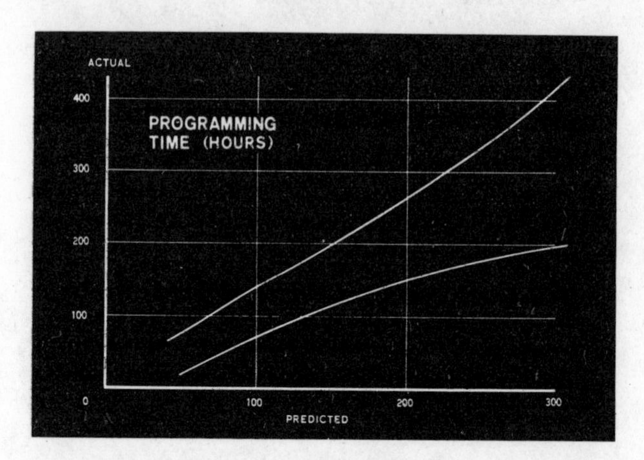

Figure 4

The above graph shows the possible error which can occur in estimating programming time utilizing linear regression modeling techniques. As the number of programming hours increases significantly beyond the average of the models analyzed, then the possibility of predicted programming hour error increases.

Module Testing

The basic equation used to develop programming time for electronic modules (which are not digital) was developed utilizing linear regression modeling techniques from over 400 modules characterized by factors easily determined by the operator. These factors are:

T = Number of active element groups

C = Complexity factor

P = Number of active connector pins plus test points

The level of effort expended on programming these modules was to the limited diagnostic level. PA, PB, and PC represent the programming time for functional testing, limited diagnostic testing and piece part diagnostic testing respectively. The TCP equation shown below is therefore defined as PB. PA is considered to be 80% of PB. PC is found by multiplying PB times a factor M which is the ratio of extending the limited diagnostic program to the piece part level. Factor M was derived by curve fitting to actual experience achieved on the 5500 ATS during the past year's operation.

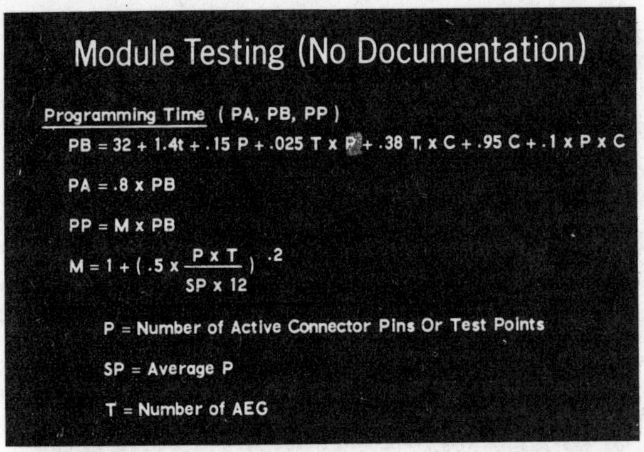

Module Testing (No Documentation)

Programming Time (PA, PB, PP)

$$PB = 32 + 1.4t + .15\,P + .025\,T \times P + .38\,T \times C + .95\,C + .1 \times P \times C$$

$$PA = .8 \times PB$$

$$PP = M \times PB$$

$$M = 1 + \left(.5 \times \frac{P \times T}{SP \times 12} \right)^{.2}$$

P = Number of Active Connector Pins Or Test Points

SP = Average P

T = Number of AEG

Figure 5

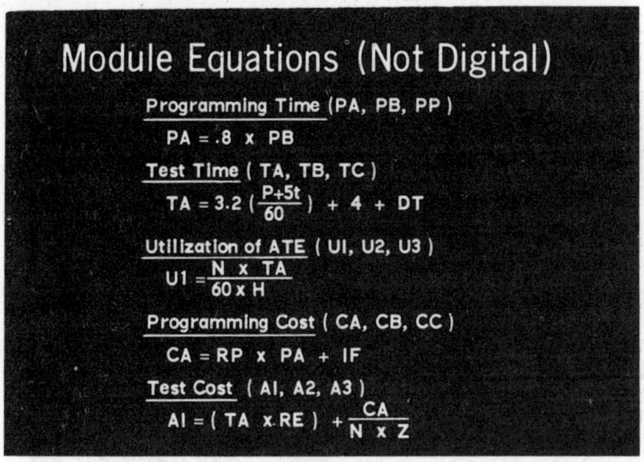

Module Equations (Not Digital)

Programming Time (PA, PB, PP)

$$PA = .8 \times PB$$

Test Time (TA, TB, TC)

$$TA = 3.2 \left(\frac{P+5t}{60} \right) + 4 + DT$$

Utilization of ATE (UI, U2, U3)

$$U1 = \frac{N \times TA}{60 \times H}$$

Programming Cost (CA, CB, CC)

$$CA = RP \times PA + IF$$

Test Cost (AI, A2, A3)

$$AI = (TA \times RE) + \frac{CA}{N \times Z}$$

Figure 6

Figure 6 shows the development of test costs for functional testing on ATE. In a similar fashion, A2 and A3, test costs for limited diagnostic and piece part diagnostics can also be estimated. Test time utilizes previously defined factors and includes also a delay time. This delay time is computed for servo and pneumatic testing where significant delays are experienced even in automatic testing. Provisions are also included in the program for operator input of significant manual delays anticipated during automatic testing.

Programming cost is shown to include a factor IF which is the interface cost. In the 5500, this figure is $200 and is increased significantly if testing is of a microwave subsystem.

Digital SRA

The equations shown in figure 7 are used to estimate programming time and test time for the three levels of testing. These equations are intended for use on digital pattern testing where documentation does not exist for functional or diagnostic test patterns. The factor NC is the number of logic elements or integrated circuits in the module and L is the maximum logic depth of sequential circuits in the module.

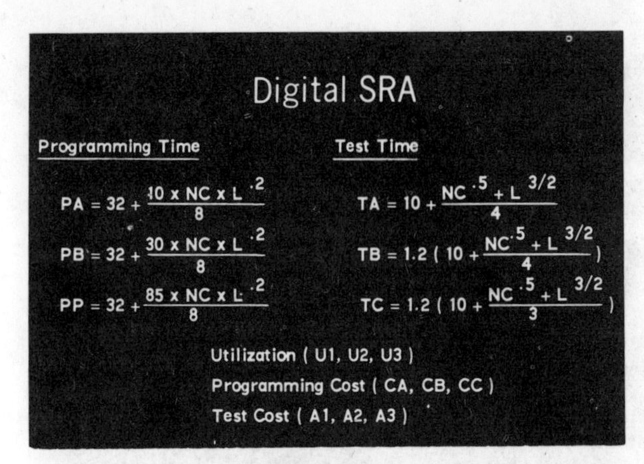

Figure 7

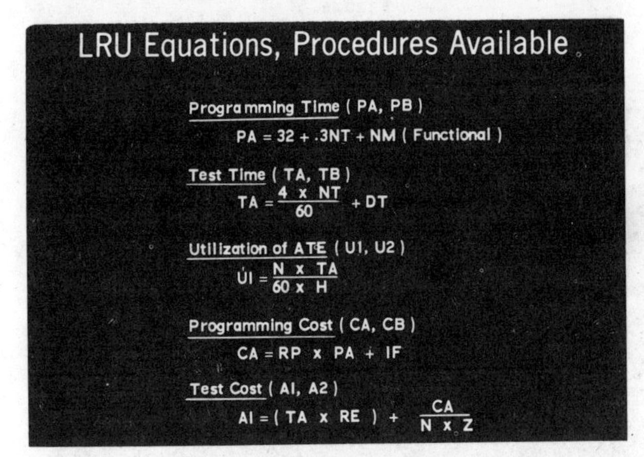

Figure 8

It should be noted that in all test time equations, a significant amount of time is added to the computed value for set up of cables and interface hardware. In cases where high quantity production testing will occur, these times may be higher than actually experienced.

ATE utilization, programming cost and test cost are computed using equations of the form shown in figure 6.

LRU Equations

End items are defined in the program as LRU's (Line Replaceable Units) or WRA's (Weapon Replaceable Assemblies) and are considered to be units which are repaired by module replacement as opposed to component replacement. If procedures are not available, then equations similar to the module equations are utilized in developing functional and limited diagnostic programming time and test costs. Where procedures are available, the operator is requested to define the number of tests (NT) and the number of modules (NM) in the UUT. The equations shown in figure 8 develop the test time, utilization, programming costs, and test cost for functional testing. Similar equations extend these numbers to the limited diagnostic level.

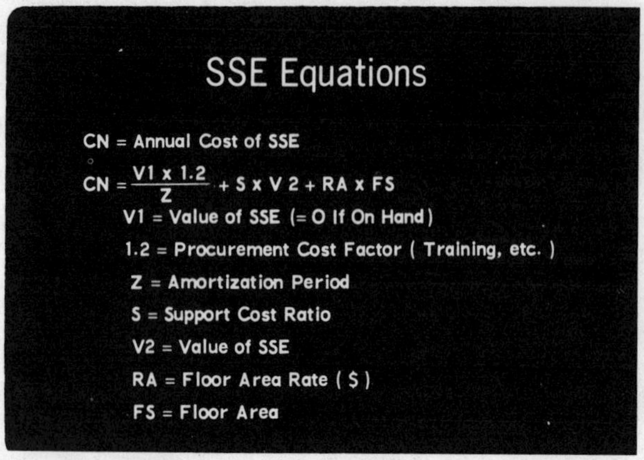

SSE Equations

CN = Annual Cost of SSE

$$CN = \frac{V1 \times 1.2}{Z} + S \times V2 + RA \times FS$$

V1 = Value of SSE (= O If On Hand)

1.2 = Procurement Cost Factor (Training, etc.)

Z = Amortization Period

S = Support Cost Ratio

V2 = Value of SSE

RA = Floor Area Rate ($)

FS = Floor Area

Figure 9

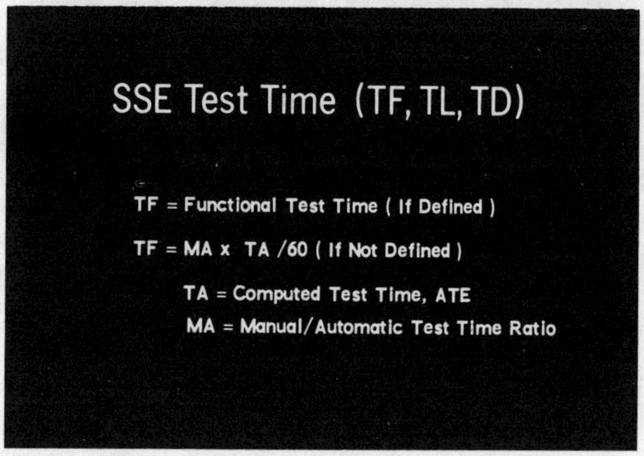

SSE Test Time (TF, TL, TD)

TF = Functional Test Time (If Defined)

TF = MA x TA /60 (If Not Defined)

TA = Computed Test Time, ATE

MA = Manual/Automatic Test Time Ratio

Figure 10

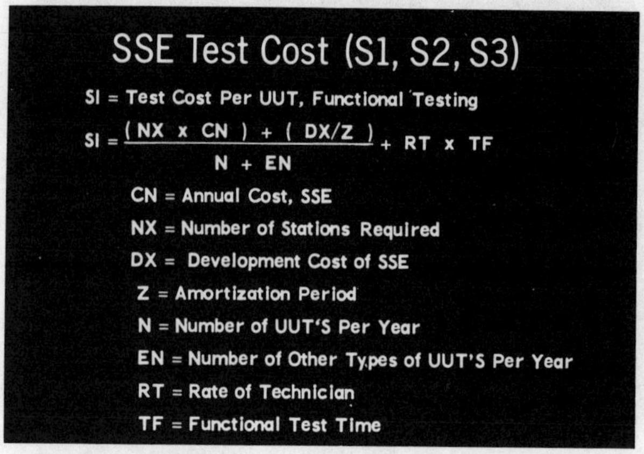

SSE Test Cost (S1, S2, S3)

SI = Test Cost Per UUT, Functional Testing

$$SI = \frac{(NX \times CN) + (DX/Z)}{N + EN} + RT \times TF$$

CN = Annual Cost, SSE

NX = Number of Stations Required

DX = Development Cost of SSE

Z = Amortization Period

N = Number of UUT'S Per Year

EN = Number of Other Types of UUT'S Per Year

RT = Rate of Technician

TF = Functional Test Time

Figure 11

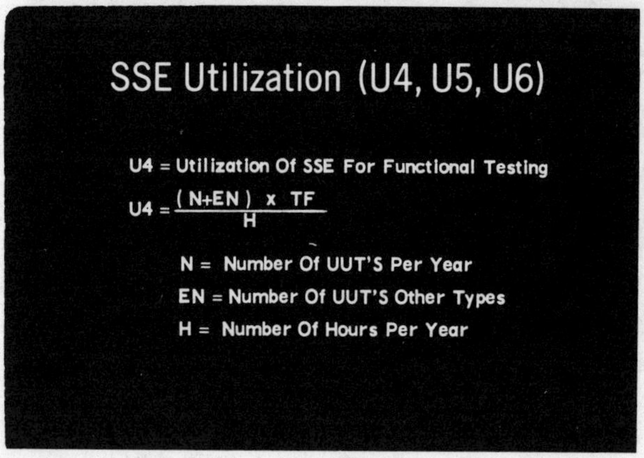

SSE Utilization (U4, U5, U6)

U4 = Utilization Of SSE For Functional Testing

$$U4 = \frac{(N+EN) \times TF}{H}$$

N = Number Of UUT'S Per Year

EN = Number Of UUT'S Other Types

H = Number Of Hours Per Year

Figure 12

SSE and GPTE Computations

The development of annual costs, test times, test cost, and utilization of SSE and GPTE are straight forward and are included here to illustrate their general form for the functional test level only. The same form of equation is utilized for limited diagnostic and piece part diagnostic testing where this capability has been defined by the operator.

Figures 9, 10, 11, and 12 show the basic form for the SSE equations.

Figures 13, 14, 15, and 16 show the general form of the equations for general purpose test equipment.

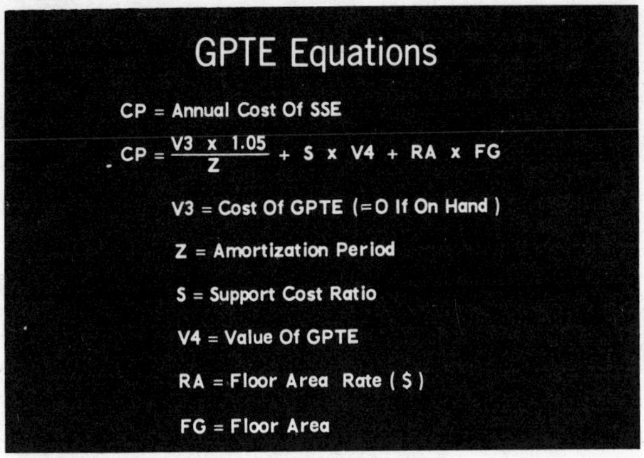

GPTE Equations

CP = Annual Cost Of SSE

$$CP = \frac{V3 \times 1.05}{Z} + S \times V4 + RA \times FG$$

V3 = Cost Of GPTE (=0 If On Hand)

Z = Amortization Period

S = Support Cost Ratio

V4 = Value Of GPTE

RA = Floor Area Rate ($)

FG = Floor Area

Figure 13

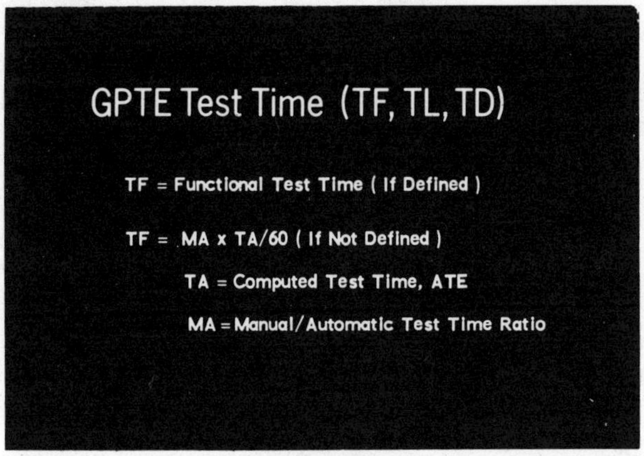

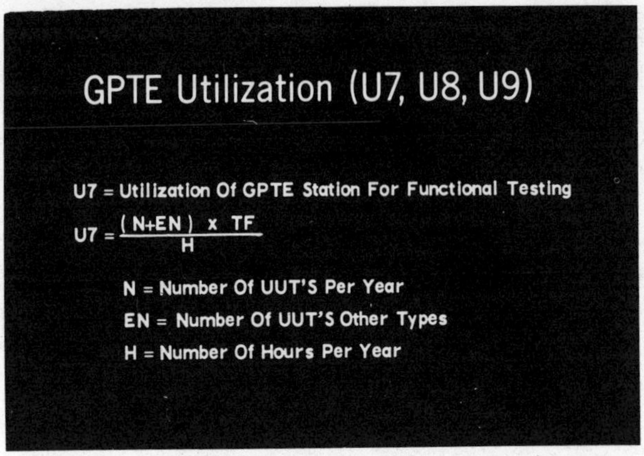

GPTE Utilization (U7, U8, U9)

U7 = Utilization Of GPTE Station For Functional Testing

$$U7 = \frac{(N+EN) \times TF}{H}$$

N = Number Of UUT'S Per Year

EN = Number Of UUT'S Other Types

H = Number Of Hours Per Year

Figure 15

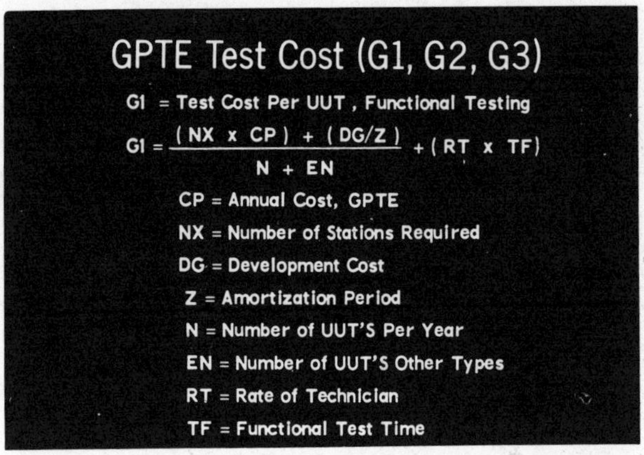

GPTE Test Time (TF, TL, TD)

TF = Functional Test Time (If Defined)

TF = MA x TA/60 (If Not Defined)

TA = Computed Test Time, ATE

MA = Manual/Automatic Test Time Ratio

Figure 14

GPTE Test Cost (G1, G2, G3)

G1 = Test Cost Per UUT, Functional Testing

$$G1 = \frac{(NX \times CP) + (DG/Z)}{N + EN} + (RT \times TF)$$

CP = Annual Cost, GPTE

NX = Number of Stations Required

DG = Development Cost

Z = Amortization Period

N = Number of UUT'S Per Year

EN = Number of UUT'S Other Types

RT = Rate of Technician

TF = Functional Test Time

Figure 16

Conclusion

The 5500 Automatic Test System, presently being utilized at the seven NARF's for depot support of avionics systems, was procured as a shop aid primarily to reduce maintenance costs. The manual to automatic test time ratios presently being experienced through use of this equipment indicate a 20 to 1 saving in test time. After amortization of the hardware costs and with full consideration to annual support cost of the ATE (12% of procurement cost per year), a production work load which results in excess of 10% utilization of the ATE will result in reduced maintenance costs.

Through use of the Support Analysis Program, it will be possible for each NARF to quickly evaluate candidates for support on the 5500 ATS and determine those which result in significant test cost savings compared to alternate support possibilities. Through use of this estimating tool, all factors relating to an economic trade-off analysis can be quickly evaluated, permitting rapid assignment of UUT's to be supported on the 5500 ATS resulting in a maximum return on investment.

TECHNICAL MANAGEMENT TECHNIQUES FOR LARGE SCALE AUTOMATIC TEST SYSTEMS ENGINEERING

SHERMAN N. MULLIN

LOCKHEED-CALIFORNIA COMPANY
A DIVISION OF LOCKHEED AIRCRAFT CORPORATION

ABSTRACT

Large scale automatic test systems (ATS) are defined and potential technical management problem areas are outlined. Critical technical management objectives for ATS projects are listed. Technical management techniques useful in the following areas are discussed: organization, supervisory and technical staffing, establishment of technical, schedule, and cost objectives, product technical quality assurance, and project control. An annotated bibliography is included as well as an appendix containing twenty-five widely believed myths of ATS engineering and management. The point of view reflected throughout the paper is that ATS technology has moved from an adolescent to a mature art rather recently and that similarly mature ATS management techniques have been evolved but are not being vigorously and consistently applied.

INTRODUCTION

Large scale automatic test systems (ATS) have emerged as a necessary but difficult to implement technology. The term large scale is used deliberately, since many of the problems and management techniques related to large scale ATS projects are not applicable to small projects. To be more specific, characteristics of large scale ATS taken as a class are listed in Figure 1. Systems of this size and complexity present unique technical management problems. Further expansion of the number and potential impact of the technical management problems results from large number of personnel involved, the multi-disciplinary staffing requirements, involvement of large number of organizations, and multiple geographic location of organizations involved. From a technical viewpoint, a large scale modern ATS is highly integrated and success is achieved when the total conglomerate of equipment, computer programs, data, and people function satisfactorily in the end-user operational environment. This success is only achieved by rigorous application of appropriate technical management techniques throughout the project life cycle. Many of the applicable techniques are not unique

1. AUTOMATIC TEST EQUIPMENT (ATE) — 20 - 100 UNITS
   - NUMBER OF EQUIPMENTS BUILT — $ 2 - $50 MILLION
   - DEVELOPMENT PROJECT COST — $. 1 - $3 MILLION
   - RECURRING UNIT COST — $ 1 - $50 MILLION
2. AUTOMATIC TEST SYSTEM APPLICATIONS
   - NUMBER OF DIFFERENT UNITS UNDER TEST (UUT) TYPES — 50 - 1,000
   - COMPLEXITY OF UNITS UNDER TEST: BOXES: 10 - 40 LARGE MODULES; MODULES: 50 - 1,000 INTEGRATED CIRCUITS CHIPS OR EQUIVALENT; MICROELECTRONIC COMPONENTS: 10 - 3,000 DISCRETE CIRCUIT EQUIVALENTS
   - NUMBER OF APPLICATIONS DEVELOPMENT GEOGRAPHIC LOCATIONS: — 2 - 20
   - TOTAL SUPPORT DEVELOPMENT COST PER UNIT UNDER TEST: (HARDWARE, SOFTWARE AND DOCUMENTATION) — $5 - $500 THOUSAND
   - TOTAL APPLICATIONS DEVELOPMENT COST (ALL UNITS UNDER TEST): — $10 - $200 MILLION
   - NUMBER OF END USER LOCATIONS: — 5 - 50 DIFFERENT GEOGRAPHIC LOCATIONS CONCURRENTLY
   - TOTAL ANNUAL END USER COSTS: (PRIMARILY OPERATIONS AND MAINTENANCE) — $ 5 - 30 MILLION
3. SYSTEM LIFE CYCLE: (DEVELOPMENT INITIATION THROUGH END USER PHASEOUT). — 10 - 25 YEARS
4. TOTAL LIFE CYCLE COSTS (NOT INCLUDING UUT'S). — $65 MILLION- $1.5 BILLION

FIGURE 1: LARGE SCALE **A**UTOMATIC **T**EST **S**YSTEMS: CHARACTERISTICS

Reprinted from *IEEE Auto. Support Systems Symp. Rec.*, Nov. 1973.

to ATS engineering management nor are they novel or difficult to comprehend. However fundamental and well known these techniques may be, there is considerable evidence that they are not being systematically applied to large scale ATS projects.

This paper addresses a number of controversial issues which are not necessarily amenable to simple conclusions. Notwithstanding this fact, the bulk of the statements set forth are singular to the point of being dogmatic. This is done intentionally and provocatively in most cases. Further, only several major areas of the total spectrum of large scale ATS management are discussed. This selectivity is based on the following assertions:

(1) Management of electrical and mechanical design of equipment (electronic hardware) is mature and not generally a problem area.

(2) Total system technical integration management is almost always a problem area.

(3) Software (computer program) development and sustaining management is difficult and still not a mature art.

(4) Configuration management and control, particular software configuration control, is a mature and well understood technique but is rarely practiced adequately.

(5) Sustaining engineering for the total large scale ATS (including all applications software and hardware) is always a major problem area.

Successful large scale ATS projects require continuous, vigorous attention to the following classic areas, in the following priority order:

Staffing: selection and development of ATS management, supervisory, technical, and administrative personnel.

Organizational Structure: clear, logical, well documented organization functions and responsibilities, frequently realigned and restructured to meet the specific needs of the current phase of the ATS project.

Systems and Procedures: a practical, concisely documented set of policies and procedures which set standards for all routine activities required for large scale ATS development and support.

MAJOR POTENTIAL PROBLEM AREAS

One significant role of ATS engineering management, as with any other engineering management, is to anticipate and preclude the occurrence of major problems which will adversely affect meeting the ATS technical, schedule, or cost objectives. The following discussion covers several major potential problems in any ATS project, selected on a highly subjective basis.

ATS requirements, design and detailed technical interface definition: Are the requirements explicitly defined and fully coordinated with the end user? Are system design performance and first article test requirements at the ATE/UUT interface defined in a technically detailed interface control document? Are no changes permitted in the interface control document without written consent of all users? Is the interface specified on a conservative (wide tolerance) basis to include circuits, mechanical design, and parts selection which will minimize quantity, size, complexity and cost of UUT adapters (interface devices)?

System software functional performance, documentation, and testing: Is all software to be developed specified in the same level of technical detail as the hardware? Is the complete package of software documentation to be prepared fully defined? Is a separate organization (whether within the ATS development organization or external, preferably external) chartered to prepare software test plans and procedures, conduct software acceptance tests and prepare formal software test reports? Has a detailed software configuration management, control, and accounting plan and procedure been prepared? Is adherence to this plan audited in-depth periodically?

UUT/ATE Interface: Have large, conservative margins been specified for all circuit interfaces between the ATE and UUT? Are detailed grounding and shielding conventions fully delineated in the applications reference manual? Are allowable amplitudes and spectra for noise on all ATE output lines fully specified?

SOME SELECTED FUNDAMENTAL ATS TECHNICAL MANAGEMENT PRINCIPLES

The fundamental principles of ATS engineering management are so simple as to be not given sufficient daily emphasis on large scale ATS projects:

Objectives and plans: establish well defined, achievable technical, schedule, and cost objectives. Prepare detailed plans to meet these objectives. Accept the fact that planning and re-planning is a continuous function of ATS engineering management, not just a one-time activity performed early in a new ATS project.

Unit Under Test (UUT) design for ATS applications: Prepare detailed, specific ATS independent technical specifications and conduct frequent on-site technical design reviews during UUT design, development, documentation, and testing. Avoid two classic pitfalls: (1) assuming that the specific ATE performance specification is a valuable tool in achieving UUT design for ATE support, and (2) also assuming that preparation and analysis of special documentation related to "proving" ATS support compatibility in any way assures such compatibility. The result is a pile of paper that is no substitute of frequent informal, detailed technical UUT design reviews by competent ATS applications specialists.

Centralize supporting operations: Data libraries, keypunching, data processing machine operations, adapter (interconnecting device) modifications, UUT maintenance and control, ATE maintenance, calibration and modifications, engineering drafting support, and many other similar activities are a necessary part of any large scale ATS project. It is nearly always most economic to establish central support groups or individuals to perform these activities on a project-wide support basis. Further, much better engineering management control is obtained by organizing in this manner. Finally, ATE engineer/programmers and ATE test program engineers usually do not perform support tasks such as those cited above in an economic or highly disciplined manner.

Establish and maintain a vigorous formal and informal management and technical communications activity: continuous formal and informal written and oral communication between ATS development management and end-user (or intermediate user) management is essential to project. Candor is essential. Good communications have to be planned and nurtured on large scale ATS projects; they are not necessarily the natural result of the fact that various individual and organizational interrelations happen to exist on a particular ATS project.

ORGANIZATION AND STAFFING

The inherent organizational structure of large scale automatic test systems projects are depicted in Figure 2. The term inherent is used due to the fact that all of these major activities must take place and be managed. The management problem is always simpler when all of these activities take place within a single organization at a single geographic location, but that is not the problem being addressed here. Unfortunately, more often these activities take place in many different locations within many different organizations.

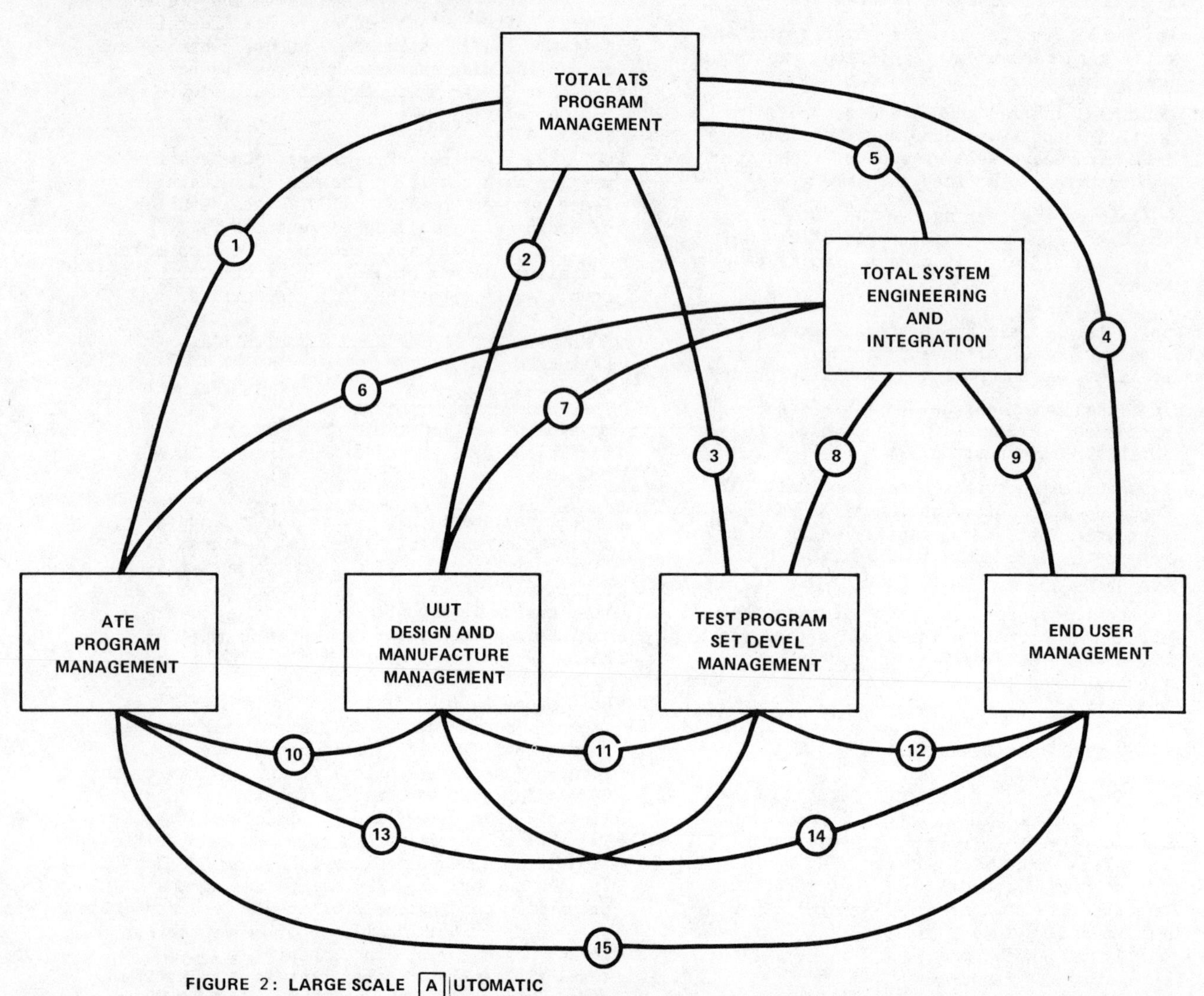

FIGURE 2: LARGE SCALE AUTOMATIC TEST SYSTEMS: INHERENT ORGANIZATION STRUCTURE/INTERFACE

Whether explicit or implicit the fifteen organizational interfaces listed on Figure 2 are the absolute minimum number which will exist in any large scale ATS project. Achievement of project objectives normally requires establishment of formal procedures and controls for each of these interfaces, as well as continuous management attention.

Test Program development is a major element of all large scale ATS projects. All of the major functions involved are shown on Figure 3. Figure 3 includes a recommended organizational structure for a test program development organization. This structure is established to provide the following features: centralization of systems engineering and applications in a separate group, centralization of all non-engineering support and control functions in a separate systems operations group, and test and evaluation of the product by a separate laboratory organization which does not report to the test program design management. These features are significant in enhancement of development efficiency, improved management controls, and achievement of good product quality assurance.

Careful management and technical staffing should always be one of the major concerns of anyone involved in ATS management. Proper staffing is a continuous activity, not simply a major task to be performed once early in the organization of a new large scale ATS project. Knowledge of UUT technology should be given major consideration in selection of both management and technical personnel at all levels. Intimate knowledge of both general UUT technology and specific UUT's is critical to test program development. Knowledge of older support systems, computer programming, or other secondary disciplines should not be given undue weight in the staffing process.

Test and diagnostic technology, particularly digital test and diagnostic technology, is beginning to emerge as a mature technical discipline. This technology is used in many areas other than automatic test systems. There now exists a substantial cadre of engineers with both management and technical experience in this discipline. The probability of meeting all large scale ATS project objectives can be significantly improved by staffing to a maximum extent from the resources of this professional cadre, many of whom do not have explicit ATS experience.

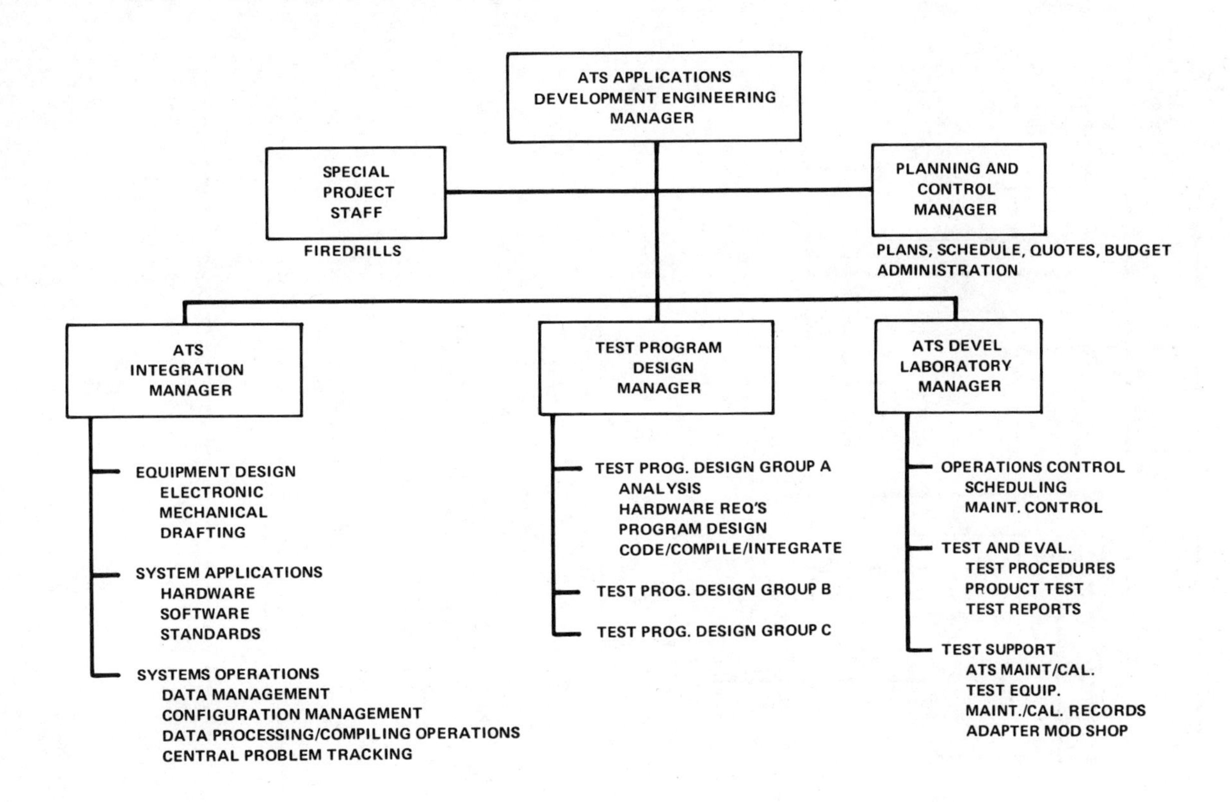

FIGURE 3: LARGE SCALE ATS (AUTOMATIC TEST SYSTEMS): TEST PROGRAM DEVELOPMENT ORGANIZATION

Test program development is a difficult job. Liguori [4] has carefully and accurately analyzed what type of engineer does this job most successfully. Major compromises in technical personnel selection usually results in grave consequences not favorable to the career of an ATS Manager.

Carefully selected new engineering college graduates can play a major role in large scale ATS project. Modern digital hardware and software technology is now a major element of most large scale ATS. Most current college graduates with a major in electronics engineering are well grounded in the fundamentals of this technology and develop rapidly into capable technical contributors. The engineering management of any ATS project should give careful attention to continuous revitalization of the organization by periodic addition of new or recent graduates.

MANAGING THE CRITICAL COST ELEMENT

Systems integration testing at several levels, culminating in the total ATS system integration test, is the largest and most critical element of cost in nearly every ATS project. The major tests required are shown in Figure 4. Obviously, a key engineering management objective is to minimize the manhours spent accomplishing this important and required activity.

Total ATS system development cost will be minimized by reaching total system integration testing at the earliest possible date. Often, the preceding tests required (see Figure 4) are assumed to be more critical and valuable than they actually are in the total project context. Further, meeting project objectives dictates that end-user operations and maintenance personnel be engaged in system testing operations at an early date. Involving these personnel will insure proper emphasis on detection and elimination of end-user operating problems. Also, system tests will be heavily influenced to be more practical in nature than would be the case if only specialized ATS development engineering personnel are involved.

The total cost of system integration testing is heavily a function of the number of on-station ATE system integration test hours required. For example, test program set integration testing has the following major cost elements: multi-shift operations control and direction, multi-shift ATE maintenance and calibration activity, engineering test programmer debugging and test activity, support equipment maintenance and calibration, general technician support, directly related program recompilation support, technical supervision and clerical support. Once a project decision is reached, usually of necessity, to conduct development testing on a single or multi-shift basis,

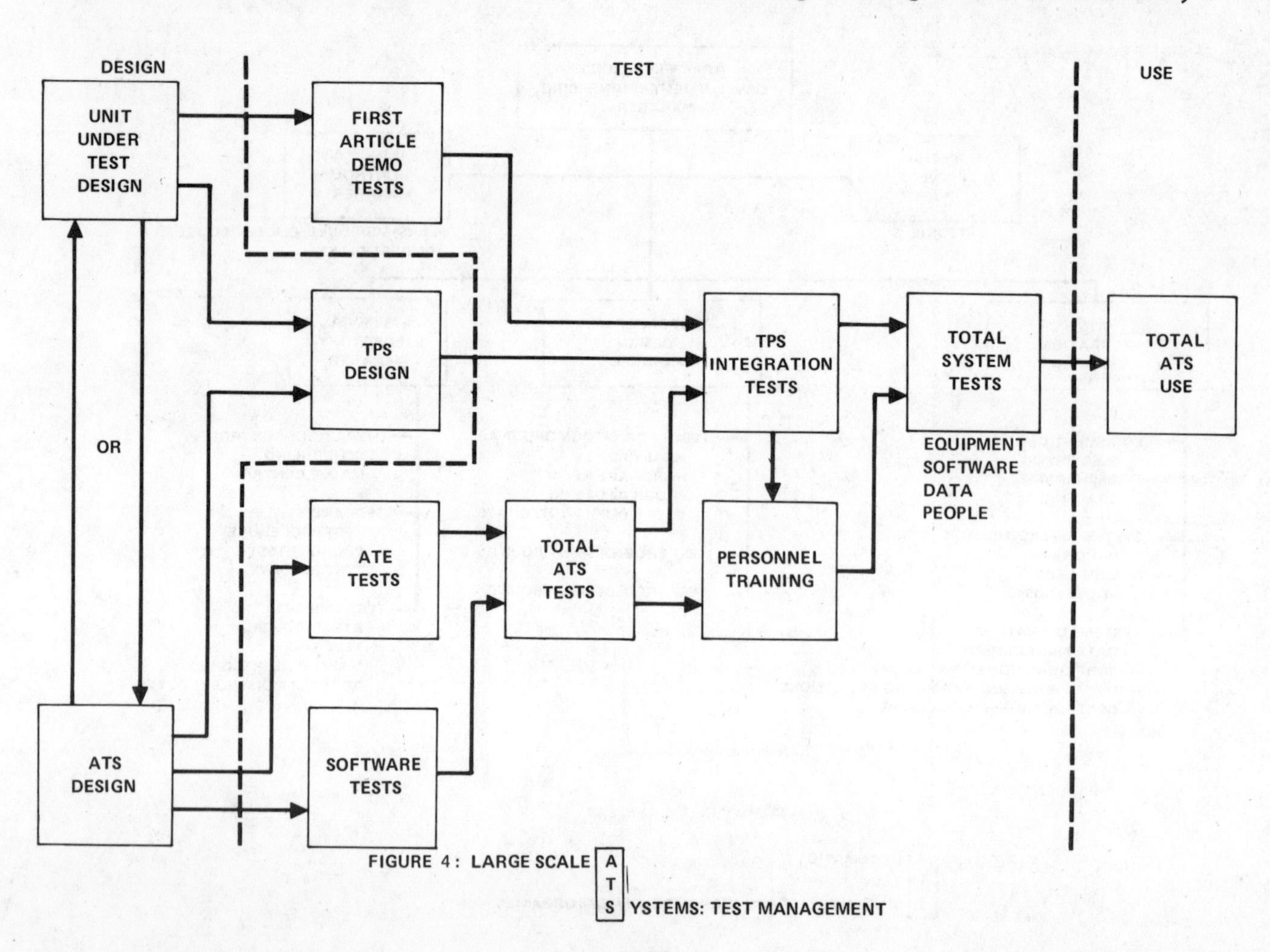

FIGURE 4: LARGE SCALE ATS YSTEMS: TEST MANAGEMENT

the operating costs per unit time are relatively constant. Minimal total cost obviously depends on finding successful methods of minimizing the number of ATE operating hours (calendar time) required. In order to do this, two significant areas merit major technical management attention:

1. Hardware and software debug aids, of which by far the most important is the ability to modify source language programs on a quasi real-time basis directly on a particular ATE.
2. Use of an operational control technique in development laboratories which affords both of the seemingly conflicting management objectives of tight control on the one hand and minute-by-minute flexible resource allocation on the other.

The first of these is implementable with current ATS technology if properly addressed at project inception. The second is a technical management technique which deserves further discussion.

THE QUEUING METHOD OF SCHEDULING AND RESOURCE ALLOCATION IN LARGE SCALE ATS DEVELOPMENT LABORATORIES

Figure 5 depicts the major variables in an involved typical ATS development laboratory. This problem has a large number of dimensions:

1 = A Automatic Test Equipment Units Involved in Development Use

2 = B Adapters (Interconnecting Devices)

3 = C Different types of Units Under Test

4 = D Development Configurations of Test and Diagnostic Computer Programs

5 = E Test Program Development Engineers and Supporting Personnel

6 = F Different types in Quantities of Special Support Equipment Required for Development Use

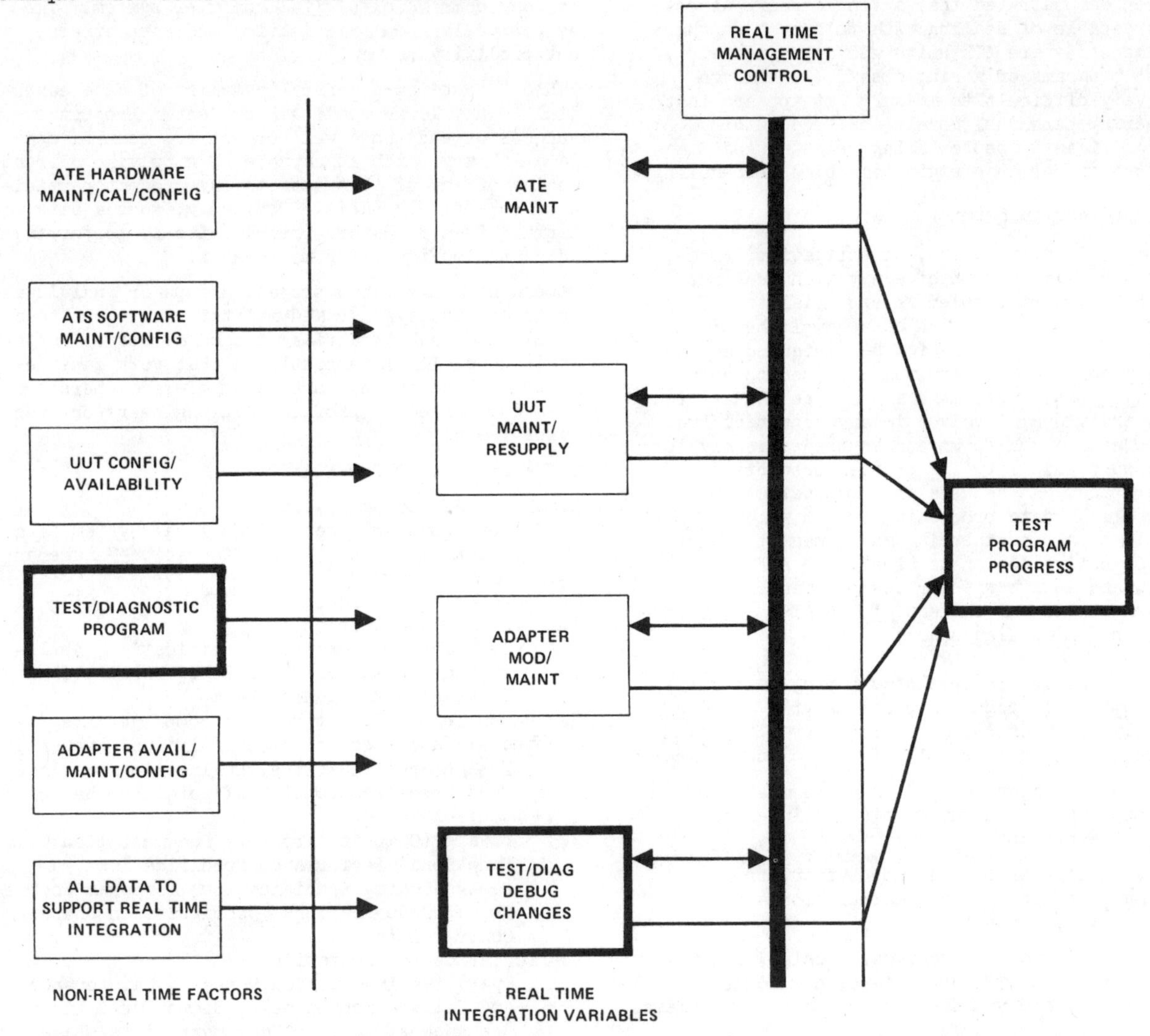

FIGURE 5. LARGE SCALE [A]UTOMATIC [T]EST [S]YSTEMS: MANAGING THE CRITICAL COST ELEMENT

Experience has been obtained on the S-3A VAST Development Program which clearly demonstrates that the Queuing method of control outlined below maximizes effective use of resources.

One shift director, unaided by any automation, makes all allocation decisions in real time. He is supplied with Queue Chits for each test program/test programmer immediately available to perform development testing activities. He is also supplied with up-to-the-minute information on maintenance, calibration and configuration status of all ATE units and UUT's available. Further, he has the same sort of information on interconnecting devices and other support equipment required. Using this method, no one is scheduled for development testing on a rigid basis except at the initiation of the first shift each morning. Data obtained (of a statistically significant quantity) clearly shows that this method is significantly superior to rigid time block type scheduling of development test activities. It has been demonstrated that a single shift director is capable of dealing with this problem where the dimensions are ATE units and approximately 25 test programs on a single shift. Although it is very difficult to measure, it appears that engineering personnel morale and productivity are significantly better using this Queuing technique rather than rigid time block scheduling.

TECHNICAL PROBLEM CONTROL

Systematic identification, investigation, resolution and verification of every technical and technical support problem is essential to ATS project success. This can be accomplished in a practical and timely fashion by designing a single problem sheet form that is used to document all types of problems and can be handwritten by any individual involved in any aspect of the ATS project. Such forms can be used directly to assign responsibility for problem resolution. Further, they can be used as input data to a fairly simple data processing system which is valuable to ATS engineering management to assure that corrective action is timely, to spot serious trends, and as a basis for design changes and improvements to any element of the ATS. This process is shown on Figure 6.

In a large scale ATS development program a very large number of technical problems will arise (possibly 5,000 to 20,000). These problems can be expected in at least all of the following areas:

Compiler: Object Code, diagnostic messages, applications documentation, maintenance documentation

ATE: Design errors, documentation errors, maintenance actions, calibration actions, modification actions

Test and Diagnostic Programs: Design, coding, on-station ATE, operational procedures, timing, and validation/verification problems.

Adapter (Interconnecting Device): Drafting errors, integration modifications, maintenance actions

UUT: Data deficiencies, configuration anomalies, maintenance actions

DOCUMENTATION AND DATA MANAGEMENT

Accurate, comprehensive, centrally controlled, readily available data is critical to any large scale ATS project. If any one item of data is critical, it is to have a near perfect ATE applications engineering reference manual. This manual should contain integrated hardware/software detailed technical applications data. It should communicate to the user both the capabilities and, most important, all of the ATE hardware and software limitations. It should be organized to be readily updated to accommodate the inevitable configuration changes in ATE hardware/software. Note that a design or test specification is totally unsuitable for use as an applicable reference manual. For a convincing demonstration of the immaturity of ATS technology in this critical area, compare the "applications reference data" for any large scale ATS with that of the applications manual for any commercial digital computer system. The objective is good organization and compactness, not prolixity.

Table 1 contains a generalized list of data essential to any large scale ATS project. Project success depends on a well organized, centralized data library which is capable of supplying current copies of any of this data to any project participant on a timely basis. Obviously, such a data library is a mandatory prerequisite to performing adequate configuration management.

Dependence on various project groups or individual engineers to organize and maintain their own technical data bank is a risky method at best. Further, it is generally more costly in that work capable of being done by well motivated clerical personnel is accomplished by professional engineers or even supervisors.

TABLE 1: ATS DOCUMENTATION

1. Total ATS Documentation
   Total System Description, Total System Data Index, Total System Configuration Management Data Summary
2. Automatic Test Equipment
2.1 System Documentation
   Technical Performance Specification, Applications Engineering Reference Manual, Configuration Management Data
2.2 Hardware Documentation (For complete integrated ATE and also for each individual subunit)
   Design Specifications, Engineering Drawings, Test Specifications, Configuration Change/Control Data
2.3 Software (Computer Program) Documentation
   Functional Performance Specifications, Design/Coding Specifications, Program Source/Object Code Listings, Configuration Change/Control Data
2.4 Logistics Documentation
   Operating Instruction Manuals, Maintenance and Calibration Manuals, Spare Parts Lists/Breakdowns, Support Equipment Lists/Usage Procedures
3. System and Support Software

3.1 Compiler/Assembler/Interpreter Programs
Function Performance Specifications, Design/Coding Specifications, Program Source/Object Code Listing, Configuration Change/Control Data, Applications Reference Manual

3.2 All Other Support Software Packages
(same as 3.1 above)

4. Units Under Test
Functional Performance Specifications, Design Specifications, Engineering Drawings, Test Specifications and Procedures, Operating Procedures (where applicable), Configuration Change/Control Data, Maintenance Manuals

5. Test Program Sets (One per unique unit under test)

5.1 Test and Diagnostic Software (Computer Programs)
Design Specifications, Flow Charts, Source/Object Code Listings, Configuration Change/Control Data

5.2 Adapter (Interface Device) Hardware
Design Specifications, Engineering Drawings

5.3 Logistics Documentation
Training Manuals, Operating Instruction Manuals, Integrated Functional Test Diagrams, Adapter Maintenance Manuals/Spare Parts Lists

AUTOMATIC TEST SYSTEMS ENGINEERING MYTHS

Twenty-five widely believed myths are contained in Appendix A. Belief in any of these by someone involved in ATS management will likely lead to major technical, schedule or budget problems on any specific ATS project. It may be asserted that successful ATS engineering management consists in large part of evolving a project plan in which none of these myths plays any role.

SUMMARY AND CONCLUSIONS

Large scale automatic test systems have become an essential element of many advanced maintenance systems, rather than merely an option conditioned by technical, cost, or schedule constraints. Management techniques have emerged (and deserve intensive further development) which are essential to the success of large scale ATS projects. These techniques are adaptations of classic engineering management techniques to the specific requirements and potential problems of large scale ATS projects.

REFERENCES AND BIBLIOGRAPHY

[1] Arthur D. Hall, A Methodology for Systems Engineering. Princeton: D. Van Nostrand, 1962. Anyone managing large scale ATS engineering should own a copy and re-read it every 90 days. Particularly valuable as a source of sound methods for the early phases of ATS project engineering and management: system synthesis, analysis and configuration selection, economic justification and ramifications, how to do good systems engineering, and much more.

[2] Nicholas A. Bond, Jr. Some Persistent Myths About Military Electronics Maintenance. HUMAN FACTORS, 1970, 12(3), 241-252 A stimulating constructive, and somewhat devastating series of blows at the folklore of military electronics maintenance. Valuable reading for the new or experienced ATS manager as an aid in remaining connected to reality.

[3] Gerald M. Weinberg, The Psychology of Computer Programming. New York: Van Nostrand Reinhold, 1971. No one managing ATS engineering can avoid the far from solved problem of computer programming management. A stimulating and wide-ranging book, with an extensive bibliography. If it doesn't aid solving software management problems, it at least gives considerable insight into why such problems permeate most ATS projects.

[4] F. Liguori, "ATE Test Designer: Programmer or System Engineer?", Proc. 1967 Automatic Support Systems Symposium for Advanced Maintainability. Good advice on what type of people are critically needed in an ATS engineering organization and why.

[5] W. L. Alexander and J. K. B. Woods, "Procurement Considerations for Computer-Controlled Test Equipment", Proc. 1967 Automatic Support Systems Symposium for Advanced Maintainability. Some valuable thoughts on the subject and quite comprehensive, but little consideration of the UUT data/configuration problem or that of test program development and maintenance. Must reading if you are managing an ATS project based on purchased ATE.

[6] A. M. Greenspan and E. H. Schmuhl, "Allocation of Responsibility for Program Preparation on ATE Supported Systems", Proc. 1969 Automatic Support Systems Symposium for Advanced Maintainability. Should the ATE manufacturer, UUT manufacturer, or ATS user prepare the test programs? A logical paper without a fully convincing conclusion. Ignores the UUT supply and maintenance problem to support test program development. Underestimates the critical problem of UUT knowledge essential to success.

[7] D. H. Lord, "ATE - Can It be Managed?", Proc. 1969 Automatic Support Systems Symposium for Advanced Maintainability. Not as general as the title suggests. Addresses how to introduce the Versatile Avionics Shop Test (VAST) system into operational Navy use for support of several types of aircraft, with full recognition of the complex management problems involved.

[8] D. L. Wood, "Automatic Test Equipment Software Configuration Management". Proc. 1972 Automatic Support Systems for Advanced Maintainability. An excellent and detailed paper on how it should be done but rarely is. The formidable problems of management implementation and operating costs are not addressed.

[9] B. W. Boehm, "Software and Its Impact: A Quantitative Assessment". DATAMATION, May 1973, pp. 48-59. ATS management is increasingly a software (computer programming) management task, a trend certain to continue. This article addresses many key issues, including the increasing and formidable costs of software integration and testing.

APPENDIX A: AUTOMATIC TEST SYSTEMS ENGINEERING MYTHS

The idea of documenting myths for a hopefully constructive purpose was stimulated by Bond's paper 2 . For those who are statistically oriented, it should be stated that these myths may well be true 10% of the time. However, it is the other 90% of the instances which are likely to be long remembered.

MYTH 1: Subunits of ATE are designed for thorough test and unambiguous diagnosis by the ATS of which they are a part. Other UUT's are usually not designed to be compatible and meet these objectives. Stated in other terms, ATS engineers intimately understand how to design for automatic test and diagnosis, whereas the typical UUT designer is usually ignorant or uninterested in this aspect of equipment design.

NOTE: The term software is used here as a synonym for a single computer program or a set of interrelated computer programs and directly related program documentation.

MYTH 2: The same ATS system software should be used for operational (factory and field shop) ATS use and for UUT test and diagnostic software development laboratory purposes.

MYTH 3: Automatic test system applications programs (ATS test and diagnostic programs) are software and are best prepared by professional computer programmers.

MYTH 4: ATS programming languages should be, can be, and are being readily designed for use by average test and maintenance technicians in factory and field maintenance shops.

MYTH 5: ATS are simply the latest generation of support systems and therefore ATS projects should be staffed primarily with very seasoned support systems personnel.

MYTH 6: ATS/Unit Under Test (UUT) interface circuit incompatibilities are usually the result of willful engineering violation of well defined, practical, realistic specification requirements for interface circuits, which normally include very realistic tolerances.

MYTH 7: Test and diagnostic program design is a mature engineering discipline widely supported by well documented, thoroughly tested computer programs. Accordingly, integration testing is usually brief and routine and extensive verification tests of diagnostic programs are neither required nor cost effective.

MYTH 8: An automatic test system should be designed to be as universal as possible in order to satisfy all known and probable applications requirements. This approach minimizes costs.

MYTH 9: System software design, development and testing is a mature technical discipline which has developed a solid body of generally applicable knowledge over the last two decades. Since this knowledge and the skilled practitioners are readily available and widely used to support ATS projects, ATS system software development and application is generally routine.

MYTH 10: Since large scale automatic test systems have a large and diverse group of users, comprehensive applications and reference manuals are always available and continuously updated.

MYTH 11: Interconnecting device (UUT adapter) design, documentation, and manufacture is a very minor aspect of total ATS development in terms of cost, technical scope and complexity.

MYTH 12: Design reviews are the primary product quality assurance technique for ATS applications test programs.

MYTH 13: The large scale ATS user and/or customer carefully analyzes "his system" requirements, clearly documents these requirements, and then seeks proposals from qualified ATS development and/or applications development organizations.

MYTH 14: The above requirements are systematically reviewed in depth with all levels of management in the organization which will use, operate, maintain, benefit from, and be dependent upon the large scale ATS to be developed.

MYTH 15: Programmers involved in ATS projects are competent, objective, thorough testers and evaluators of the programs they have produced.

MYTH 16: Software development organizations supporting ATS projects are continuously aware of the system reliability requirements of the large scale programs they are developing; this awareness guides their technical performance in all phases of their work.

MYTH 17: Large scale automatic test systems are highly flexible and need not be frozen in configuration until late in the development cycle. Further, most engineers and programmers understand this need for flexibility late in the development cycle, design to accommodate it, and participate very competently in implementation of late development changes.

MYTH 18: Designers of computers, peripheral equipment, and related hardware are very interested in the problems of total automatic test system design and, accordingly, eagerly participate in definition and documentation of total system requirements.

MYTH 19: Large scale automatic test system design is normally based on rigorous, mature synthesis methods, supported by rigorous analyses and/or simulation early in the development process to validate all key parameters of the total system design.

MYTH 20: Most academic devotees of large scale automatic test system engineering have designed several such systems and are well aware of synthesis problems and the numerous potential practical pitfalls to be avoided.

MYTH 21: It is nearly impossible to systematically plan and schedule large scale ATS development, largely due to the fact that all computer programming is a highly creative activity, comparable to composing a symphony, writing classical poetry, or doing original mathematical research.

MYTH 22: Large scale automatic test system engineering is so radically different from other large engineering projects that nearly all of the classical technical management techniques and procedures are either inapplicable or of marginal value.

MYTH 23: Digital computer hardware and software technology (technical performance, reliability, and cost) is the main constraint in designing advanced large scale automatic test systems.

MYTH 24: Computer peripheral equipment (data links, printers, magnetic tape units, cassette-based magnetic tape systems, input/output keyboards, magnetic discs, etc.) is an extremely mature and high reliability technology, so that little attention to this area is necessary in the design of ATS.

MYTH 25: The costs of developing ATS test and diagnostic software and adapter hardware are largely independent of the particular ATE hardware/software. Accordingly, the ATE should be selected on the basis of cost.

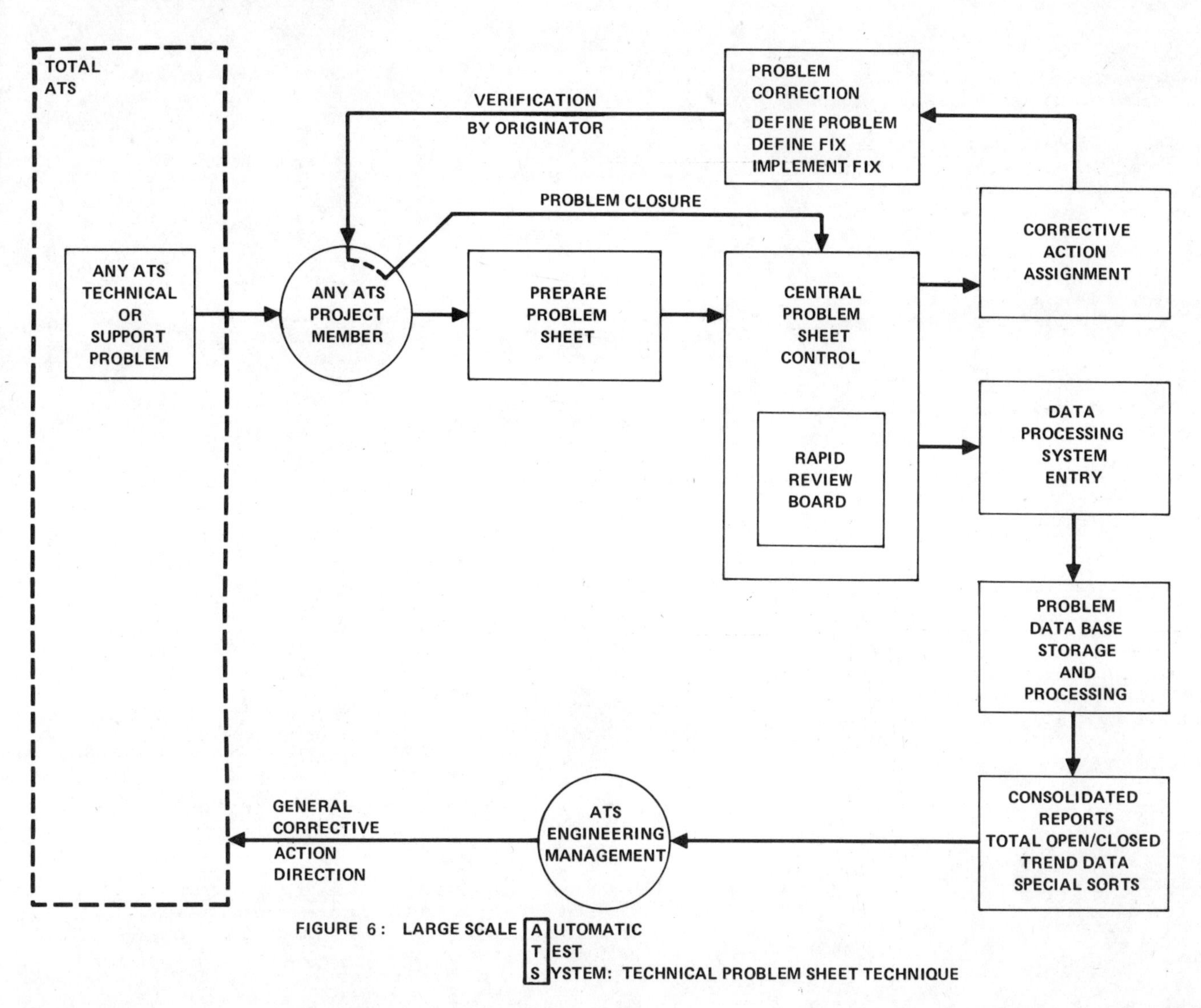

FIGURE 6: LARGE SCALE AUTOMATIC TEST SYSTEM: TECHNICAL PROBLEM SHEET TECHNIQUE

PART VII: Bibliography

A. Conference Records Devoted to Automatic Testing

[1] *Proc. Automatic Support Systems Symp. Advanced Maintainability*, sponsored by the St. Louis, Mo., Section of the IEEE, June 7-9, 1965.

[2] *Proc. Automatic Support Systems Symp. Advanced Maintainability*, sponsored by the St. Louis, Mo., Section of the IEEE, Nov. 7-9, 1966.

[3] *Proc. Automatic Support Systems Symp. Advanced Maintainability*, sponsored by the St. Louis, Mo., Section of the IEEE, Nov. 7-9, 1967.

[4] *Proc. Automatic Support Systems Symp. Advanced Maintainability*, sponsored by the St. Louis, Mo., Section of the IEEE, Nov. 12-14, 1968.

[5] *Proc. Automatic Support Systems Symp. Advanced Maintainability*, sponsored by the St. Louis, Mo., Section of the IEEE, Nov. 3-5, 1969.

[6] *Symp. Rec., Automated Support Systems for Advanced Maintainability*, sponsored by the IEEE Aerospace and Electronic Systems Society and the St. Louis, Mo., Section of the IEEE, IEEE Publ. 70 C52-AES, Oct. 19-21, 1970.

[7] *Symp. Rec., Automated Support Systems for Advanced Maintainability*, sponsored by the IEEE Aerospace and Electronic Systems Society and the Philadelphia, Pa., Section of the IEEE, IEEE Publ. 72 CHO 699-9-AES, Nov. 13-15, 1972.

[8] *Symp. Rec., Automated Support Systems for Advanced Maintainability*, sponsored by the IEEE Aerospace and Electronic Systems Society and the Ft. Worth, Tex., Section of the IEEE, IEEE Publ. 73 CHO 804-5-AES, Nov. 5-7, 1973.

[9] *Proc. Joint Conf. Automatic Test Systems* sponsored by the Institution of Electronic and Radio Engineers, Univ. Birmingham, London, England, Apr. 14-17, 1970.

[10] *Automatic Testing '73 Conf. Rec.*, sponsored by Network, 84 High St., Newport Pagnell, Bucks, England, Oct. 27-29, 1973.

[11] WESCON '69 Tech. Papers, Session 21, "Computer aided testing, management and implementation," Western Electron. Show and Conv., San Francisco, Calif., Aug. 19-21, 1969.

[12] WESCON '70 Tech. Papers, Session 5, "Military user views of ATE," Western Electron. Show and Conv., Los Angeles, Calif., Aug. 1970.

[13] WESCON '71 Tech. Papers, Session 3, "Automated testing of MOS integrated circuits," Western Electron. Show and Conv., San Francisco, Calif., Aug. 24-27, 1971.

[14] WESCON '71 Tech. Papers, Session 14, "Present and future of automatic test languages," Western Electron. Show and Conv., San Francisco, Calif., Aug. 24-27, 1971.

[15] WESCON '71 Tech. Papers, Session 17, "Instrumentation for automatic dynamic test systems," Western Electron. Show and Conv., San Francisco, Calif., Aug. 24-27, 1971.

[16] WESCON '71 Tech. Papers, Session 18, "Automatic manufacturing," Western Electron. Show and Conv., San Francisco, Calif., Aug. 24-27, 1971.

[17] WESCON '71 Tech. Papers, Session 21, "Commercial applications of automatic test equipment," San Francisco, Calif., Aug. 24-27, 1971.

[18] WESCON '73 Tech. Papers, Session 10, "Computer aided test design for automatic test equipment," Western Electron. Show and Conv., San Francisco, Calif., Sept. 11-14, 1973.

B. Major Works and Special Reports Devoted to Automatic Testing

[1] "Automation in electronic test equipment," vols. I and II, edited by D. M. Goodman, Director, Project SETE[1] (Secretariat for Electronic Test Equipment), School Eng. Sci., New York Univ., Aug. 1966. Hardbound volumes of technical papers by various authors involved in design courses given at NYU in the summer of 1960 and 1961.

[2] "Automation in electronic test equipment: Built-in test and continuous monitoring," vols. III, IV, and V, edited by D. M. Goodman, Director, Project SETE (Secretariat for Electronic Test Equipment), School Eng. Sci., New York Univ., Apr. 1967. Hardbound volumes of technical papers by various authors involved in design courses given at NYU in the summer of 1966.

[3] "Automation in electronic test equipment: Apollo-Saturn," vol. VI, edited by D. M. Goodman, Director, Project SETE (Secretariat for Electronic Test Equipment), School Eng. Sci., New York Univ., June 1968. Hardbound volume of technical papers by various authors involved in design courses given at NYU in the summer of 1967.

[4] "Automation in electronic test equipment: Factory and depot," vol. VII, edited by D. M. Goodman, Director, Project SETE (Secretariat for Electronic Test Equipment), School Eng. Sci., New York Univ., May 1969. Hardbound volume of technical papers by various authors involved in design courses given at NYU in the summer of 1968.

[5] V. Mayper, Jr., "A survey of programming aspects of

[1] Project SETE has been relocated to the following address:
Project CETE Director
c/o Officer-In-Charge
FMSAEG Annex (Code 862)
Naval Weapons Station, Seal Beach
Corona, Calif. 91720
ATTN: GIDEP Admin. Office

computer controlled' automatic test equipment," vols. I and II, Project SETE Doc. 210/78, June 1964.

[6] V. Mayper, Jr., and D. M. Goodman, "Problems and pitfalls in automatic test computer programming," Project SETE Doc. 210/86, May 1965.

[7] J. Lustig and D. M. Goodman, "Trends in the development of automatic test equipment," Project SETE Doc. 210/106, June 1973.

[8] "Program design handbook for automatic test equipment," RCA Doc. ATE-15 (CR-68-588-30), June 25, 1968. Handbook of test program design experience prepared by RCA engineers under contract to the U. S. Naval Air Development Cen., Warminster, Pa., approximately 200 pages.

[9] F. Liguori, "Environmental model program improved programming procedures," RCA Doc. EMP66, Mar. 16, 1967. Report on one-year evaluation of test programming techniques and costs related to ATE. Develops cost relationships and techniques for test program cost estimating. Approximately 100 pages.

[10] "Systems support," RCA, Camden, N. J., Nov. 1962. Collection of selected articles on ATE previously published by RCA authors, reprinted in magazine form.

[11] "Automated support systems," RCA, Burlington, Mass., 1968. Collection of selected articles on ATE previously published by RCA authors, reprinted in magazine form.

[12] F. Liguori, A. M. Greenspan, *et al.*, "Commercial automatic testing systems, lecture notes revision 3, vol. I and II," published by the Center for Professional Advancement, Nov. 1973. Lecture notes bound in two volumes, consisting of selected reprint papers and original lecture notes, 320 pages. Provided to students enrolling in the the course "Commercial Automatic Testing Systems" which is offered twice annually by the Center for Professional Advancement, P. O. Box 997, Somerville, N. J. 08876.

[13] *Proc. Programming/Software Short Course for Automatic Test Equipment*, Naval Electron. Syst. Command, Washington, D. C., Mar. 1-5, 1971. Softbound volume of technical papers presented by invited speakers at the short course.

[14] *Proc. Navy Sponsored, Government/Industry Workshop on Automated Testing and Monitoring Equipment*, Naval Electron. Lab. Cen., San Diego, Calif., Mar. 1-3, 1972. Softbound book of 197 pages containing individual technical papers by members of industry and government as well as the transcript of the government panel session.

[15] "Automatic test equipment (ATE) phase I audit and appraisal, vol. I, Preliminary analysis of ATE data," Naval Electron. Lab. Cen., San Diego, Calif., Sept. 15, 1971.

[16] "Automatic test equipment (ATE) phase I audit and appraisal, vol. II, Detailed data from ATE survey," Naval Electron. Lab. Cen., San Diego, Calif., Sept. 15, 1971. The combined volumes I and II comprise Tech. Doc. 138 which reports on an industry survey of systems and related management aspects of ATE. Volume I and volume II consist of 123 and 148 pages, respectively.

C. Specifications Related to Automatic Testing

[1] Abbreviated test language for avionics systems (ATLAS)," Aeronautical Radio, Inc. Series of specifications defining the ATLAS test language designed to document manual or automatic test procedures. Latest version and supplements issued periodically by Aeronautical Radio, Inc., 2551 Riva Rd., Annapolis, Md. 21401.

[2] "General requirements for versatile avionic shop test system/avionics system compatibility," Naval Air Syst. Command, Dep. Navy, Washington, D. C., Requirement AR-8A, Mar. 1969. Specifies requirements for designing avionics systems to be compatible for testing on a U. S. Navy ATE system known as VAST.

[3] "General requirements for operational test program sets," Naval Air Syst. Command, Dep. Navy, Washington, D. C., Requirement AR-9A, May 1970. Specifies requirements for development, test, documentation, configuration management, and delivery of test program sets for avionic units to be tested by the U. S. Navy ATE system known as VAST.

[4] "General requirements for maintainability of avionics equipment and systems," Naval Air Syst. Command, Dep. Navy, Washington, D. C., Requirement AR-10A, Jan. 1969. Specifies maintainability and built-in test requirements for avionic equipment to be tested by the U. S. Navy ATE system known as VAST.

[5] "Definition of terms for automatic electronic test and checkout," Dep. Defense, Washington, D. C., Military Standard MIL-STD-1309A, Apr. 12, 1972. Defines 22 pages of terms related to ATE.

[6] "Preparation of test requirements document," Dep. Air Force, Wright Patterson AFB, Military Standard MIL-STD-1519 (USAF), Ohio, Sept. 17, 1971. Establishes requirements for preparation and control of test requirements documents (TRD's) used to specify testing requirements for avionics units. TRD's become source documents for preparing test program sets for ATE.

D. Chronological Bibliography of Items Referenced by Papers in This Book

[1] F. Lesh, "Methods of simulating a differential analyzer on a digital computer," J. Asso. Comput. Mach., vol. 5, no. 3, 1958.

[2] G. Terborgh, "Business investment policy," Machinery and Allied Products Inst., Washington, D. C., 1958.

[3] D. B. Dobson and L. L. Wolff, "Automatic test equipment checks missile systems," *Electronics*, July 15, 1960.

[4] *Automation, It's Impact on Business and People*. Buckingham, England: Mentor, 1961.

[5] A. D. Hall, *A Methodology for Systems Engineering.* Princeton, N. J.: Van Nostrand, 1962.

[6] F. Everhard, "DEE . . . digital evaluation equipment . . . for multi-missile checkout," Syst. Support, RCA, Camden, N. J., Nov. 1962.

[7] D. B. Dobson and L. L. Wolff, "The DEE concept—A family of test systems," Syst. Support, RCA, Camden, N. J., Nov. 1962.

[8] "General specification for systems readiness/maintainability; Avionic systems design," U. S. Government Printing Office, Washington, D. C., Military Spec. MIL-S-23603 (WEP), Mar. 1, 1963.

[9] "A tabulation, automatic and semi-automatic checkout and test equipment," *Ground Support Equipment Magazine*, May/June 1964.

[10] D. J. Seaman, "Automatic test equipment: A brief review of principles and background," *Ground Support Equipment Magazine*, May/June 1964.

[11] V. Mayper, "A survey of programming aspects of computer-controlled automatic test equipment," vols. I and II, Project SETE, New York Univ., June 1964.

[12] M. V. Wilkes, "Lists and why they are useful," in *Proc. ACM 19th Nat. Conf.*, Aug. 1964.

[13] R. D. Brennan and R. N. Linebarger, "A survey of digital simulation: Digital analog simulator programs," *Simulation*, vol. 3, Dec. 1964.

[14] B. L. Ryle, "The Atoll checkout language," *Datamation*, Apr. 1965.

[15] V. Mayper, "Programming for automated checkout, p. I," *Datamation*, Apr. 1965.

[16] J. W. Cooley and J. W. Tukey, "An algorithm for the machine calculation of complex fourier series," *Math. Comput.*, vol. 19, Apr. 1965.

[17] V. Mayper, "Programming for automated checkout: p. II," *Datamation*, May 1965.

[18] V. Mayper, Jr., and D. M. Goodman, "Problems and pitfalls in automatic test computer programming," Project SETE Doc. 210/86, May 1965.

[19] R. C. Miller, "Simplifying the use of automatic test equipment with a compiler," in *Proc. 1965 Automatic Support Systems Symp. Advanced Maintainability*, sponsored by the St. Louis, Mo., Section of the IEEE, June 7-9, 1965.

[20] W. Metcalf, "Progress report on GPATS," in *Proc. 1965 Automatic Support Systems Symp. Advanced Maintainability*, sponsored by the St. Louis, Mo., Section of the IEEE, June 7-9, 1965.

[21] B. J. Evanzia, "Automatic test equipment: A million dollar screwdriver," *Electronics*, Aug. 23, 1965.

[22] R. D. Brennan and R. N. Linebarger, "An evaluation of digital analog simulator languages," in *1965 Proc. IFIP*, vol. 2, 1965.

[23] J. J. Clancy and M. S. Findeberg, "Digital simulation languages: A critique and a guide," in *AFIPS Conf. Proc.* vol. 27, p. I, 1965.

[24] B. H. Scheff, "Simple user oriented compiler source language for programming automatic test equipment," *Commun. Ass. Comput. Mach.*, Apr. 1966.

[25] B. L. Ryle, "An engineering approach to configuration management," in *Proc. Automatic Support Systems Symp. Advanced Maintainability*, St. Louis, Mo., Nov. 7-9, 1966.

[26] R. V. Jacobsen, "Digital simulation of large scale systems," *AFIPS Conf. Proc.*, vol. 28, 1966.

[27] W. M. Syn, "DSL/90—A digital simulation program for continuous system modeling," *AFIPS Conf. Proc.*, vol. 28, 1966.

[28] P. Z. Ingerman, A Syntax-Oriented Translator. New York: Academic Press, 1966.

[29] J. S. Bendat and A. G. Piersol, *Measurement and Analysis of Random Data.* New York: Wiley, 1966.

[30] B. H. Liebowitz, "The technical specification—Key to management control of computer programming," *AFIPS* Conf. Proc., vol. 30, Apr. 1967.

[31] L. V. Searle and G. Neal, "Configuration management of computer programs by the Air Force," *AFIPS Conf. Proc.*, vol. 30, Apr. 1967.

[32] M. V. Ratynski, "The Air Force computer program acquisition concept," *AFIPS Conf. Proc.*, vol. 30, Apr. 1967.

[33] W. T. Cochran *et al.*, "What is the fast Fourier transform?" *IEEE Trans. Audio Electroacoust.*, vol. AU-15, pp. 45-55, June 1967.

[34] F. Liguroi, "ATE designer—Programmer or systems engineer," in *Proc. Automatic Support Systems Symp. Advanced Maintainability*, sponsored by the St. Louis, Mo., Section of the IEEE, Nov. 7-9, 1967.

[35] R. J. Meyer, "Computer controlled GPATS," in *Proc. Automatic Support Systems Symp. Advanced Maintainability*, sponsored by the St. Louis, Mo., Section of the IEEE, Nov. 7-9, 1967.

[36] W. L. Alexander and J. K. B. Woods, "Procurement considerations for computer controlled test equipment," in *Proc. Automatic Support Systems Symp. Advanced Maintainability*, sponsored by the St. Louis, Mo., Section of the IEEE, Nov. 7-9, 1967.

[37] E. A. Helfert, "Techniques of financial analysis," Richard D. Irwin, Inc., Homewood, Ill., 1967.

[38] R. Lucey and D. Newth, "When to inspect all incoming semi-conductors," *Electron. Equipment Eng.*, Mar. 1968.

[39] T. A. Ellison and L. S. O'Neill, "ATLAS—A standard compiler input language for commercial airlines," in *Proc. 1968 Automatic Support Systems Symp. Advanced Maintainability*, sponsored by the St. Louis, Mo., Section of the IEEE, Nov. 12-14, 1968.

[40] R. G. Fulks and J. Lamont, "An automatic computer-controlled system for the measurement of cable capacitance," *IEEE Trans. Instrumen. Measurement*, vol. IM-17, pp. 299-303, Dec., 1968.

[41] D. S. Ammer, *Manufacturing Management and Control* New York: Appleton, 1968.

[42] P. Wegner, *Programming Languages, Information Structures, and Machine Organization.* New York: McGraw-Hill, 1968.

[43] R. L. Mattison and R. T. Mitchell, "UTEC—A universal test equipment compiler," in *Proc. Automatic Support Systems Symp. Advanced Maintainability,* sponsored by the St. Louis, Mo., Section of the IEEE, Nov. 12-14, 1968.

[44] "A guide to ATLAS for specification test writers," Aeronautical Radio, Inc., ARINC Rep. 418, May 15, 1969.

[45] G. D. Bergland, "A guided tour of the fast Fourier transform," *IEEE Spectrum*, vol. 6, pp. 41-52, July 1969.

[46] R. A. Grimm, M. D. Ewy, and S. C. Shank, "Automated testing," *Hewlett-Packard J.*, Aug. 1969.

[47] D. A. Bobroff, "Avoid pitfalls in computerized testing," *Electron. Design*, Aug. 16, 1969.

[48] M. L. Fichtenbaum, "A computer controlled system for testing digital logic circuits," *NEREM Rec.*, 1969.

[49] A. M. Greenspan and E. H. Schmuhl, "Allocation of responsibility for program preparation on ATE supported systems," in *Proc. Automatic Support Systems Symp. Advanced Maintainability*, sponsored by the St. Louis, Mo., Section of the IEEE, Nov. 3-5, 1969.

[50] D. H. Lord, "ATE—Can it be managed?" *Automatic Support Systems Symp. Advanced Maintainability*, sponsored by the St. Louis, Mo., Section of the IEEE, Nov. 3-5, 1969.

[51] "IC test equipment buyer's guide," *Solid State Technol.*, Mar. 1970.

[52] D. T. Ross, "Fourth-generation software—A building block science replaces hand-crafted art," *Comput. Decisions*, Apr. 1970.

[53] P. R. Roth, "Digital Fourier analysis," *Hewlett-Packard J.*, June 1970.

[54] A. Holt and A. Stoughton, "Guidelines for the purchase of memory testing equipment," *Evaluation Eng.*, July/Aug. 1970.

[55] R. G. Loughlin and F. McCoy, "The introduction of VAST," WESCON Tech. Papers, Session 5, Los Angeles, Calif., Aug. 1970.

[56] M. T. Ellis, "VITAL—A general-purpose multiple-dialect automatic test language," in *Symp. Rec., Automatic Support Systems Symp. Advanced Maintainability*, sponsored by the St. Louis, Mo., Section of the IEEE, Oct. 19-21, 1970.

[57] L. W. Wagner and J. W. Edwards, "Evaluation of automatic test equipment through Monte Carlo simulation," in *Symp. Rec. Automatic Support Systems Symp. Advanced Maintainability*, sponsored by the St. Louis, Mo., Section of the IEEE, Oct. 19-21, 1970.

[58] E. J. Fuller and J. A. Procter, "The evolution and development of a digital test capability for VAST," *Symp. Rec., Automatic Support Systems Symp. Advanced Maintainability*, sponsored by the St. Louis, Mo., Section of the IEEE, Oct. 19-21, 1970.

[59] R. K. Gundal, and J. F. King, "A central time-sharing mini-computer—Provides data collection and automatic control over a variety of remote test systems," NEREM '70 Tech. Appl. Papers, Nov. 1970.

[60] "Abbreviated test language for avionic systems," British Aircraft Corp., IPG/TS/126, Stevenage, Herts., England, Nov. 13, 1970.

[61] N. A. Bond, Jr., "Some persistent myths about military electronics/Maintenance," *Human Factors*, 1970.

[62] H. Chang, E. Manning, and G. Metze, *Fault Diagnosis of Digital Systems.* New York: Wiley, 1970.

[63] M. A. Robinson, "A critique of MOL/LSI testing," *Electronics*, Feb. 1, 1971.

[64] "Abbreviated test language for avionic systems," Aeronautical Radio, Inc., 2551 Riva Road, Annapolis, Md., ARINC Spec. 416, Mar. 1, 1971.

[65] L. Curran, "Readers reply on MOL/LSI testing," *Electronics*, Mar. 1, 1971.

[66] W. R. Johnson, Jr., "Proving out large PC boards" *Electronics*, Mar. 15, 1971.

[67] J. J. Bartik, "Common computer interfaces," in *Proc. IEEE Comput. Soc., Mideastern Area*, Cherry Hill, N. J., Mar. 1971.

[68] J. Litzinger, "Hardware/software characteristics of a computer-controlled test system," in *Proc. IEEE Comput. Soc., Mideastern Area*, Cherry Hill, N. J., Mar. 1971.

[69] P. R. Roth, "Effective measurements using digital signal analysis," *IEEE Spectrum*, vol. 8, pp. 62-70, Apr. 1971.

[70] R. G. Bennetts and D. W. Lewin, "Fault diagnosis of digital systems—A review," *Computer*, vol. 4, July/Aug. 1971.

[71] A. Friedman and P. R. Menoa, *Fault Detection in Digital Circuits.* Englewood Cliffs, N. J.: Prentice-Hall, 1971.

[72] W. J. Kahn, "How to establish an automated testing facility, problems and solutions," Industrial and Scientific Management, Inc., 1971.

[73] G. M. Weinbert, *The Psychology of Computer Programming.* New York: Van Nostrand-Reinhold, 1971.

[74] R. M. McClure, "Fault simulation of digital logic utilizing a small host machine," in *Proc. ACM–IEEE Design Automation Workshop*, June 1972.

[75] "Basic concepts in data base management systems," *Datamation*, July 1972.

[76] D. L. Wood, "Automatic test equipment software configuration management," in *Proc. Automatic Support Systems Symp. Advanced Maintainability*, sponsored by the Philadelphia, Pa., Section of the IEEE, Nov. 13–15, 1972.

[77] S. A. Szygenda, "Implementation of synthesized techniques for a comprehensive digital design, verification, and diagnosis system," in *Proc. 5th Hawaii Int. Conf. Systems Sciences*, 1972.

[78] M. A. Breuer, *Design Automation of Digital Systems*, vol. I. Englewood Cliffs, N. J.: Prentice-Hall, 1972.

[79] B. W. Boehm, "Software and it's impact: A quantitative assessment," *Datamation*, May 1973.

[80] E. G. Cromer, "Fault isolation and repair techniques: Current strategies and equipment," *Electron. Packag. Prod.*, vol. 13, June 1973.

Author Index

Editor's Biography

Fred Liguori, (S'57–M'58–SM'67) graduated from the Electrical Engineering College, Tufts University, Medford, Mass., and received the Master's degree from Hofstra University, Hempstead, N. Y. He has completed graduate and professional level courses at Washington University, St. Louis, Mo., Adelphi College, Garden City, N. Y., and the Massachussetts Institute of Technology, Cambridge.

He is currently Head of the ATE Branch, Naval Air Engineering Center, Philadelphia, Pa. Prior to accepting this appointment in 1971, he was Manager of the Software Systems Engineering Laboratory, Emerson Electric Company, with earlier ATE experience at RCA. He has over 20 years experience in electronic testing, more than half of which have been as an Engineer and Manager of ATE design and related software projects. He has actively contributed to the annual IEEE Automatic Support Systems Symposium since its founding in 1965 as author, speaker, session chairman, committeeman, registrar, conference organizer, and chairman for 1972. He has published extensively on the subject of ATE at other conferences including government sponsored seminars, the 1969 and 1971 ACM National Conventions, AIAA meetings, and the first ATE conference held in the United Kingdom. Other publications include papers in *Computers and Automation* and *Software Age* and contributions to two books on ATE and software. He was session chairman for the ATE sessions at WESCON '70, '71, '73. He is currently Course Director and Instructor for a course on commercial automatic test systems given by the Center for Professional Advancement, Somerville, N. J.

Mr. Liguori is the chairman of the IEEE Technical Committee on Automated Instrumentation and an associate fellow of the AIAA for which he served on the National Technical Committee on Computer Systems. He is a member of the National Civilian Administrators Association and a Registered Professional Engineer in the Commonwealth of Massachusetts.